4th Engineering Science and Technology International Conference (ESTIC 2018)

MATEC Web of Conferences Volume 248 (2018)

Padang, Indonesia
28 - 29 August 2018

Editors:

Yovial Mahyoedin
Ariadi Hazmi
Rika Ampuh Hadiguna
Hendra Suherman
Reni Desmiarti

ISBN: 978-1-5108-7703-0

Printed from e-media with permission by:

Curran Associates, Inc.
57 Morehouse Lane
Red Hook, NY 12571

proceedings.com

Some format issues inherent in the e-media version may also appear in this print version.

For additional information, please contact EDP Sciences – Web of Conferences at the address below.

EDP Sciences – Web of Conferences
17, Avenue du Hoggar
Parc d'Activité de Courtabœuf
BP 112
F-91944 Les Ulis Cedex A
France

Phone: +33 (0) 1 69 18 75 75
Fax: +33 (0) 1 69 28 84 91

contact-edps@webofconferences.org

Additional copies of this publication are available from:

Curran Associates, Inc.
57 Morehouse Lane
Red Hook, NY 12571 USA
Phone: 845-758-0400
Fax: 845-758-2633
Email: curran@proceedings.com
Web: www.proceedings.com

TABLE OF CONTENTS

The 4th Engineering Science and Technology International Conference (ESTIC) 2018

Padang, Indonesia
August 28-29, 2018

The Axana Hotel Padang
http://theaxanahotel.ambacang.com/
Address: Jl. Bundo Kandung No 14-16 (25119) Padang, Indonesia
Telp: (+62) 751 7530456 or (+62) 751 39966

GIFU UNIVERSITY

Universiti Kebangsaan Malaysia
The National University of Malaysia

MATEC Web of Conferences

Scopus

Preface

It is our great pleasure to introduce the proceedings of 2018 4th Engineering Science and Technology Conference (ESTIC 2018) held in Padang, West Sumatera, Indonesia from August 28-29, 2018. ESTIC 2018 is a scientific meeting aimed to promote mutual exchange between scientists and also experts, to discuss innovative ideas in scientific research and to find potential opportunities for international cooperation. This conference is dedicated to issues related to mechanical, chemical, industrial, electrical, information, civil and environmental engineering

The theme for this conference is "Applied Technology for Sustainable Development". Applied engineering was progressing rapidly and enables us to reveal and comprehend how this universe works. I believe that this theme has challenges for all engineers how to develop of sustainable industry process and energy, innovative material, conceptual and process design. However, the aim of the research is how to create the better life for the future without neglect the negative impact due to the actives incurred.

We would like to express our gratitude to keynote speakers and invited speakers for active participant in this conference. We would also like to thank the reviewers, who spared their valuable time, for their advice. They have certainly helped improve the quality, accuracy, and relevance of each paper selected for the conference program and for publication. We also wish to thank all the authors who have contributed to this conference, as well as the member of organizing committee, who has worked very hard to run the conference and all the conference participants for their support for ESTIC 2018.

Padang, August 28, 2018

Yovial Mahyoedin, PhD.

4th Engineering Science and Technology International Conference (ESTIC 2018)

INTRODUCTION.

The Engineering Science and Technology International Conference organized by the Faculty of Industrial Technology, Bung Hatta University, has been held since 2011, in the 2018, it was already its 4th volume

This event is aimed at promoting engineering technology research activities in Indonesia and overseas, in the hope of building and strengthening networks and collaborations. The upcoming conference theme is "Applied Technology *for Sustainable Development*".The conference is open to all researchers in applied engineering technology and science.

ORGANIZER:

Faculty of Industrial Technology, Bung Hatta University, Padang , West Sumatera, Indonesia

LOCATION AND DATE:

The Axana Hotel, Padang, 28th – 29th August 2018

CONFERENCE WEBSITES:

http://www.esticfti.bunghatta.ac.id

PROCEEDINGS EDITOR:

Yovial Mahyoedin, PhD.
Department of Mechanical Engineering, Faculty of Industrial Technology, Bung Hatta University, Padang, West Sumatera, Indonesia.
Email : estic2018@bunghatta.ac.id

INTERNATIONAL SCIENTIFIC COMMITTEE

KEYNOTE SPEAKERS

Prof. Yoshinori Itaya (Gifu University, Japan)

Prof. Dr. Ir. Andi Isra Mahyudin (Bandung Institute of Technology, Indonesia)

Assoc. Prof. Dr. Abu Bakar Sulong (National University of Malaysia)

INVITED SPEAKERS

Assoc. Prof. Hendra Suherman, MT. (Bung Hatta University, Indonesia)

Assoc. Prof. Dr. Eng. Reni Desmiarti (Bung Hatta University, Indonesia)

Dr. Ardiansyah Syahrom (Universiti Teknologi Malaysia)

Dr. Inna Kholida Sari (Bung Hatta University, Indonesia)

Dr. Hidayat, ST., MT (Bung Hatta University, Indonesia)

ORGANIZING COMMITTEE

Prof. Dr. Azwar Ananda, MA (Rector of Bung Hatta University)

Assoc. Prof. Dr. Hendra Suherman (Vice Rector of Bung Hatta University)

Dr. Hidayat, MT (Dean of the Faculty of Industrial Technology, Bung Hatta University)

Supervisor	: Assoc. Prof. Ir. Drs. Mulyanef, M.Sc
Chairperson	: Yovial Mahyoedin, PhD.
Secretary	: Ir. Burmawi, MT
Treasury	: Ir. Wenny Marthiana, MT

Secretariat Office of ESTIC 2018

Engineering Science and Technology International Conference
Department of Mechanical Engineering
Faculty of Industrial Technology
Kampus Proklamator III Bung Hatta University
Jl. Gajah Mada No.19 OloNanggalo, Padang, West Sumatera, Indonesia.
Website : http://www.esticfti.bunghatta.ac.id
Email : estic2018@bunghatta.ac.id

GIFU UNIVERSITY

MATEC Web of Conferences

edp sciences

Web of Conferences

Statement of Peer review

In submitting conference proceedings to MATEC WEB of Conference, I certify to the publisher that I adhere to the Policy on Publishing Integrity of the journal in order to safeguard good scientific practice in publishing.

1. All articles have been subjected to peer review administered by the proceedings editors.
2. Reviews have been conducted by expert referees, who have been requested to provide unbiased and constructive comments aimed, whenever possible, at improving the work.
3. Proceedings editors have taken all reasonable steps to ensure the quality of the materials they publish and their decision to accept or reject a paper for publication has been based only on the merits of the work and the relevance to the journal.

Title, date and place of the conference

4th Engineering Science and Technology International Conference (ESTIC 2018), August 28th - 29th ,2018
The Axana Hotel , Padang, West Sumatera - Indonesia

Proceedings editor(s):

Prof. Dr. Eng. Ariadi Hazmi
Prof. Dr. Rika Ampuh Hadiguna
Assoc. Prof. Dr. Hendra Suherman
Assoc. Prof. Dr. Eng. Reni Desmiarti

Date and editor's signature

10.11.2018

Effect of Thickness Reduction on Cold Rolling Process to Microstructure and Brass Hardness

Agung Setyo Darmawan[1,*], *Bambang Waluyo* Febriantoko[1], and *Agus Dwi* Anggono[1], *Tri Widodo Besar* Riyadi[1], *Abdul* Hamid[2]

[1]Department of Mechanical Engineering, Faculty of Engineering, Universitas Muhammadiyah Surakarta, Jl. Ahmad Yani, Tromol Pos 1 Pabelan, Surakarta 57162, Indonesia
[2]Department of Mechanical Engineering, Faculty of Engineering, Universitas Batam, 29464 Batam, Indonesia

Abstract. Brass is an alloy between copper (Cu) and zinc (Zn) where the majority of the elements are copper. Brass that undergoes a cold rolling process will experience changes in microstructure and mechanical properties. This study aims to find out the effect of cold rolling process on changes in microstructure and hardness of brass. The composition of brass is analyzed by using Energy Dispersive Spectroscopy technique, while microscope optic is used to obtain microstructure photograph of brass. The hardness of brass is measured before and after cold rolling process. The test results show that cold rolling process causes the increase of hardness on brass, change of grain shape, and the appearance of deformation bands on brass.

1 Introduction

Metals have various useful properties such as hard, ductile, and formable. Given these properties, a variety of products are made of metal. The product is produced through a series of process including moulding, forming, machining, welding, heat treatment and surface treatment [1].

One of metal processes is forming. The forming process is classified into a hot forming and cold forming process. The hot forming process is carried out above the recrystallization temperature and the cold forming process is carried out under the recrystallization temperature. The property of that utilized in the forming process is the ductility. Ductility is important for structural materials. Changing the shape is performed by giving plastic deformation [2]. An example of the forming process is rolling.

Large deformations occur in the hot forming process, while small deformations are obtained from cold forming. Metal work is often conducted by heat forming process which is then continued with cold forming. The cold formed metal surface is smoother than the hot formed metal surface.

The cold forming process can make the metal undergo some changes in mechanical properties, namely getting stronger, getting harder, and getting brittle [3,4]. The larger the

* Corresponding author: Agung.Darmawan@ums.ac.id

plastic deformation, the greater the strength, hardness and brittleness of the metal. The crack will occur in metal once excessive brittleness is reached. A product has a requirement for mechanical properties, so the process of metal forming must also refer to the requirements of that property [5]. In this case it is necessary to have data on the level of deformation and its relation to the mechanical properties of the material. Therefore it is necessary to test the various levels of deformation and changes in its mechanical properties.

The cold working processes also provide a way to directionally influence the micro structural changes [6-8]. Several researchers have been investigated the effect of plastic deformation to microstructure of material. Li et.al investigated the combination effect of plastic deformation and phase transformation on the evolution of microstructure and property of NiTi alloy [9]. Ren et.al studied grain refinement and mechanical response of U-5.5 wt.% Nb alloy produced by cold rolling and heat treatment [10]. Kawalko et.al investigated microstructure evolution of CP titanium during deformation in KoBo process followed by cold rolling [11].

This study aims find out the effect of cold rolling process on changes in microstructure and hardness of brass. Hardness, one of mechanical property that important to consider, is a measure of a material's resistance to localized plastic deformation. Hardness is an indicator of wear resistance [12]. Brass is widely used as an industrial material, clinical and scientific application because of its excellent characteristics such as high corrosion resistance, non-magnetism, good plasticity deformability, and machinability [13,14].

2 Materials and Methods

Several tests are conducted on brass test specimen to reach the objective of this research. The composition of Brass are 62.702 % weight Cu and 37.298 % weight Zn. The phases of the material are α and β'. The tests are as follows:

1. Composition test
2. Cold rolling process
3. Hardness test
4. Metallographic.

The composition of brass is analyzed by using Energy Dispersive Spectroscopy technique. In this way, the element composition can be obtained qualitatively or semi-qualitatively on a small area on the surface of the specimen. The cold rolling process is carried out at a temperature below the recrystallization temperature to provide a plastic deformation of the brass specimen in the form of a reduction in the thickness of the brass plate. TBlitz type 2035 roll machine was used in this research. Hardness and metallographic tests are performed at each stage of reduction. Hardness is measured by using Vickers Hardness tester and metallographic test is performed by using microscope optic.

Percent reduction in thickness (% *RT*) is defined as

$$\% \, RT = \ln \frac{t_o}{t_i} \quad (1)$$

where t_o is the initial thickness of brass specimen and t_i is the instantaneous thickness.

3 Results and Discussions

The phases shown in Fig. 1 are α (dark) and β' (light) phases. This corresponds to the Cu-Zn phase diagram. There is also a banded structure, a second phase structure that resembles fiber [15]. This structure occurs because alloys have experienced the rolling process and experience segregation of Zn composition.

Fig. 1 also shows the colour difference of the α phase (a phase that looks dark and is indicated by B and C). Position B shows a slightly lighter colour than position C. This is due to the α phase at position B has more Zn element than that at position C. The atomic number of Zn is greater than Cu atomic number. The composition of Cu and Zn atoms is shown in the Table 1.

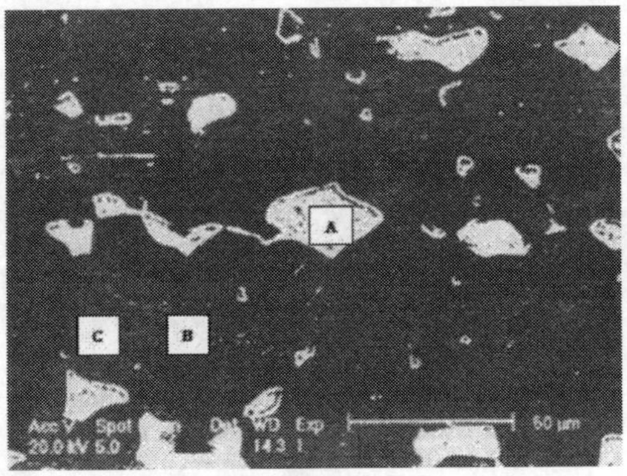

Fig. 1. Scanning Electron Microscopy of Brass

Table 1. The Composition of Cu and Zn in brass.

Position	% wt Cu	% wt Zn	% at Cu	% at Zn
A	56.565	43.435	57.261	42.739
B	64.570	35.430	65.217	34.783
C	65.185	34.815	65.827	34.173

Photographs of metallographic results are shown in Fig. 2 and Fig. 3. Fig. 2 shows microstructure of brass before cold rolling process. The microstructure consists of α phase and β' phase. The annealing twin due to previous hot rolling process appears on the microstructure.

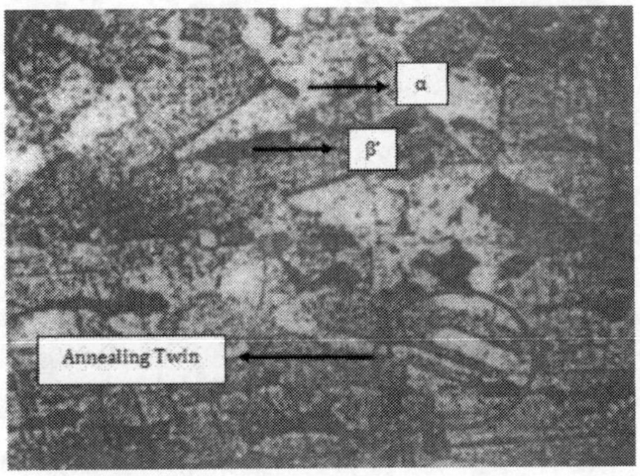

Fig. 2. Initial microstructure photograph of Brass

Fig. 3 shows the microstructure changes because of cold rolling process. It can be seen that the higher the reduction, the grains will be flatter because of deformation. Annealing twin on the reduction in thickness of 10.8% is deformed so that the shape is slightly curved as can be seen in Fig. 3a. There is no mechanical twinning on the microstructure because the brass has face-centered cubic (FCC) unit cell.

At a reduction in thickness of 32.9 % to a reduction in thickness of 113.55 % (Fig. 3b to Fig. 3f), there is a deformation band in a flat-shaped item. The deformation band is formed because SFE (Stacking Fault Energy) value for brass is low (SFE brass is smaller than 10mJ / m2).

Fig. 4 shows the effect of reduction in thickness to the hardness. It can be shown that the higher the reduction, the higher the hardness [16]. In this research, hardness increase from 94.3 HV to 211.7 HV. Increasing of hardness is due to the occurrence of strain hardening phenomenon. The strain hardening phenomenon occurs as dislocation dislocations become more difficult to move due to the doubling of dislocations.

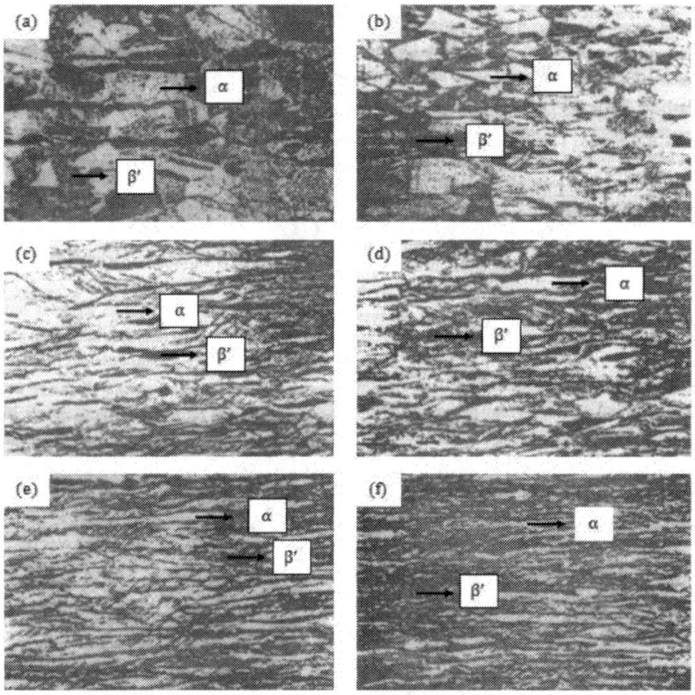

Fig. 3. Microstructure photograph of Brass with reduction in thickness of (a) 10.8 %, (b) 32.9 %, (c) 57.8 %, (d) 78.9 %, (e) 94.7 % and (f) 113.5 %.

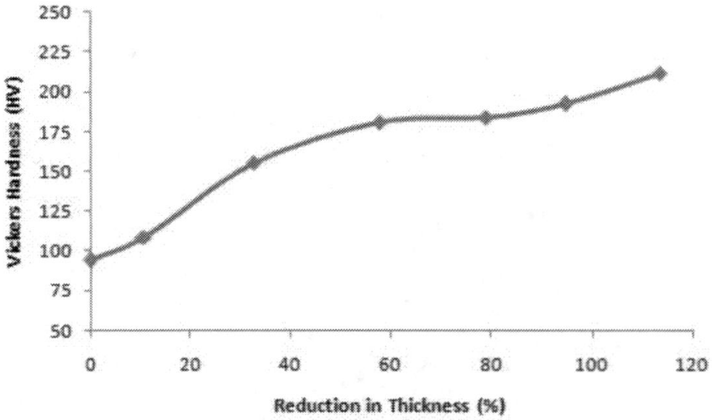

Fig. 4. The effect of reduction in thickness to the hardness.

4 Conclusions

From the result in this research, it is can be concluded as follows:

1. The cold rolling process causes an increase in hardness of brass. In this research, hardness increase from 94.3 HV to 211.7 HV.
2. The cold roll process causes the formation of the deformation band on the brass

The authors gratefully acknowledge Mechanical Engineering Laboratory of Universitas Muhammadiyah Surakarta for cold rolling process.

References

1. A.S. Darmawan, W.A. Siswanto, T. Sujitno, Advanced Materials Research, **789**, 347 (2013).
2. Y. Lian, J. Huang, J. Zhang, C. Zhao, W. Gao, Z. Zhang, M. Ma, Materials Science & Engineering A, **712**, 663 (2018).
3. A.S. Darmawan, A.D. Anggono, A. Hamid, MATEC Web of Conferences, **154**, 01105 (2018).
4. Y. Bai, T. He, Y. Liu, Materials Characterization, 137, 142 (2018).
5. A. Hamid, I. Baba, W. Sani, International Journal on Advanced Science, Engineering and Information Technology, **7 (2)**, 446 (2017).
6. C. Lan, Y. Wu, L. Guo, F. Chen, Materials Science and Engineering: A, **690**, 170 (2017).
7. S. Wang, L. Niu, C. Chen, Y.L. Jia, M.P. Wang, Z. Li, Z.H. Zhong, P. Lu, P. Li, Y.C. Wu, Fusion Engineering and Design, **125**, 510 (2017).
8. M. Eskandari, M.A. Mohtadi-Bonab, J.A. Szpunar, Materials and Design, **90**, 618 (2016).
9. Y. Li, J.Y. Li, M. Liu, Y.Y. Ren, F. Chen, G.C. Yao, Q.S. Mei, Journal of Alloys and Compounds, **653**, 156 (2015).
10. Z. Ren, R. Ma, G. Hu, J. Wu, Z. Wang, C. Luo, Journal of Nuclear Materials, **494**, 72 (2017).
11. J. Kawałko, P. Bobrowski, P. Koprowski, A. Jarzę̨bska, M. Bieda, M. Łagoda, K. Sztwiertnia, Journal of Alloys and Compounds, **707**, 298 (2017).
12. A. S. Darmawan, W. A. Siswanto, T. Sujitno, Applied Mechanics and Materials, **315**, 700 (2013).
13. J. Li, Y. Ren, M. Ibrahim, Songbai, K. Yang, Materials Letters, **202**, 162 (2017).
14. H. Imai, Y. Kosaka, A. Kojima, S. Li, K. Kondoh, J. Umeda, H. Atsumi, Powder Technology, **198**, 417 (2010).
15. J. Ryś, G. Cempura, Materials Science and Engineering: A, **700**, 656 (2017).
16. J. Hou, M. Zhang, S. Ma, P.K. Liaw, Y. Zhang, J. Qiao, Materials Science & Engineering A, **707,** 593 (2017).

Air flow Simulation in correlation of the Outside Engine Temperature Change with Variation of the Grill

Rahmat Riza*, *Paryana* Puspaputra and *Wajib* Haryono

Universitas Islam Indonesia, Yogyakarta, Indonesia

Abstract. This research is objected to study the effect of air volume variation in the cooling process for the outer wall of vehicle engine. The results were expected to be consideration in Grill's engineering process in term of better efficiency which is the discussion nowadays. The simulation was executed on a model rebuilt from a rough estimate of one of the existing urban vehicles in Indonesia. This model was performed in computational fluid dynamics software. Four condition of hole's grill of the vehicle engine were prepared as the main parameters to experiment the air flow which would receive the heat from the outer wall of the engine. These parameters were study to investigate the outer wall temperature of the engine. The results showed that reducing 25% of hole's area of the grill would increase temperature around 7.4%, the decreasing of 50% of the hole had made the increasing temperature around 28%. The last condition of the decreasing of 75% of the hole had caused the increasing temperature around 400%.

1 Introduction

Internal combustion engine works in the optimum system in the consideration of many parameters which referring to the maximum power output produced in the lowest fuel consumption without neglectful safety and cost of maintenance[1]. One of the parameters is the optimum temperature of the combustion.The change of the temperature, outside of the optimum range, will be ensured to implicate the engine in the worse direction. Whenever the temperature of ignition does not reach the optimum point, the process is facing some problem like imperfect burning which producing less power and more pollutants for our environment [1]. The process is called overheating whenever the temperature exceeding the optimum value. In the overheating condition, the engine also produces many problems such as pre-ignition which cause less power and reducing the lifetime of the engine component [2]. This is regarding to the process where most of energy generated by fuel burning is used to drive piston and some of them is exhausted as energy loss. The loss will cause temperature increasing around the engine and released to the environment by heat transfer which is known as the cooling system. This system is an important process to maintain the combustion process in the optimum temperature. The importance of cooling system in internal combustion engine had been shown by many experiment conducted [1–6]. They showed that

* Corresponding author: rahmatriza@uii.ac.id

cooling system take an important position in the system. These results also showed that the cooling system influenced the performance of the engine overall.

The cooling system for the vehicle engine naturally takes place by conduction which is transferring the heat from the inner wall of the engine to the outer wall of the engine. Then, the heat in the outer wall of the engine is removed by convection to the environment. The heat is also removed by fluid, called as coolant, whenever the heat in the combustion room is too high to remove to the environment in nature. The coolant is utilized to distribute the area of heat releasing to the environment[1] within more alternatives. For the latest vehicle, temperature in this outer wall is approximated in the range 90 to 115°C [2]. This temperature is maintained in order to prevent the overall components of the engine from the defect [7]. This situation can be obtained by supplying as much as possible the number of air in the environment to the engine room. This is parallel with formula [2]:

$$q = m.C_p.\Delta T \tag{1}$$

where q referred to the number of heat transferred, m is number of mass involved in the process and ΔT symbolized temperature difference between low temperature and high temperature and C_p is specific heat. This showed that the heat released to the environment from the engine will be determined by the amount of air in the outer wall of the engine. However, the air that can be provided in the engine room is limited by the space for the safety, design and the efficiency. Thus, it was crucial to recognized the effect of air variation which supplied to engine room to the change of outer temperature in order to adopt different many variations in the design and safety consideration.

2 Model Development

Many factors that was determined how much the air can delivery to the engine room. One of them is grill design. Grill of vehicle has multiple function. The first function is designed as the protector of the engine components from the large size material from the road. The second function is used to determine number of air entering the engine room. From its shape, the volume of outside air could be controlled to delivery to the engine room. The other function was to develop attractive and aerodynamic shape of the vehicle [8]. In some of condition, all the function cannot adequate each other. Therefore, variation of grill design was used to conduct this experiment. It was performed by following steps, namely: 1. model determination, 2. Model simulation and 3. Results and Analysis.

This work was implemented for Honda Brio Satya which is manufactured in 2014. The grill and engine room of this sample were modelled as the grill was assumed as given in Figure 1(a):

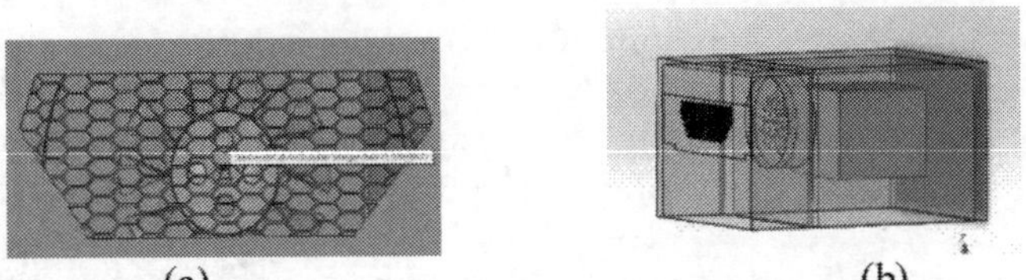

(a) (b)

Fig. 1. (a) Recondition of Grill Design of Honda Brio Satya which manufacturing in 2014; (b) Simulation Model of Vehicle Engine Room.

This grill design was re-produced by measuring the original one and drawn in Computer Aided Drawing (CAD) with the approximated dimension. However, the difference should be accepted since the objective of this experiment was to observe the trend of the change of the air volume entering the engine room in relation with the change of the outer wall temperature of the engine. The engine in this model was assumed as the black box which generated heat. Its dimension is only approximated from the real condition. Based on this approaching, the model for the simulation in this work was given in Figure 1(b).

In order to obtained different of air volume, this work was committed by varying the area of hole in the grill. Three configurations were examined in order to explore how the volume of air entering the engine room influence the temperature of outer wall of the engine. The configurations were determined for 25% closed area of hole, 50% closed area of hole and the last one was 75% closed area of hole.

The simulations were implemented within environment temperature of 32°C which referring to higher level of the temperature in the D.I. Yogykarta and wind velocity around 2,1 m/s. These data were obtained from Climatology station D.I. Yogyakarta (references). This condition was combined with condition vehicle run at 1500 RPM which temperature observation show to 85°C. This temperature was taken by running the engine for five minutes in 1500 RPM. The measurement was executed by turning off the cooling fan where the instrument was able to measure till 400°C.

3 Results and Discussion

Simulation was performed by using software for computational fluids dynamics (CFD). Number of Meshing for the calculation was set up automatically by the software assumption. The computational results of CFD software for the condition of the original grill as the comparison with the experimental condition were given on Fig. 2.

Fig.2. Air Flow Simulation Result of the Engine Cooling System for the Original Grill Design

Temperature distribution in Fig. 2. described that air from the outside entering room engine with temperature 32°C had absorbed heat from the engine. This process was brought the air temperature in the room engine increase to the highest of 36,54°C.

The first variation in this experiment was performed by modifying the aperture of the hole in the grill of the engine. This modification was intent to obtain 75% of the fresh air from the surrounding flowing to the engine room. It had been implemented by modifying the hole's diameter of the grill thereby the area of the hole reduced for 25% from its original size. It was modelled as given in Fig. 3.(a). The second one was conducted on the grill with 50% of hole's area reduced. The last modification was carried through 75% of closed hole's area.

(a) (b) (c)

Fig.3. (a) Grill model with 25% of the hole's area was closed; (b) Grill model with 50% of the hole's area was closed; (c) Grill model with 75% of the hole's area was closed

All of those experiment configurations were designated in order to explore the effect of air flow to cool the outer wall of the engine. The results for the first experiment were given in Fig.4, where Fig.5 refer to the results of air flow simulation for second experiment. And the outputs of the last condition of the experiment were illustrated in Fig.6.

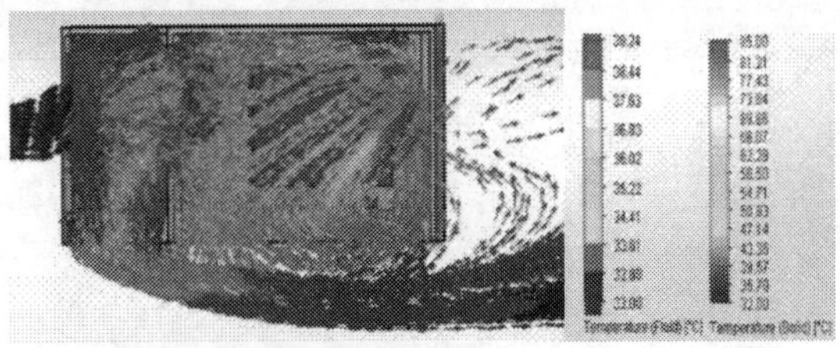

Fig.4. Air flow simulation for 25% closed area of hole in y-axis.

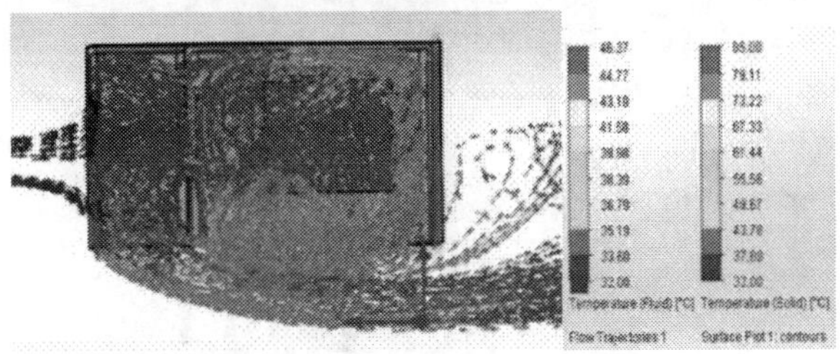

Fig.5. Air flow simulation for 50% closed area of hole in y-axis.

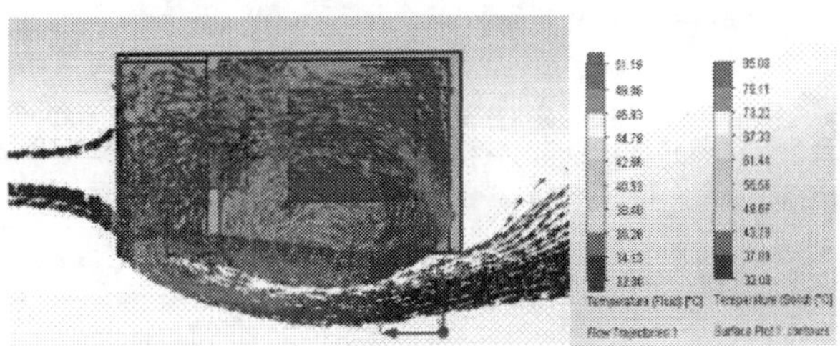

Fig.6. Air flow simulation for 75% closed area of hole in y-axis.

Overall, the results for all configurations showed that air flow in the room engine had started to develop turbulent flow. This condition was highlighted by the pattern of the arrows highly fluctuating. The turbulent condition was normally developed when inertia forces greater than viscous forces as determined by Reynold number which formulated [9]:

$$Re_x = \frac{\rho u_\infty x}{\mu} \tag{2}$$

Where ρ refers to air density, x is the length of heat transfer processes, μ is viscosity of the air, and u_∞ is the velocity of air entering the engine room. Based on this equation, the air velocity influences the turbulent development of the airflow. It depends on several condition, especially the area that is gone through by the air itself. The surrounding air enters the engine room will pass the grill. On this situation, the velocity of the surrounding air would have surface resistance from the grill as discussed by Crowe, et al. [10] described as in Figure 7:

When the surrounding air pass the grill, some of the air would hit the grill, some of the air would tangent with the surface's hole of the grill and the rest could pass through the grill freely. The part of air which hit the grill would yield drag. This drag would influence the air in the engine room and form drop pressure in the back of the grill which is located in the engine room. This pressure would cause most of the air restrained near in the grill as figured in the simulation results as in Fig. 4 to Fig. 6. Only few of them could directly reach the engine wall. This meant that the air volume which received heat from the outer wall of the engine reduce and caused the air in the engine room would take extra heat from the engine surface. The small of grill hole would led the small volume of the surrounding air to reduce temperature of the outer wall of the engine. This is indicated from the results illustrated by Fig. 4 to Fig.6 which show that the variation of the hole's area of the vehicle grill will influence the wall temperature of the engine

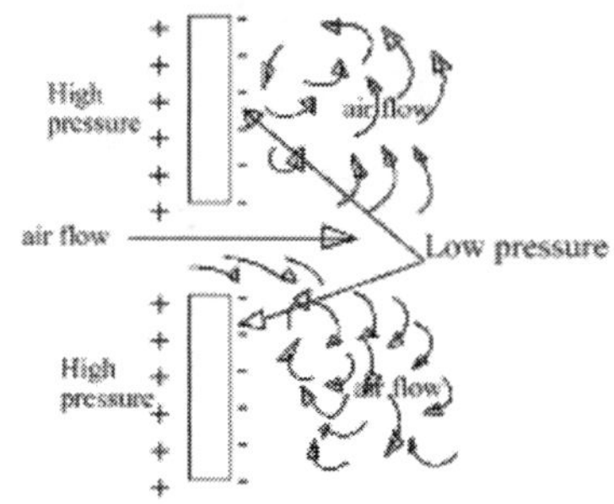

Fig.7. Air flow illustration in the area of grill

4 Conclusion

Air flow simulation results showed that the change of the hole's area of the grill vehicle had bought difference heat transfer from the outer wall of the engine. Diminishing 25% of grill area could increase the engine temperature around 2.7°C. However, reducing the diameter of the hole was not linier with increasing temperature where 50% of diminishing diameter had increased the engine's temperature around 10.23°C and 75% of diameter reducing had brought the engine's wall temperature to be 51,19°C.

References

1. Y. Singh and N. K. Singh, J. Energy South. Afr. **24**, 7 (2013).
2. M. M. A. Morad and J. Alrajhi, J. Mech. Eng. Autom. **4**, 55 (2014).
3. B. Yang, S. Han, and H. Cho, Int. J. Appl. Eng. Res. **11**, 9106 (n.d.) (2015).
4. C. Wilhelmsson, A. Vressner, P. Tunestål, B. Johansson, G. Särner, and M. Aldén, in (2005).
5. C. Wang, Q. Sun, and L. Xu, J. Electr. Comput. Eng. **2017**, 1 (2017).
6. A. K. Hossain, D. I. Smith, and P. A. Davies, J. Sustain. Dev. Energy Water Environ. Syst. **5**, 46 (2017).
7. G. R. Safakish, Adv. Mech. Eng. **4**, 931584 (2012).
8. L. Liu, C. Jing, and X. Hu, in Atlantis Press, (2015).
9. F. P. Incopera, D. P. Dewitt, T. L. Bergman, and A. S. Lavine, *Fundamentals of Heat and Mass Transfer*, Sixth John Wiley & Sons (Asia) Pte Ltd., (2007).
10. C. T. Crowe, D. F. Elger, and J. A. Roberson, *Engineering Fluid Mechanics*, Eighth John Wiley & Sons, Inc., (2005).

Performance Study on Solar Hybrid Air-Conditioning System for Residential Water Heating

Kaidir[1,*], Mulyanef[1] and Burmawi[1]

[1]Department of Mechanical Engineering, Bung Hatta University, Padang, Indonesia

Abstract. The Government of Indonesia has done some efforts to overcome energy crisis nationwide. Among them by issuing a policy as the Foundation for the development and improvement of the capacity of providing energy. The Government has issued a presidential instruction number 10 of the year 2005 on energy savings and last Presidential Instruction No. 2 of the year 2008 about saving energy and water. One of the implementation efforts of this energy crisis is overcome with effort saves energy consumption on the air conditioning with develop hybrids air conditioning, namely as the air conditioner at the same time to the water heating. This research is a new technology for energy saving of electricity. The purpose of this research is to produce a prototype of a hybrid energy efficient refrigeration machine that can be function simultaneously as the engine room air conditioner at the same time water heaters. Specific research targets to be achieved is the realization of a prototype of a hybrid air conditioning engine capacity 1.5 PK which can save energy around 60%. The method of research is the planning, creation, testing the performance of air conditioning machines of compression cycle steam hybrid.

1 Introduction

Hot water can be produced by using the waste heat from an air conditioner. In the places with a year round air conditioning requirement, the heat rejected in the condenser is waste, unless the heat is recovered by using a heat exchanger. This can contributed to energy conservation, and be economically viable. The use of a desuperheater for heating water deserves more attention in the ASEAN countries than in places where there is winter, because a desuperheater does not recover waste heat in winter, but rather uses the energy intend for heating. Normally, heating and cooling systems are widely used in comfort air conditioner and industrial applications. However, the share of the energy for heating and cooling purposes in total energy consumption increases. Due to the economic benefits resulting from high coefficient of performance (COP) values, mechanical heat pump systems become convenient devices for heating and cooling purposes [1], [2]. Heat pump is an apparatus or machine that moves heat from the heat source at a lower temperature to the heat sink at a higher temperature by means of mechanical work or a high temperature heat source [3]. The difference between a conventional a conditioner and a heat pump

* Corresponding author: irkaidir@bunghatta.ac.id

is that a heat pump can be used to provide heating or cooling. However, heat pump is still uses the same basic refrigeration cycle for working. It can be easier to say that a heat pump can change which coil is the condenser and which the evaporator by using a reversing valve. So, in cooling conditions, it is common to require heat pumps that are designed only to provide heating. Now a day, the cost of energy continues to rise and it becomes an imperative to save energy and improve overall energy efficiency. A key idea for improving the energy efficiency of many industries is to recover every possible sources of waste heat and convert this energy to a useful output [4]. Moreover, many efforts tried to increase the heat pump performance. However, this study is aimed to investigate the hot water making potential by using the waste heat that released from condenser of the air conditioner. The advantages of this system are as follows: 1) Save energy due to use of the waste heat to produce hot water 2) Without electric short circuit and 3) Fast to produce hot water and 4) Reduce greenhouse gases [5]. So, some recent researches focused on this area are summarized as follows: Roongutai et al. [6] studied the warm water making from air-conditioning system by using of the waste heat that released from the air conditioner. A pressure switch was used to activate both of the condensers, which are automatically controlled. Their results indicated that the highest temperature of the water in the reservoir is 49^{0}C. Saisanit et al. [7] design and construct the prototype of the hot water making machine using waste heat released from a common air conditioner system.

Two types of the hot water system such as “submerged coil” and “flow through” are used in their study. Solenoid valves were used to control the flow direction of the refrigerant. Air conditioner with cooling capacity of 3,51 KW and working with R-22 was used in this study. The result indicated that the hot water making machine with submerged coil type is more appropriate to use than the flow through type. As mentioned before, the purposes of this study are to investigate the potential of hot water making by using a conventional air conditioner as an air-water heat pump and then to compare the COP of the system between conventional air conditioner cycle and heat pump cycle. The energy saving potential for making of 120 L hot water is also presented.

Some work on the use of a heat exchanger for domestic water heating has been investigated by a number of authors Xingxing Zhang et al [8], however, are interested in characterization of a solar photovoltaic/loop-heat-pipe heat pump water heating system. Wonseok Kim et al [9], performance analysis of hybrid solar-geothermal CO2 heat pump system for residential heating were tested

Pradeep Bansal et al [10] Status of not-in-kind refrigeration technologies for household space conditioning, water heating and food refrigeration. Boonrit Prasartkaew et al [11] interested in a study of the Experimental study on the performance of a solar-biomass hybrid air conditioning system. Anjali at al [12] studied the performance analysis of a solar hybrid air conditioner with waste heat recovery and re-use using evacuated tube collector. Aziz et al [13] studied the potential use of heat energy wasted in condenser ac central for water heating to save energy.

2 Experimental Apparatus

A Panasonic Eco smart Under-Ceiling split room air - conditioner had a rated capacity of 3.51 kW (12,000 Btu/h). The air conditioner had a hermetic type compressor and a thermostatic expansion device. A Helix heat exchanger was installed between its compressor discharge and condenser inlet. A schematic of the system is shown in Figure 1. The hot water storage tank had a capacity of 120 liters, and was located beside the air-conditioner. The system was tested in Machine Performance of room, which was at 27° C. During the test, no hot water was consumed, and the water temperature in the storage tank increased from 27° C to 62.1^O C in 240 minutes. Consider the simple vapor-compression refrigeration cycle that forms part of an air conditioning plants as shown schematically in Figure 1.

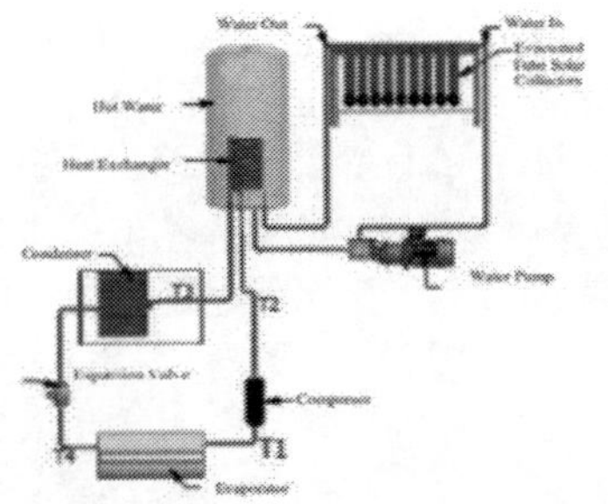

Fig. 1. Schematic of a hybrid air-conditioner

3 Energy Consideration

In order to study the hot water making potential of the air conditioning system and then compare of the COP between common cycle and air conditioning cycle, the important parameters can be calculated from following equation.

Cooling capacity at evaporator (Q_{Evap})

$$Q_{Evap} = Q_e/\dot{m}\,r = h_1 - h_4 \quad (1)$$

Compressor work (WComp)

$$W_{Comp} = wc/\dot{m}\,r = h_2 - h_1 \quad (2)$$

Heating Capacity at Condensor, (Q_{Cond})

$$Q_{Cond} = Qk/\dot{m}\,r = h_2 - h_3 \quad (3)$$

Coefficient of performance (COP_A)

$$COP_A = Q_{COn}/W_{Comp} \quad (4)$$

Coefficient of performance Carnot (COP_C)

$$(COP_C) = T_{Cond}/(T_{Cond} - T_{Eva}) \quad (5)$$

Coefficient of performance Refrigeration, (COP_{Ref})

$$COP_{Ref} = (h_1 - H_4)/(h_2 - h_1) \quad (6)$$

Coefficient of performance Hybrid, (COP_{Hyb})

$$COP_{Hyb} = (Q_{kond} + Q_{eva})/\ W_{comp} \quad (7)$$

4 Result and Discussion

Figure 4.1 shows that the air humidity level of an average of 75.1 percent at the beginning of the air conditioning is operated by room air humidity 90.5% within 120 minutes of operation there is a decrease in air humidity of 22.5% so that air humidity reaches 68%. The initial temperature of the conditioned room was 26.5 ^{o}C after operating for 120 minutes with a temperature drop of 8 ^{o}C resulting in a temperature of 18.5 ^{o}C. Voltage electrical energy is a constant 220 volt compressor and a large amperage of electrical energy used for hybrid air refrigeration machines constant 5 amperes. The initial temperature in a tank with a capacity of 100 liters is 27.5 °C, within 120 minutes the temperature of hot water reaches 45.7 °C.

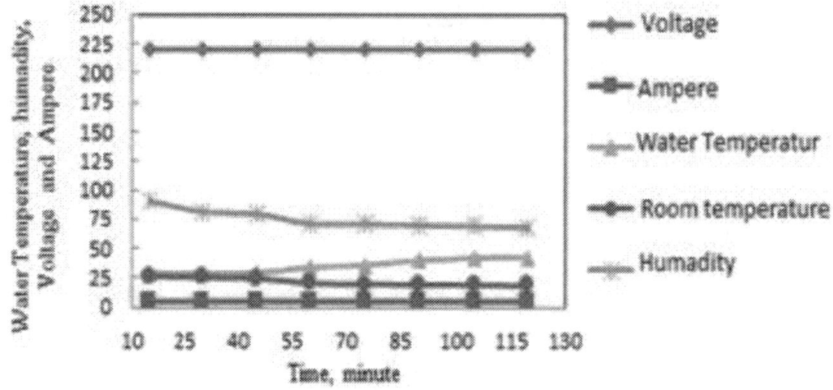

Fig. 4.1 Temperature, Voltage, Amphere vs Time Operation Hybrid Air Conditioning

The variation in the capacity of hot air in the tank which results in the operation of the hybrid air conditioning machine (hybrid AC) within 120 minutes is shown in Figure 4.2 below. The air variation in the tank starts from 20 liters, 35 liters, 40 liters, 50 liters, 60 liters, 70 liters, 85 liters to a maximum of 100 liters of air. The maximum temperature of water produced in the tank is 55.5 OC at a 50 liter tank capacity. In the research of air conditioning machines carried out for 6 hours, the maximum temperature of hot air obtained reached 88.5 OC.

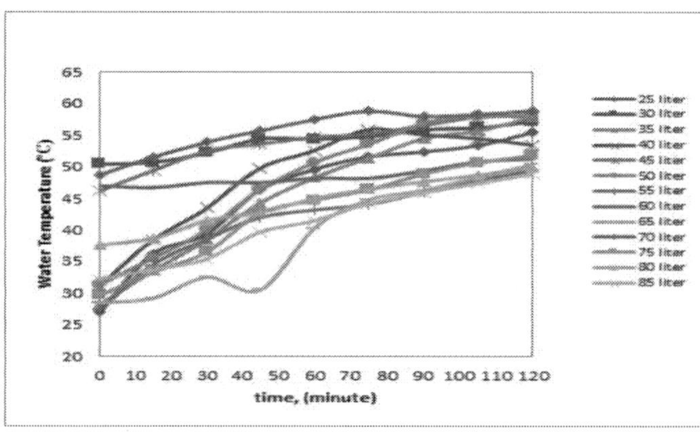

Fig. 4.2 Water Temperature, OC vs Time Operation Hybrid Air Conditioning

5 Conclusion

Hybrid vapor compression air conditioning machines as air conditioners for residential buildings or hotels can be used to produce water for bathing and washing purposes. In this research, the refrigerant used as working fluid R22. In this study using a standard air conditioning machine with a capacity of 12000 BTU, this refrigeration machine was modified by adding a heat exchanger unit. This heat exchanger is installed between the compressor and condenser. Exchanges planned and made are spring shaped helic heat exchangers made from quarter inch copper pipe material. The results of the research that have been done are obtained by the hybridization engine perfusion coefficient, as follows, the average value of the actual hybrid air refraction engine coefficient of performance $(COP)_A$ = 3.35. Average Rankine hybrid air

refraction (COP) machine performance coefficient value R = 8.46. And the average value of the Carnot hybrid air refraction machine (COP) C performance coefficient = 9.85.
The production of hot water using heat from refrigerant heat out the compressor of the air conditioning hybrid engine studied produces hot water with water capacities varying from 20 liters to 100 liters. For the operating time of the hybrid refrigeration machine for 120 minutes, the temperature of hot water was 55.5 ^{O}C. At a tank capacity of 100 liters of water with the operating time of the air conditioning machine for 6 hours the maximum hot water temperature is 88.5 ^{O}C.

I would like to thank Directorate General of Higher, Ministery Research and Higher Education, Republic of Indonesia, DIPA 2017 DIPA- 042.01.1.655462/2017

References

1. Ulku, S., Adsorption heat pumps. Heat Recovery System, pp. 277-284, (1986)
2. Ulku, S., The Netherlands: Martinus Nijkoff Publishers, 1987 Systems and Equipment volume of the ASHRAE Handbook, ASHRAE, Inc., Atlanta, GA, (2004)
3. Chua, K.J., Chou, S.K. and Yang, W.M., Applied Eergy,. **87,** 3611-3624, (2010)
4. Roongutai, P., Sangsuwan, S., Channaen, S. and Lakchan, A., King Mongkut's University of Technology, (1999)
5. Saisanit, P., PattanaIm, N., Hachanont, P., and Chanpueng, W., Research report at Rajamangala University of Technology Thanyaburi, (2008).
6. Saisanit, P., PattanaIm, N., Hachanont, P., and Chanpueng, W., Research report at Rajamangala University of Technology Thanyaburi, (2008).
7. Zhang.X., Zhao.X .,Xu.J., and Yu.X,, Applied Energy **102**, 1229– 1245, (2013)
8. Kim.W, Choi.J, and Cho.H,, Renewable Energy **50** 596-604, (2013).
9. Bansal.P, Vineyard.E, and Abdelaziz.O, International Journal of Sustainable Built Environment **1**, 85–101, (2012)
10. Prasartkaewa.B, S. Kumar B., Renewable Energy **57** 86-93. (2013)
11. S. Nithya, Lalitha Shree, Kiruthika and Krishnaveni, International Journal of Electronics and Communication Engineering and Technology, **8**(2)75–80. (2017)
12. Aruna. S. Ningade and Rudranna Nandihalli, International Journal of Electrical Engineering & Technology (IJEET), **5** 191-198 (2014)

Performance Experimental Study of Solar Still With Reflector To Produce Fresh Water and Salt

Mulyanef [1]*, *Duskiardi*[2], *Kamaruzzaman* Sopian[3], *Kaidir*[4], and *Zulfika* Rahman[5]

[1,2,4,5]Department of Mechanical Engineering, University of Bung Hatta, 25143 Padang, Indonesia
[3] Solar Energy Research Institute, Universiti Kebangsaan Malaysia, 43600 Bangi, Selangor, Malaysia

Abstract. This paper presents a study of experimental performance of solar still using reflector to produce fresh and salt water in the climate of Padang city, Indonesia. The solar collector is equipped with a reflector placed on the top of the cover with slope of 30°. Test results show that using reflector can increase temperature in the basin. Freshwater productivity also increases and speeds up time in producing salt. Average daily freshwater productivity increased by 16.8% during the testing period by adding reflector compared to solar still without reflector

1 Introduction

Solar energy is alternative energy that never runs out, clean and available almost all over the world. Indonesia is a maritime country, which has vast seas and the longest beaches in the world. People who live along the coast and small islands still have difficulty in getting fresh water for their daily needs. This problem can be solved by using simple solar still to process sea water into fresh water. Solar still is an environmentally friendly sea water treatment process and a great technology for wider use. Solar still can also be used to process sea water into salt by heating sea water continuously to dry. There are several ways to improve the performance of solar still, one of which is by adding a reflector to the side of the collector to increase solar radiation into basin liner. Figure 1 shows the simplest device of solar still. Basin has the depth of salt water and has transparent cover of solar radiation, which can block the wavelength radiation emitted by interior surface of collector. Slanting cover is placed for condensing water vapor and flow along the slope into condensate pool. The basin consists of black absorbent plate to maximize absorption of solar radiation, and is isolated on the outer surface to minimize heat loss. Several experimental and theoretical studies have been found in literature that examines the effects of using reflectors to improve performance of solar still. Al-Hayek and Badran [1] found that single-slope solar still with internal reflector was more productive (20%) than solar still double-slope without reflector. Abdallah et al. [2] Experimental results show the thermal performance of traditional single-slope type of solar still can be improved by installing internal reflection mirror. Average freshwater productivity increases by 30% compared to the type of solar still without reflection mirror(reflector). Badran and Al-Tahaineh [3] have found the coupling effect of flat plate solar collector with solar still can increase freshwater productivity by 36%. The deeper the

* Corresponding author: mulyanef@bunghatta.ac.id

water inbasin, the less solar still productivity. Productivity of solar still is found proportional to the intensity of solar radiation.

Mulyanef et al. [4] analyzed the performance of solar still to process sea water into fresh water and salt. Using a flat plate collector with area of 140 cm x 220 cm and sea water volume of 12,000 ml in basin,we obtained 10.076 ml of fresh water and 322 grams of salt. Tanaka and Nakatake [5] analyzed the effect of external reflector inclination angle of basin-type of solar still. Productivity of fresh water from solar still can be increased by inclining external reflector 30 ° in summer. Tanaka [6] conducted a research on solar still using reflector placed at the bottom of flat plate collector. The amount of fresh water produced is increase throughout the year using flat plate base reflector.Khalifa and Ibrahim [7] reported the results of experimental investigations on basin productivity, solar still type with internal and external reflectors tilted at an angle of 0° (vertical), 10°, 20° and 30°. Daily productivity is found to be greater for solar still with larger closing angle. The benefit of vertical external reflectors in winter is found decrease at cover angle exceeding 40°. The most productive solar still in winter is solar still with 20° cover angle, and internal and external reflector tilt angle at 20°. Tanaka and Nakatake [8] concluded that increasing solar radiation absorbed by basin liner would increase daily output of solar still using internal and external reflectors. Furthermore, an increase in daily output above average throughout the year by adding an internal reflector was 22%, which increased to 48% when both reflectors were used at glass cover slope angle of 20 °. A.J.N. Khalifa [9] conducted research on the effects of internal and external reflectors on the performance of basin-type solar stills in various seasons. Daily productivity increased by 35.5% by adding internal and external reflectors. While using internal reflector only increase productivity by 19.9%. Z.M. Omara et.al [10] studied the performance of the stepped solar still using internal and external reflectors. The result shows an increase in temperature on absorbent plate, ambient temperature, water temperature in basin, cover glass temperature compared to solar still without reflector.

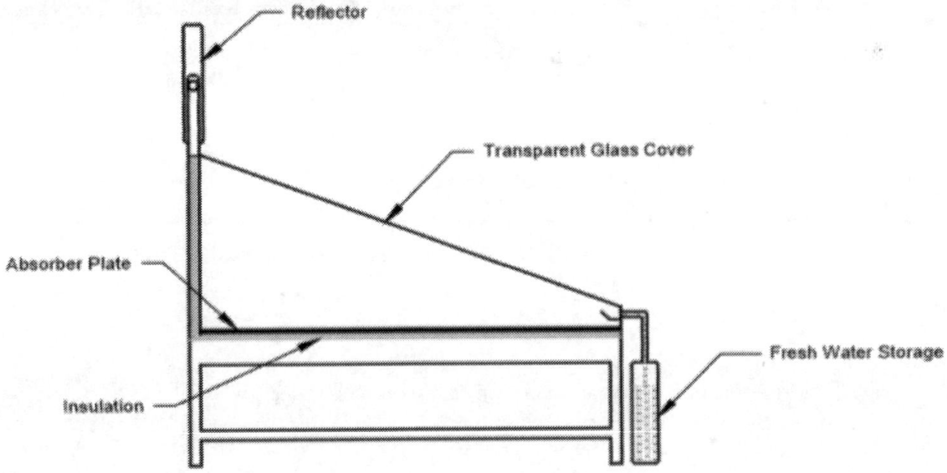

Fig.1. Schematic diagram of the solar still to produce fresh water and salt

2 The experimental setup

Solar still uses flat plate collector, which consists of absorbent plate, cover, insulation, reflector and freshwater reservoir. Absorbent plate with area of 1.96 m2, made of aluminum sheet and given black paint. The bottom and side of the collector are installed with glass wool insulation with a thickness of 2 cm, and in other parts covered by galvanized sheets. At the top of the cover is a reflector with a size of 180 cm x 42 cm. The flat plate collector has a cover with tilt angle of 30^0 made of glass. To drain condensate (fresh water) produced, at the bottom of the cover is placed condensate collection channel. Condensate channels are connected with plastic pipes to storage. Logger datais used to record the absorbent plate temperature,water temperaturein basin, evaporation temperature, outer glass cover temperature, ambient temperature, wind speed and solar intensity. Data is recorded every one hour from 9:00 am to 6:00 pm Padang time, Indonesia. To produce salt, the test is carried out continuously until sea water in the basin dries and salt is harvested. Seawater is reinserted into the basin for subsequent testing and fresh water (condensate) and salt are produced.

3 Results and discussion

Figure 2 shows different temperature variations of solar still with reflectors in the experiment conducted on April 17, 2017. It can be seen that the temperature of absorbent plate is the highest and is followed by evaporation temperature, water temperature in basin, glass temperature, and ambient temperature. The graph shows temperature relationship with time, when the temperature from 08.00 to 13.00 rises and then decreases until 17.00. The temperature of solar still collector with reflector is closely related to solar intensity. The higher the solar intensity, the higher the temperature in collector. This result is similar to the findings of many studies [1], [2], [3], [4].

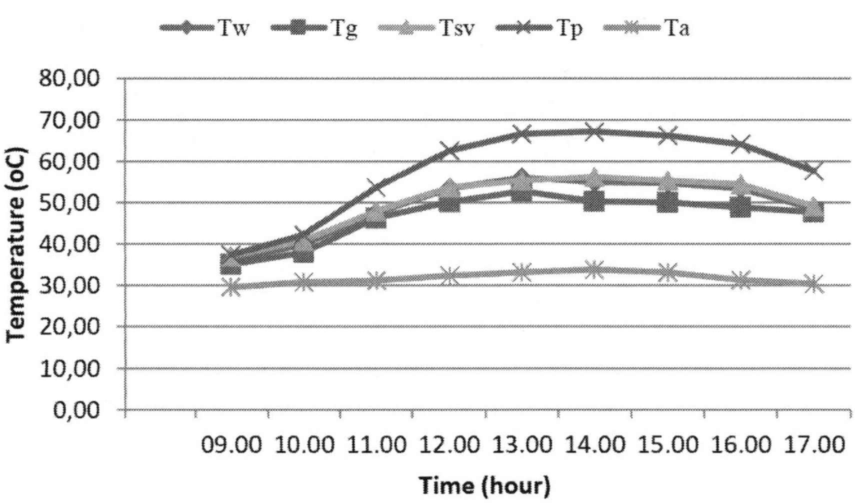

Fig 2. Different temperature variations on solar still with reflector

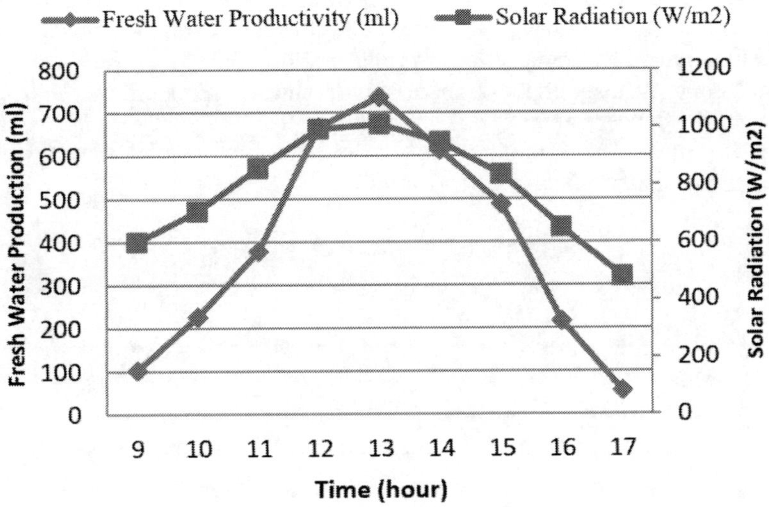

Fig 3. Relationship among freshwater productivity, time, and solar intensity

Figure 3. Showing the relationship among freshwater productivity, time and solar intensity in first test using reflectors (April 17, 2017). At 09.00, fresh water production had begun and continued to rise until 12:30 and after that the productivity of fresh water decreased to 17.00. In the graph, the productivity of fresh water is closely related to solar intensity. The higher the solar intensity the higher the productivity of fresh water produced. From 12 liters of sea water which was put into the collector's basin, 10.944 ml of fresh water and 371 grams of salt were obtained. Test time required is 4 days to produce salt with average solar intensity of 696 W/m^2.

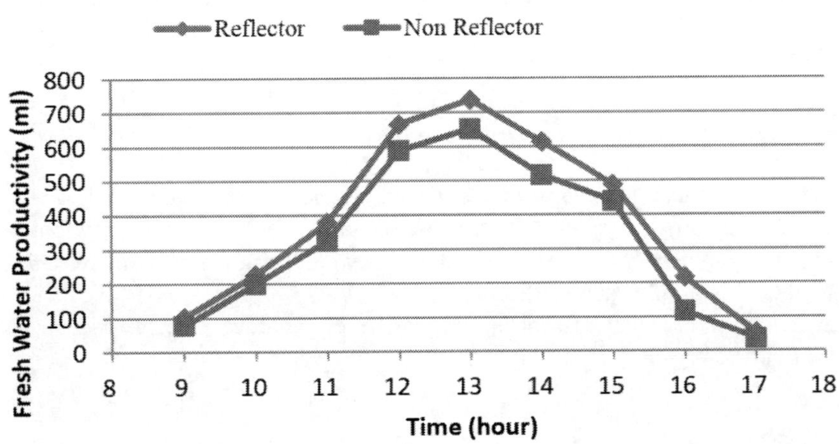

Fig 4. Comparison of freshwater productivity with time for solar still with reflector and non-reflector

Figure 4 shows a comparison of fresh water productivity for solar still using reflector and non-reflector. Freshwater productivity using reflector is higher than the type of non-reflector. Fresh water productivity is very dependent on solar intensity. The higher the solar energy, the more fresh water produced. The effect of solar radiation on condensate productivity is shown in Figure 4. For solar still using reflector with 12 liters of sea water and average solar intensity of 787 W / m2, the amount of fresh water (condensate) on the first day was 3440 ml / day. While the solar still without reflector produces 2945 ml / day of fresh water. Reflectors on solar still have significant effect on increasing freshwater productivity (18.6%), which is associated with the concentration of solar radiation reflected into the basin. H. Tanaka and Y. Nakatake [8] and A.J.N. Khalifa et.al [9] found the same results, that reflector is very influential in increasing productivity of solar still.

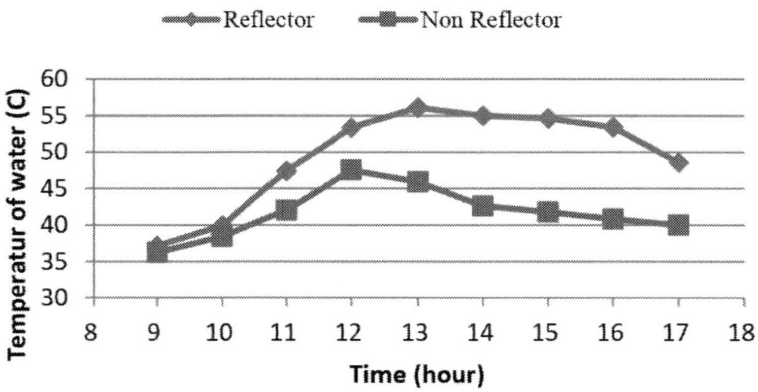

Fig 5. Relationship between water temperature in basin and time

Picture 5 displays relationship between water temperature in basin and time for solar still using reflector and non-reflector. High temperature for both solar still sizes is produced at 1:00 p.m. compared to 09.00-12.00 and 14.00-17.00. This is because solar intensity in the morning is still low, with time elapsed, intensity gets higher and after passing 12.00-13.00solar intensity drops again. For solar still using reflector, water temperature in basin is higher compared to non-reflector. The results of this test are the same as those conducted by Z.M. Omara et.al [10], that the temperature of water in the basin is higher for solar still with reflector.

4 Conclusion

The experimental results that have been carried out can be summarized as follows:

- The use of reflectors in simple solar still basins can increase solar still productivity.
- Productivity of solar still using reflector above collector cover increased 16.8%, compared to solar still without reflector.
- The use of reflector in solar still basin can increase temperature in basin and can accelerate the time to produce salt..

References

1. I. Al-Hayek and O. Badran, The effect of using different designs of solar stills on water distillation, Desalination J, **169** 121–127. (2004)
2. S. Abdallah, O. Badran and M. Abu-Khader, Performance evaluation of a modifi ed design of a single slope solar still, Desalination J, **219** 222–230. (2008)
3. O.O. Badran and H.A. Al-Tahainah, The effect of coupling fl at plate collector on the solar still productivity, Desalination J, **183** 137–142. (2005)

4. Mulyanef, Duskiardi and Hidayat Rahmad, Performance Analysis of Sea Water Solar Still To Produce Fresh Water and Salt, Matec Web of Conferences, Proc. **159**, 02013 (2018)
5. H. Tanaka and Y. Nakatake, Effect of inclination of the external fl at plate refl ector of basin type still in winter, Solar Energy J, **81** 1035–1042. (2007)
6. H. Tanaka, Optimum inclination of still and bottom reflector for tilted wick solar still with flat plate bottom reflector, Desalination and Water Treatment J, **51** 6482–6489. (2013)
7. A.J.N. Khalifa, H.A. Ibrahim, Effect of inclination of the external reflector of simple solar still in winter: An experimental investigation for different cover angles, Desalination J, **264** 129–133. (2010)
8. H. Tanaka and Y. Nakatake, Theoretical analysis of a basin type solar still with internal and external reflectors, Desalination J, **197** 205–216. (2006)
9. A.J.N. Khalifa and A.M and Hamood, H.A Ibrahim, Experimental study on the effect of internal and external reflectors on the performance of basin type solar stills at various seasons, Desalination and Water Treatment J, **27** 313–318. (2011)
10. Z.M. Omara, A.E. Kabeel, M.M. Younes. Enhancing xthestepped solar still performance using internal and external reflectors, Energy Conversion and Management J, **78** 876–881 (2014)

Caracterization and Compressive Strength of Biocomposite Hydroxyapatite-Borosilicate with Hight Temperature Sintering

Burmawi[1,2] ,*Novesar* Jamarun[2*] , *Syukri* Arief[2] , Gunawarman[3]

[1,2]Department of Mechanical Engineering, Bung Hatta University, Padang Indonesia
[2] Department of Chemistry, Andalas University, Padang Indonesia
[3]Department of Mechanical Engineering, Andalas University Padang Indonesia

Abstract. The use of Hydroxyapatite as bone graft material continues to increase. To utilize hydroxyapatite, it is often formed into biocomposite. The formation of hydroxyapatite biocomposite intended to address the fragile hydroxyapatite weaknesses. The added element in the formation of this biocomposite is borosilicate. Mixing of this material is arranged with certain composition using ball miling. To form a test specimen, it was printed with compacting force of 5 KN, 15 KN and 25 KN. Test results Compressive Strength.and XRD showed that addition of borosilicate did not change the compound that is hydroxyapatite, but the intensity of hydroxyapatite tends to decrease with addition of borosilicate amount, this is corroborated by FTIR testing. For visible compressive strength, the maximum value occurs at 75: 25 compositions, ie 45 MPa, 42.9 MPa and 52.3 MPa. This suggests that the addition of borosilicate in biocomposite formation does not alter hydroxyapatite compound and can generally increase the compressive strength and modulus of elasticity to 25% wt% borosilicate addition.

1 Introduction

Hydroxyapatite is a chemical compound $Ca_{10}(PO)_6$ $(OH)_2$ or HAp. It is an attractive and potential material used to repair bone, bone filler and biocomposite as well as for engineering scaffolds [1]. Hydroxyapatite can be obtained by various methods, including: precipitation, sol-gel, hydrothermal and mechanical [2]. Hidroksiapatit of bovine bone has advantages, among others, because it has physical and chemical properties that almost similar to human bones. Hydroxyapatite is also osteoconductivity, biodegradable and non-toxic.

Powder of bovine bone hydroxyapatite still has fragile mechanical properties. Therefore, to improve the mechanical properties, need to be formed into biocomposite material. Biocomposite is combination of two or more ingredients, each having a function as filler and as matrix. In this case hydroxyapatite as filler and borosilicate as a matrix. Silica can substitute phosphate compound in HAp, considering the average length of the bond between Si-O and P-O are 1.62 Å and 1.51 Å. Therefore, this substitution is likely to occur in PO4 3- with SiO44- [3]. The addition of borax material generates better inhibition of cell

* Corresponding author: novesar62@yahoo.com

development in static conditions, making it excellent for bone growth [4. 5]. Sintering at a temperature of 1000°C to 1100°C, forming a superplastic material. This happens because in sintering process there is an increase in granular density, but the strain will decrease [6].

In previous study [7], production of borosilicate-hydroxyapatite biocomposite with mold pressure of 70 kg/cm^2, borosilicate composition of 5, 10 and 15 wt% and sintering temperature of 1000 °C also, providing maximum compressive strength of 3.48 MPa. In this study, the formation of the borosilicate-hydroxyapatite biocomposite besides analyzing the characteristics occurring by the addition of borosilicate compositions and compaction forces of 5 KN, 15 KN and 25 KN, it is also analyzed mechanical properties biocomposite material.

2 Material and Methods

This study was started from mixing of bovine hydroxyapatite material with silica material and borax acid with hydroxyapatite composition of 70, 75, 80, 85 and 90% remaining weight is added with silica and borax acid. The mixing process is using ball miling with 200 rpm rotation speed, 10 mm diameter ball as much as 25 pieces and performed for 60 minutes. To form a specimen, printing is conducted with compression force of 5 KN, 15 KN and 25 KN and holding time of 10 minutes with a size of 10 mm and a height of 10 mm. The sintering process is carried out at temperature of 1000 ° C for 3 hours. To check biocomposite characteristic, XRD test was performed using X-Ray Diffraction (XRD) with XRD XPERT Powder PW 30/40 and Fourier Transform Infra Red (FTIR) using FTIR Spectrometer Frontier with Perkin Elmer brand. The mechanical property test is a compressive strength test. Test of compressive strength is performed using universal press test machine with standart ASTM D3410.

3 Results and discussion

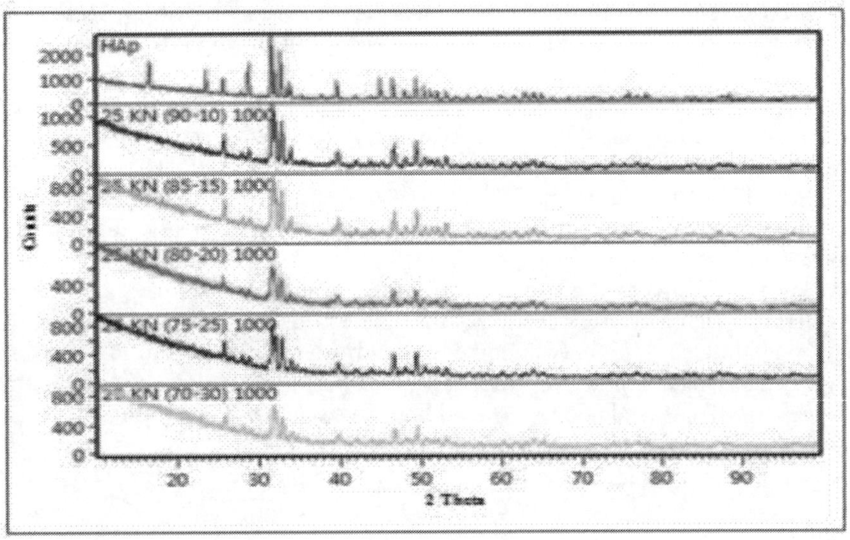

Fig 1. XRD test results with a variety of Borosilicate compositions

XRD testing is conducted to determine the compounds formed on the specimen. In Figure 1, it is seen that in pure HAp material, there is still PO4 NH4 pollutant. After adding borosilicate material with composition ranging from 10% and sintering treatment at a temperature of 1000 ° C, it is seen that the pollutant on hydroxyapatite begin to disappear, and after examination with JCPDS, it turns out the compound formed in biocomposite Hydroxyapatite-Borosilicate is in the form of hydroxyapatite compound as well. This is seen at diffraction angle of 2θ, it is not very different ie ranging from 31.7 °, 32.2 ° and 32.9 ° [8]. Similarly, other diffraction peaks formed the same chemistry of hydroxyapatite and this is an easy substitution of SiO2 in hydroxyapatite structure. The XRD profile was compared to bovine hydroxyapatite with HAp-Borosilicate biocomposite, and according to JCPDS 9-432, it showed similar hydroxyapatite [9]. However from this test, the more borosilicate, the less the intensity of hydroxyapatite compounds. This is affected by borosilicate intensity in bovine hydroxyapatite.

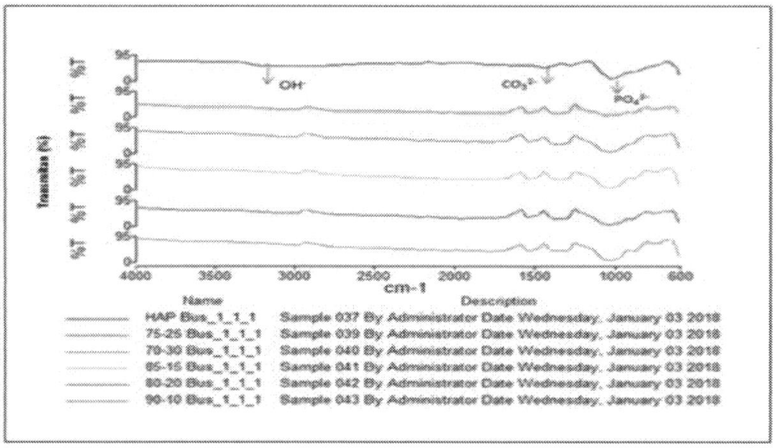

Fig. 2. FTIR test results with a variety of borosilicate compositions

The FTIR test is useful to know the functional group of compounds present in biocomposite of HAp-borosilicate. From infrared spectrum of hydroxyapatite, we saw functional groups of CaO, PO_4 and OH^-. The wavelength formed gives information that the functional group of PO_4 is formed at wavelength of 1015 cm-1, 1061 cm-1, 1026 cm-1 and 1030 cm-1. The OH^- functional group is seen in almost all compositions at the wavelengths of 2214 cm-1, 2026 cm-1, and 2987 cm-1. Wavelengths of 1267 cm-1, 1264 cm-1 and 1262 cm-1 prove that SiO_2 is present in this biocomposite material [10]. Meanwhile, the wavelength of 1528 cm-1 is formed. This indicates that a CO_3-functional group [11] is formed as the effect of borosilicate addition on the formation of borosilicate hydroxyapatite-biocomposite

Compressive strength is used to determine the ability of biocomposite materials in accepting vertical loading to cross-section of the object. This is related to material application on human bone graft. [12] We have conducted a study on hydroxyapatite-PCL with addition of 10% w/v PCl, giving compressive strength yield of 4.20 ± 0.05 MPa and elasticity modulus of 6.10 ± 0.03 MPa. In addition, [13], adding silica to tricalcium phosphate provides compressive strength value of about 2 MPa.

In the study [14], PMMA-cement provides high compressive strength above 100 MPa, but its press elasticity modulus is quite low at around 2.5 Gpa. This means the material is quite brittle. Likewise, in the study [15], HA / PSSM composites with PSSM compositions of 5, 10 and 15%, had tensile strength value of 16.20 ± 1.73 MPa to 23.96 ± 2.56 MPa and modulus of elasticity of 626.96 to 1026 MPa. When compared with the results of this study, it is noticed that a given mold compaction force varying 5 KN, 15 KN and 25 KN gives promising results. The graph show the borosilicate composition given to hydroxyapatite has a significant effect on the compressive strength of biocomposite, in which the borosilicate functioned as a matrix with hydroxyapatite filler. Figures 3, 4 and 5 show that generally the more borosilicate composition, the higher compressive strength of the material. At a pressurized force of 5 KN, the maximum compressive strength of HAp-borosilicate biocomposite material is 45 MPa at 75:25 compositions. The compacting force of 15 KN gives maximum compressive strength of 42.9 MPa also in the 75: 25 compositions. The compacting force of 25 KN gives maximum compressive strength of 52.3 MPa in 75:25 compositions

In this research, the result of this compressive strength of biocomposite is higher than PCL-hydroxyapatite composite, hydroxyapatite-PSSM as well as from Composite Tricalcium phosphate-silica but lower than Hydroxyapatite-PMMA biocomposite. It means that the material is more resilient than hydroxyapatite-PMMA. And when compared to previous studies [7], with the same material but different processes, the results obtained in this study are much higher. This can be seen from the compressive strength value obtained by 3.48 MPa versus 32 MPa. According to the results of others' research [16] for cortical bone application with compressive strength of 30 - 160 MPa, it turns out that the compressive strength produced has met the range listed in the table and can be used.

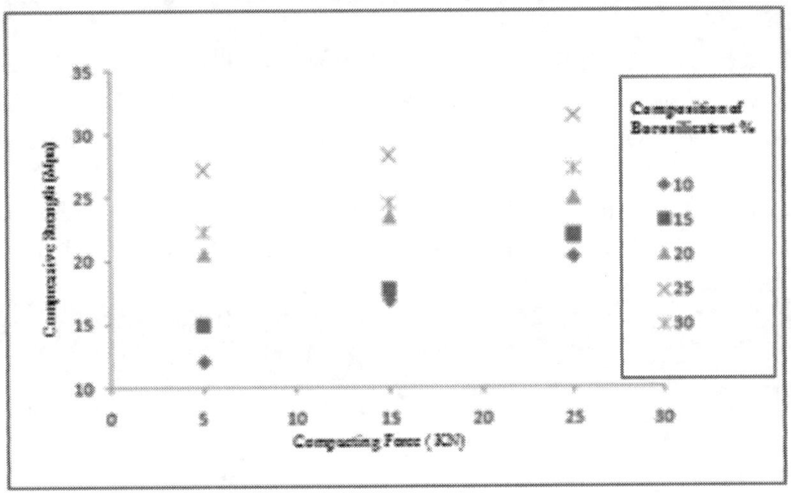

Fig. 3. HAp-Borosilicate biocomposite Compressive Strength with Compacting force of 5 KN, 15 KN and 25 KN

4 Conclusion

1. Addition of borosilicate to the formation of biocompositeHAp-borosilicate generally does not change the chemical formula of hydroxyapatite.
2. Maximum compressive strength is obtained in Borosilicate composition of 25 wt% for compacting force styles of 5 KN, 15 KN and 25 KN at sintering temperature of 1000 ° C with values of 32 MPa.

Thanks to the Ministry of Research and Technology - Higher Education, Republic of Indonesia, which has helped finance this research

References

1. P. Ducheyne. S. Radin , L King. Journal Dissolution,J.Biomed.Mater.27,25- 34(1993)

2. N. Jamarun, Z.Azharman, S.Arief, T.P.Sari, A.Asril, S.Elfina, Rasayan Journal Chemical, 8, 1, 133-137 (2015)
3. A. El Yacoubi. A. Massit, M.Fathi, B.Chafik, El Idris , K. Yamni, IOSRJAC, 7, 11,24-29 (2014)
4. R.F. Brown. R.F, M.N.Rahman, Agatha B, Dwilewicz, W. Huang, Wiley Periodical, Inc. (2008)
5. K. Itatani. K. Tsuchiya, Y. Sakka, Davies. I.J, S. Koda, 31. 2641-2648 (2011).
6. S. Ramesh., K.L.Aw, R.Tolouei, M.Amiriyan, C.Y.Tan, M.Hamdi, J.Purbolaksono, M.A.Hasan, W.D.Teng, Ceramic International, 39, 111-119 (2013)
7. Burmawi, N. Jamarun, S.Arief and Gunawarman, Oriental Journal Of Chemistry, 33, 2, 920-924 (2017)
8. S. Amjuvan. S, H. Aspersa, Bull Material Science, 35, 6, 1031-1038 (2012)
9. E. Kusrini, A..R, Pudjiastuti, S. Astutiningsih and S. Harjanto, ICBEE'2012 (2012).
10. N.Fariba., M.A.S. Sadjadi, S.J.Fateami, M.K.Mobarakeh and R.M.Afshar, 32 3, 1639-1647(2016).
11. A.Khelendra, G Singh, D. Puri, S. Prakash, biomedical Application, 10, 727-734(2011)
12. K.J.U.Sharon and Nazmin S, Contact angle, ARPN, 11, 13686-13691, (2016.)
13. Lereley. M.A, R.G. Corrodeguos, L Alberto dos Santos, Material Research, 14, 475-482 (2011).
14. K.Serbetqi., F.Korkusuz, N.Hasirci, Turk J Med Sci, 30, 543-549 (2000).
15. Weng Hong Tham, M.U. Wahit, M.R.A. Kadir and T.W. Wong, , SJST, ,35, 1, 57-61. (2013).

MATEC Web of Conferences **248**, 01006 (2018) https://doi.org/10.1051/matecconf/201824801006
ESTIC 2018

Peculiarities of Injection Molding Conducting Composites

Yovial Mahyoedin[1*], *Jaafar* Sahari[2], *Andanastuti* Mukhtar[2], *Norhamidi* Mohammad[2], Iqbal[1]

[1]Department of Mechanical Engineering, Bung Hatta University, Indonesia
[2]Department of Mechanical and Material Engineering, Universiti Kebangsaan Malaysia, Malaysia.

Abstract. The investigations in this study focused on the characteristic of feedstock in an effort to understand the mechanism of injection molded in composite material. A composite, which has 75% wt. filler, consist of graphite (G), carbon black (CB) and polypropylene copolymer (PP). Twin-screw co-rotating extruder used for mixing materials. The conductivity of the molded parts measured using a four-point probe test procedure. The results showed that the injection molding conducting composites, which aggregated into larger clusters, tended to disperse unevenly into the PP, resulting in fewer particle-particle contacts and, consequently, a lower-conductivity composite in some part of the molded.

1 Introduction

As it was well explained by percolation theory, the electrical properties of polymer composite mainly depend on the dispersion condition of filler particles, particle size and aggregate structure [1]. Whilst, injection molding is one of the preferred methods for manufacturing parts of thermoplastic polymers. However, the main difficulty of using injection molding is its inability to handle materials with poor flow characteristics since the highest the viscosity the more difficult materials to be injected [2]. One characteristic for injection-molded parts is an anisotropic character in structure and, thus, properties due to molecular chains and filler particles orientation resulting from high processing velocities. If the melt is cooled down rapidly to ensure short cycle times, these orientations are frozen and remain in the molded part. In the case of conductive filler particles having a high aspect ratio, such as CNT, orientations lead to an impairment of network formation and, thus, to a shift of the percolation threshold towards higher filler contents [3, 4].
This paper reports the injection molding study on the characteristics of polymer composites based on graphite and carbon black as fillers in order to produce an electrical conducting composite material.

2 Material and Methods

An granular polypropylene (PP) random copolymer grade TitanPro SM 688 from Titan Petchem Sdn. Bhd, was selected as matrix phase (binder) for the different composite

* Corresponding author : jmahyoedin@yahoo.com

formulations. Two type of conductive filler; supplied by Asbury Graphite Mills, Inc, were investigated in this study, i.e. natural graphite (G) type Asbury 3243 and carbon black (CB) type Asbury 5303. Table 1 illustrated the properties of these materials.

Table 1. Material properties.

Material	Density ρ (g/cm3)	Conductivity σ (S/cm)	Melting Point (°C)	Particle size D_{50}
Polypropylene	0.905	8.50E-14	~170	
Carbon Black	0.918	341		11.5 μm
Graphite	1.28	33.4		56.4 μm

The sampels for this study were formed using Injection Molding machine type Arburg Allrounder 850-210-320 D. Injection nozzle with diameter of 3 mm used to ensure smooth flow of material into the cavity.

The electrical conductivity characteristics of injection-molded composite are significantly affected by the distribution of filled conducting fibers. Moreover, the particles distribution is influenced by the injection molding processing parameters. Hence, by properly controlling molding process conditions, the filler conductive polymer can be arranged to have a higher electrical conductivity value with the aim of bipolar plate. Therefore, Taguchi optimization technique was applied to identify the optimal settings of control factors which would acquire the greatest electrical conductivity value of plates.

The injection molding process of conductive composite were conducted, after optimizations on the compound material, the injection molding process (e.g. the process parameters) and the molding tool using taguchi method. Plates of 100mm×50mm with thickness 2.0 mm were produced successfully by injection molding. The process cycle time for each plate only takes between 20 and 50 s.

Finally, the electrical conductivity of plates in the in-plane direction was measuring using four point probe type Jandel Sensitive Digital Voltmeter.

3 Results and Discussion

3.1 Filler Material.

Mixture materials to injection molding shall have the uniformity and have the appropriate of electrical conductivity as well. According to percolation study, uniformity is also can improve the electrical conductivity of the mixture materials.
A combination of fine and coarse filler particles lead to higher electrical conductivity which might be attributed to the higher maximum packing fraction of the filler powders [5]. Table 2 showed distribution of particle size and particles shape of both fillers. Large values of the distribution slope Sw correspond to narrow particle size distributions and small value corresponds to broad distributions.

Table 2. Particle size distribution and cumulative particle sizes

Filler powder	Shape particle	D_{10} (mm)	D_{50} (mm) Average size	D_{90} (mm)	Sw
Natural graphite (NG 3243)	flakes	18.586	56.435	127.089	3.066
Carbon black (CB 5303)	particles	3.922	11.585	66.225	2.085

The result showed that both filler particles are suitable for injection molding process, since S_W values of both material is quite small and close enough to 2 (the easier filler powders to mold exhibit S_W value of 2). Large values of distribution slope S_W, correspond to narrow particle size distribution and small values correspond to broad distribution [6]. This result also showed that graphite powder seem more difficult to mold when compare to carbon black. From the shape particles view, graphite also seems more difficult to mold because of flakes-shaped. It is well known that spherical particles tend to easier molding and afford of high solids loading.

3.2 Injection Molding Process

An injection-molded plate of 100mm×50mm was divided into several areas for measuring electrical conductivity. Figure 1 shows the measurement areas. The samples tested were derived from the results of taguchi optimization with a composition of 20% PP-80% G/CB.

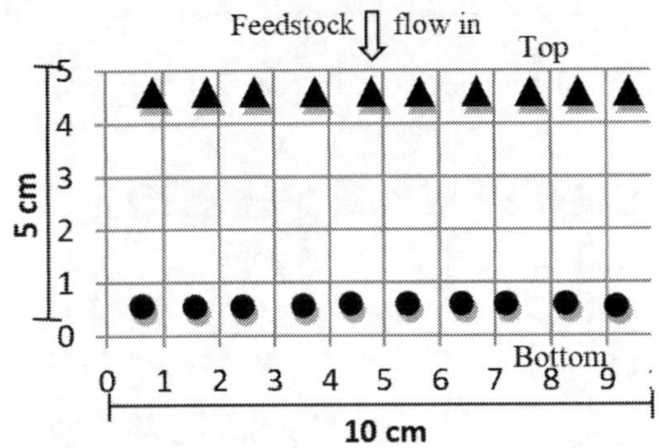

Fig. 1. a) Injection molding sample. b) Electrical conductivity measurements area

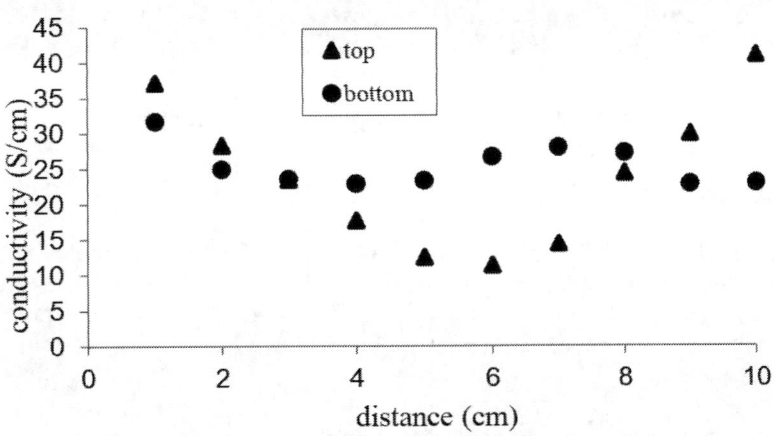

Fig. 2. Electrical conductivity of injection molding sample 20%PP:80%G/CB (90% Gnat + 10% CB)

Figure 2 shows the measured electrical conductivity on the surface of the sample with a composition of 20% PP-80% G/CB. Results show that the conductive filler tended to disperse more evenly near the wall of cavity, thereby resulting in many particle-particle contacts and, consequently, a higher-conductivity polymer composite. Whereas, the lower conductivity value was found in the center part of the sample. This suggest that there is significant anisotropy was displayed in the PP/G/CB composites (i.e. particles are randomly oriented in the matrix) thus providing a non-uniform electrical conductivity value within the plate.

However, processing of composite at low filler loading will result in filler-matrix separation when subjected to high pressure during molding. This indicates at high-filler loading the separation does not occur, thus caused non-uniform filler within the plate. Further, it seems that the flow of filler material into the cavity was not completely proper and, consequently, has great significant in predicting the electrical conductivity of final parts. Non-uniform distribution of filler also causes the electrical conductivity of the material becomes non-uniform. Reduced the amount of filler in the middle of the plate causes the electrical conductivity of materials in part also decrease as shown in Figure 3.

The viscosity of the feedstock was influenced greatly by the shear strain rate and temperature of the feedstock. The viscosity of the dispersed system filled with fine particles is also strong influenced by the degree of agglomeration of the particle in the system. As the concentration of the particles increases, in general, the ability to form agglomerates increases, thus the electrical conductivity also increase. Meanwhile, the forming of agglomerates increases the effective concentration on viscosity of the system, because the volume fraction of the liquid occluded by the agglomerates, that is, the liquid volume that is not responsible for the viscosity, increases as the size of agglomerates increases

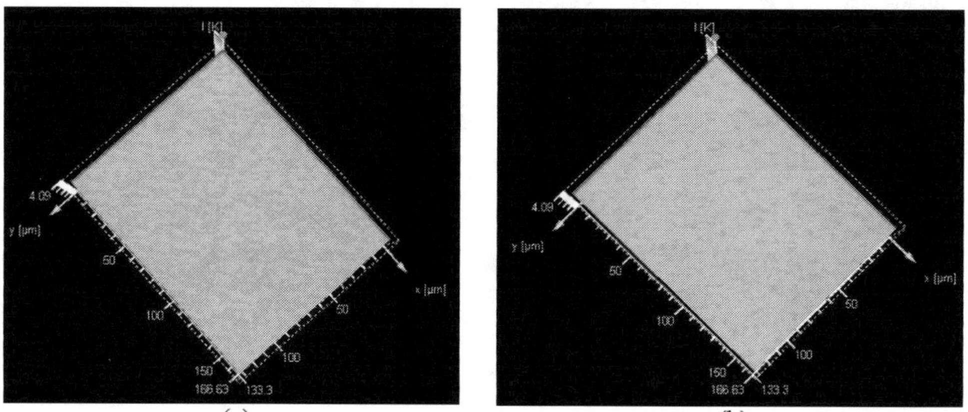

(a) (b)

Fig.3. infrared micrographs of filler material. (a) near the wall. (b) center part

4 Conclusion

In this paper, we studied the injection molding process for the production of electrically conductive composite material. High viscosity, in line with higher filler contents, which results in poor flowing of materials, are not the only drawback to using injection molding in producing polymer composite. Non uniform electrical conductivity values of final parts are common outcomes as well.

It has been stressed that homogeneous composite, which contains the maximum filler loading and without sacrificing the ease of molding, should be used in the injection molding mixing step as this will maximize electrical conductivity. Hence, from this study, composite with filler loading of 20 wt% CB was the most suitable filler loading.

The authors would like to thank The Ministry of Higher Education for their financial support of this work through Fundamental Research Grants under UKM-RF-02-FRGS 650001-2007

References

1. Jing X. Z. W., Lan L. Journal of Mat Sci Letters, **19:**377– 379 (2000)
2. Heinzel, A., Mahlendorf, F., Niemzig, O., Kreuz, C., J. Power Source, **131:** 35-40 (2004).
3. Weber M, Kamal MR. Polymer Compos. **18**(6):711–25 (1997)
4. T. Villmow, S Pegel, Petra P., Udo W. Composites Science and Technology. **68** 777–789 (2008)
5. Kalyon, D.M., Birinci, E., Yazici, R., Karuv, B. & Walsh, S. Poly Eng and Sci. **42**(7): 1606. (2002)
6. Yovial M, Jaafar S, Andanastuti M, Norhamidi M., Advanced Materials Research. **233** – 235 3057 (2011).

Effect of graphite sizes and carbon black content on flowability of the injection molded conductive composite material

Yovial Mahyoedin[1*], *Jaafar* Sahari[2], *Andanastuti* Mukhtar[2], *Norhamidi* Mohammad[2] and Suryadimal[1]

[1]Dept of Mechanical Engineering, Bung Hatta University, Padang, Indonesia
[2]Dept. of Mechanic and Materials Engineering, Faculty of Engineering & Built Environment Universiti Kebangsaan Malaysia, 43600 UKM Bangi, Malaysia

Abstract. This study is to investigate the flowability of the injection molded conductive composite material containing filler content from 70 up to 80 wt% by using spiral mold. Several moulding compounds, containing polypropylene (PP) as a matrix and graphite (G) and carbon black (CB) as conductive fillers prepared by melt compounding using twin-screw extruder. Carbon black is added as much as 10% and 20% respectively, in order to improve the electrical conductivity of composite material. Results show that the flowability of injection molded conductive composite material decreases with decreasing graphite size and with increasing filler (graphite and carbon black) content. It was found that composites containing G with particle size distribution (≤100 μm) entirely exhibits a relatively higher flowability, in the range of 10.67–6.21 cm, compared to the small size (25-60 μm), in which flowability is in the range of 5.23–3.37 cm. An attempt to combine carbon black as second filler with the PP and G found that composites containing CB showed decreased flowability of the injection molded conductive composite material, especially when the electrical conductivity formed through the resin. Results indicate that the flowability of the injection molded conductive composite material is an important design parameter to fabricate cost-effective, large, or thin composite bipolar plates.

1 Introduction

It has already known that the electrical properties of polymer composite material mainly depend on the dispersion condition of filler particles, particle size and aggregate structure [1]. High filler loading (>60 wt%) is needed to provide the required electrical conductivity [2]. Whilst, many authors identified that injection molding is the most promising technology for manufacturing parts of conducting composite [3], because it has multiple advantages over other methods of plastic molding. Not only is plastic injection molding simpler and more reliable, it is also extremely efficient. However, thermosetting polymers are generally more expensive to mold than thermoplastics and represent only about 5% of plastics processing [4].

* Corresponding author: jmahyoedin@yahoo.com

Whilst, the major difficulty in using injection molding is its inability to handle materials with poor flow characteristics [5]. Some literatures have recently compared the effect of graphite size [6] and filler on composite properties.

This study investigates the flowability effect of PP/G/CB material with different graphite contents, graphite particle sizes and the carbon black contents on the formability of the composite bipolar plate

2 Experimental

Graphite (G) type Asbury 3243 varied of 25-60 μm and ≥100 μm, and carbon black (CB) type Asbury 5303 with particle size of 30 nm from Asbury Graphite Mills Inc. used as conductive fillers. Polypropylene (PP) random copolymer grade TitanPro SM 688, in the form of granular from Titan Petchem Sdn. Bhd, selected as matrix phase for the different composite formulations.

CB/G mixture has a composition ratio, respectively were 20/80 (20% CB and 80% G) and 10/90 (10% CB and 90% G). This mixture powder were used as the conductive phase in the range of 60-80 wt.% and then mixed with PP using ThermoHaake Rheomix Polylab type 557-5501 co–rotating intermeshing twin-screw extruder with temperature of 200°C. The extrudate immediately cooled in water bath then pelletized using Centurion Guard Master. Approximately 1.5 Kg of each formulation being collected. All types of sample are tabulated in Table 1.

Tabel 1. Name code of specimens and their contents

No	Names	Content		
		Graphite size (μm)	Carbon black	Total Filler
1	CC1	100% (25-60 μm)	Non	60-80%
2	CC2	100% ≥100 μm)	Non	60-80%
3	CC3	90% (25-60 μm)	10% wt	60-80%
4	CC4	80% (≥100 μm)	20% wt	60-80%
5	CC5	90% (25-60 μm)	10% wt	60-80%
6	CC6	80% (≥100 μm)	20% wt	60-80%

Fig 1. a) Spiral mold and b) Resulting samples, generated from a variety of compositions

Test specimens formed using an Arburg Allrounder 28 ton model 221-55-250 injection molding with a 22-mm diameter single screw. The flowability was determined by means of injection molding of the constant weight of feedstcok material injected into a spiral channel mold. Spiral mold has a groove circle with a diameter of 3 mm and a depth of 1.5 mm. Figure.

1 shows the spiral mold and the actually forming shapes generated from a variety of compositions.

Injection molding process parameters are taken as follow: injection pressure varies from 100 to 150 Bar, whilst a mold temperature (30°C) and screw speed (200 rpm) were kept constant. The cooling time in the mold was 110 second, and the holding time was 8 s. The injection molded composites were produced with loadings of 70 and 80 wt% filler (i.e. mixture of G and CB). The zone temperatures varied according to the PP base composite, progressing from the solids conveying zone (220°C) to the nozzle zone (230°C).

3 Results and discussion

3.1 The effect of graphite size on the flowability of material

Figure 2 plots the flowability of conducting composite material as a function of graphite particle sizes. The flowability trend is almost linear and decreases continuously with increasing graphite content, whereas, for the size of the graphite particles shows that the smaller the particles the shorter the length of the spiral mold.

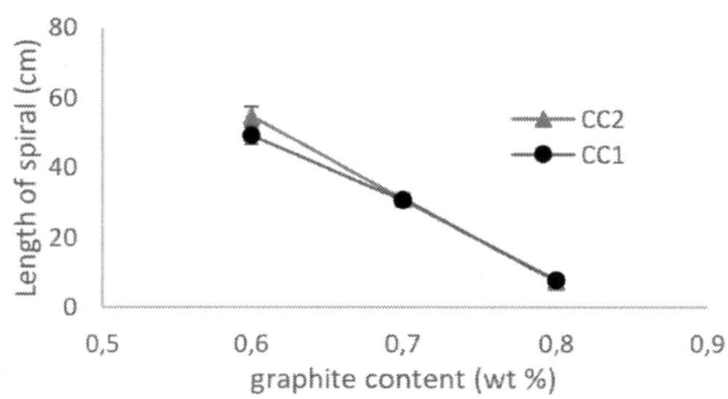

Fig 2. The flowability of conductive composite material PP/G/CB

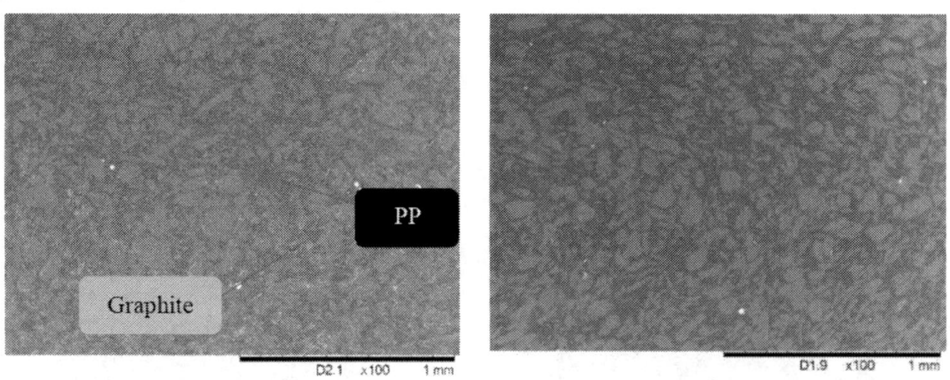

(a) (b)

Fig 3. Optical micrographs of conductive composite PP/G/CB, with different graphite sizes (a) 70% wt of ≥100 μm, (b) 70% wt of 25-60 μm.

Figure 3 (a) and (b) also shown a different size of 70% wt graphite particles, i.e 25-60 μm and ≥100 μm, respectively. At low graphite content, the particles will separate properly in the media because of the potential power of their rejection. In this case, the interaction between the particles is small and since the PP/G conductive material contain higher PP content, its serve as lubricant to decrease friction among graphite particles, leading decreased viscosity and consequently, affecting the length of spiral. At high powder loading, the particles will come into contact with one another and tend to agglomerate and to improve the interaction between particles [13]. The finer the powder particles, the easier the formation of agglomerates, the more difficult the occurrence distribution (or dispersion) of the particles. Therefore, the thermoplastic PP surrounding graphite particles cannot penetrate easily; hence, the conductive composite material becomes too sticky to flow smoothly, consequently affecting the length of spiral.

3.2 The effect of carbonblack on the flowability of material

The use of graphite powder as main electrical conductivity material and carbon black as second filler was found capable of forming a conductive network between the filler powders [7]. Carbon black also has a double percolation that can increase the electrical conductivity of the material [8]. Several researches have found that the rheological properties of CB composite are affected by CB dispersion throughout the thermoplastic. Viscosity of material increases with increasing CB content. However, the porous structure of carbon black can degrade the mechanical properties. Therefore the powder loading of carbon black as filler in a polymer matrix is limited [9].

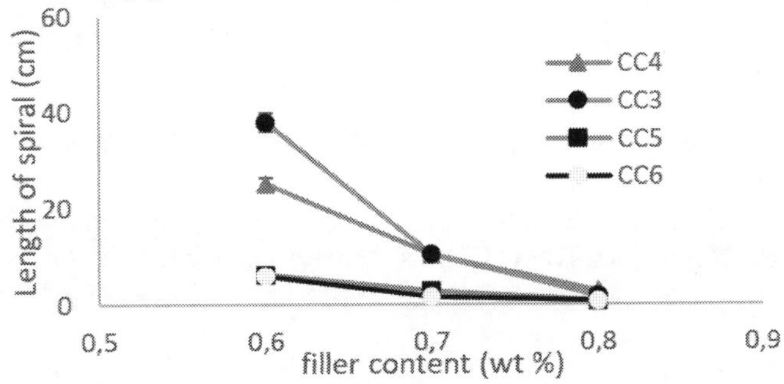

Fig. 4. The flowability of conductive composite material PP/G/CB, with different graphite particle sizes, different carbon black content and various composition of filler

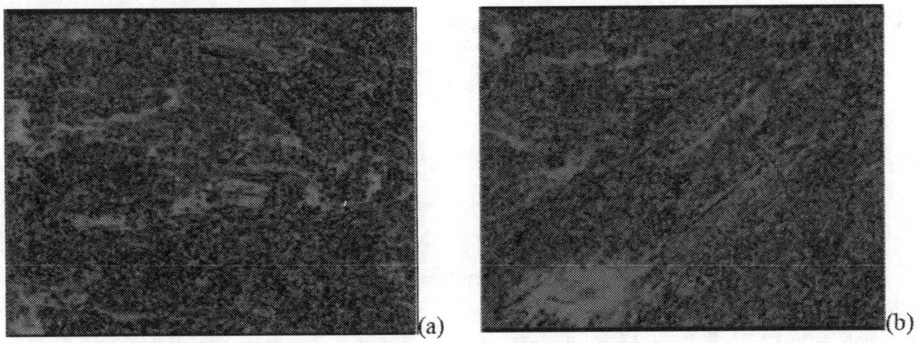

Fig 5. Infrared micrographs of conductive composite PP/G/CB, with different carbon black content. (a) 10% wt CB, (b) 20% wt CB

Fig. 4 illustrates the flowability of conductive composite material PP/G/CB with 10 and 20% wt CB as a function of filler content. The figure shows that flowability of conductive composite material decreases with increasing CB content, i.e. from 25.16 to 1.6 cm for 20% wt CB and from 37.9 to 2.7 for 10% wt CB. However, the difference flowability of conductive composite material looks not significant at very high filler content (80 wt% of filler). This implies that critical powder loading of PP/G/CB material is likely to be in the range of 80% wt of filler. However, an inhomogeneous distribution of the PP throughout the injection molded PP/G/CB samples is probably the main reason. The low PP content of the PP/G/CB composite material confirmed by infrared micrographs observations (Figure 5). One should consider some improvements in order to carry out the processing of conductive material at the very higher filler content.

4 Conclusions

The present researches have used various graphite particle sizes, graphite contents, and carbon black contents to analyze the effect of flowability on the production of conductive composite material. In this paper, experimental study of the flowability in injection molding machine is presented in term of length of a spiral molding. Although the production of polymer composite plates by means of injection molding can be troublesome, there has been success in producing plates, depending on the fillers and additives. Results indicate that the flowability of material affected by graphite particle content and size. With decreasing graphite particle size, accumulation density increases and causes the particles will come into contact with one another and tend to agglomerate, thus the thermoplastic cannot penetrate between graphite particles to form a continuous flow channel. The flowability of material also decreases with adding carbon black content. The finer the powder particles, the easier the formation of agglomerates, the more difficult the particles to flow. It has been stressed that homogeneous composite, which contains the maximum filler loading and without sacrificing the ease of molding, should be used in the injection molding mixing step as this will maximize electrical conductivity. Hence, from this study, conductive composite with amount of a second filler of 20% wt CB was the most suitable filler loading

References

1. Yovial M., Sahari J., A., Mukhtar, Norhamidi M.: Advanced Materials Research . **233** – 235 3057 (2011)
2. Minkook K., J. Choe, Jun W. L., Dai G. L. Composite Structures. **132**, P. 1122-1128 (2015).
3. Boothroyd, G., Dewhurst, P., Knight, W. in: Product design for manufacture and assembly. Marcel Dekker. (2002)
4. Heinzel, A., Mahlendorf, F., Niemzig, O., Kreuz, C. J. Power Sources, **131** 35-40 (2004).
5. H.C. Kuan, C.C.M. Ma, K.H. Chen, S.M. Chen: J. Power Sources. **134** 7 (2004).
6. Dweiri, R. & Sahari J. Composites Science and Technology. **68**:1679–1687 (2008)
7. Tchoudakov, R., M. Narkis, and A. Siegmann : Polymer Engineering & Science **44**-3:528-540 (2004)
8. Lee, Shuo-Jen, Jian-Jang Lai, and Ching-Han Huang. J. Power Sources **145**-2 362-368 (2005):
9. Barrie CL, Griffiths PC, Abbott RJ, Grillo I, Kudryashov E, Smyth C. : J. Colloid Interface Sci. **272**(1):210-7. (2004).

Experimental Investigation of Vibration Response of a Flexible Coupler In a Four Bar Mechanism Due to Varying Crank Length and Crank Speed

Rizky Arman[1,*], *Andi* Isra Mahyuddin[2], *Wenny* Marthiana[1], Iman Satria[1]

[1]Mechanical Engineering Department, Bung Hatta University, Padang, Indonesia
[2]Mechanical Engineering Department, Bandung Institute of Technology, Bandung, Indonesia

Abstract. Effort to increase productivity calls for increasingly higher operating speed. Consequently, many part of a machinery are made as light as possible to avoid the detrimental inertial effects. This condition means several of the elements are no longer rigid and tend to experience elastic deformation in operation. The existence of the deformation yields element possessing position dependent dynamic parameters. Mechanisms with time-varying parameter have a phenomenon known as parametric instability where the mechanism could becomes dynamically unstable at several frequency bands, even at relatively low excitation frequency. This research investigates the dynamic behaviour of a flexible coupler on an experimental four-bar mechanism setup. The study examines the coupler vibration response due to varying crank length and crank speed.

1 Introduction

In engineering, dynamic analysis of machines or systems with many components, which all the components are considered rigid, it can be done easily. However, in an effort to increase the productivity, the machines with high speed are needed. Therefore, some engine components are made as light as possible so that the effect of inertia is reduced. The impact of this reduction in inertial forces resulting structure is more easily deformed.

Consequently, the stiffness of the rod that changed to the position, then the dynamic behaviours of elastic rod was governed by nonlinear partial differential equations in time-varying coefficients. One characteristic of the system that has changed the equation of motion with the coefficient of time-varying is a parametric instability that occurs in various frequency bands [1-4]. Several studies on this research had been done [5] who conducted an analysis of the shaft of the mechanism with the connecting rod and crankshaft are flexible. Mahyuddin [6] investigate by analytical and experimental dynamic response and parametric instabilities in the mechanisms Cam-Follower with a flexible shaft and Schwab [7] analysed the dynamic response of the elastic mechanism. Influence of crank length on the dynamic properties of elastic rod circuited conducted by Chen [8].

* Corresponding author: rizky.arm@gmail.com

Theoretical investigation the influence of several parameters from the parametric instabilities of four-bar mechanism with flexible connecting rod has been done [6, 9].The theoretical knowledge derived through mathematical equations to investigate the parametric instability and dynamic response as a function of the change in frequency and amplitude parameters. This research aims to study the dynamic behaviours of elastic rod [10]. For that reason, a vibration response investigation using strain gage sensors employed to measure midpoint strain of elastic connecting rod on the device model test (four bar mechanism). The study conducted for a wide variety of radius R and rotational crank speed, ω. The crank radius variations are 10 to 40 mm with each interval of 3 mm, while the rotational speed variation of the crank is 180 to 1875 rpm.

2 Material and Methods

2.1 Elastic Mechanism

The elastic mechanism is defined as a mechanism that has at least one elastic component. The elastic rod will cause the mechanism can no longer be analysed as a rigid body. A general model of the elastic rod is made with the initial curvature that aims to facilitate the emergence of elastic properties that generate parametric excitation. The state equilibrium for the elastic connecting rod position before and after deflected was observed as shown as Fig. 1 [11,12].

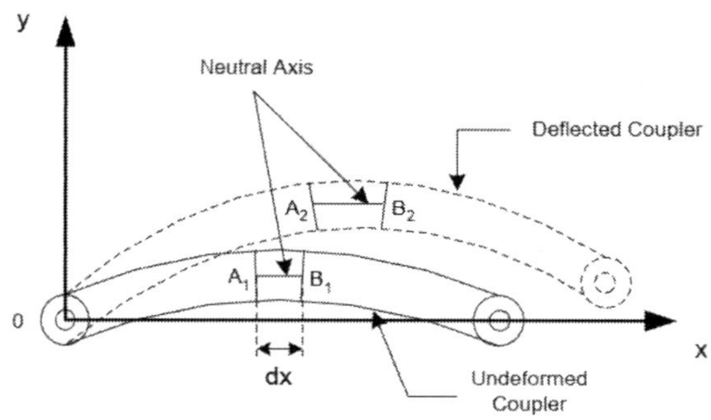

Fig. 1. Elastic rod, before and after deflected

In Fig. 2 are shown in free body diagram of *dx* elements, where M, P and Q is the bending moment, axial force and transverse force acting on the rod. Free body diagram of *dx* elements create equal balance of forces and moments.

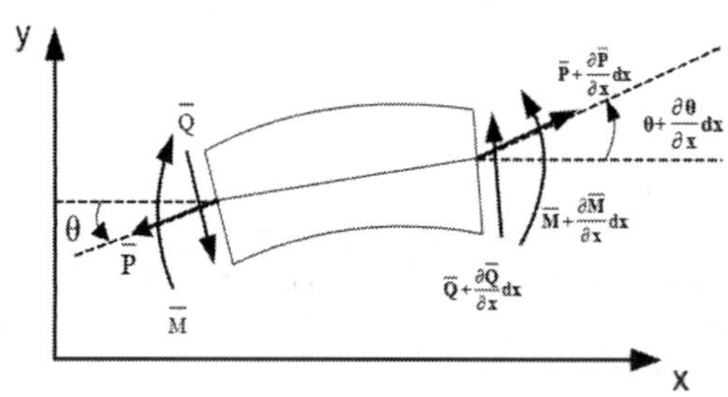

Fig. 2. Free body diagram of elastic rod

Elastic rod can be modelled as a combined system that has stiffness and damping. The damping is assumed as a linear viscoelastic [13, 14]. By using Kelvin-Voigt model, then the relationship of stress and strain can be obtained as

$$\sigma_n = E.\varepsilon_n + \eta \frac{\partial \varepsilon_n}{\partial t} \tag{1}$$

2.2 Dynamic Response and Parametric Stability of a Four Bar Mechanism

One characteristic of the system that has the equations of motion with the time-varying coefficient is a parametric instability that can occur in a variety of frequency bands [15, 16]. Many methods can be used to evaluate the stability of the steady-state solution, including Hill's method, perturbation and Floquet method. Hill's method is an analytical method that is difficult to apply to systems with many degrees of freedom, whereas the method Floquet is a numerical approach, which more easily applied to systems with many degrees of freedom.

Parametric instabilities that occur in several frequency bands can be represented as an unstable region in the field of amplitude-frequency parameters. Early emergence of parametric instability area is referred to as the critical frequency or frequency bifurcation that can describe as

$$\omega = \frac{2.\omega_i}{n} \qquad ; i = 1, 2, \ldots, k; \; n = 1, 2, \ldots \tag{2}$$

In the study conducted by Huda and Mahyuddin [6] states that the elastic rod analysis on a four bar mechanism can be obtained in the same way as on slider crank mechanism. The elastic rod Coordinates of any point of T in coordinates (*OXY*) as shown in Fig. 3, are

$$X_p = r.\cos\varphi(x+u)\cos\beta + (v + v_0)\sin\beta \tag{3}$$

$$Y_p = r.\sin\varphi(x+u)\sin\beta + (v + v_0)\cos\beta \tag{4}$$

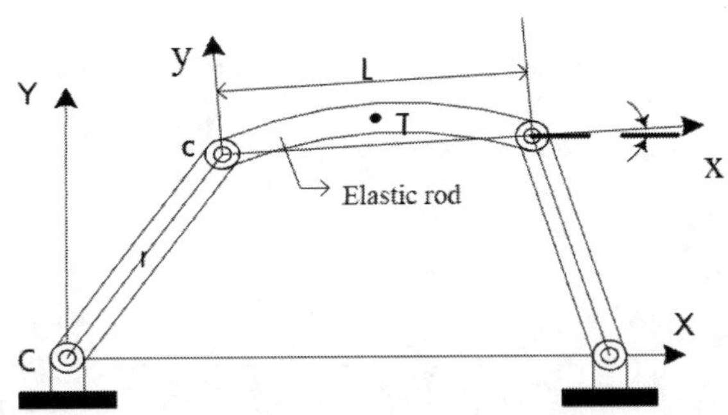

Fig. 3. Four bar mechanism

Various methods can be performed, such as using a type of filter both when doing data acquisition and data processing. The method in this research is utilizing the signal filtering on measurement devices and Matlab software for further data processing [17].

3 Results and Discussion

To plot the graph of strain versus time in frequency domain, the data in format.xls called up and programmed in Matlab. The graph obtained from strain versus time with a sampling rate of 1000 Hz still seemed 'rough', so it is still difficult to see the periodicity of the signal. We perform the method of data smoothing utilizing curve-fitting tool from *Graphical User Interface* (GUI), so that the signal curve looks more refined and periodicity can be determined. After we obtained the optimal fitting method, then it can be plotted the data of curve-fitting result as. While the data of the frequency domain and the curve-fitting result is obtained by creating a new program in Matlab command window.

From the data processing, we can determine the maximum value (*peak to peak*) response from strain versus time. Some examples of strain graph in a radius of 10, 25 and 40 mm (R10, R25 and R40), and the examples of maximum strain value is presented in Table 1. From the example shown that the rotational speed crank growing on each radius of crank R10, R25 and R40 respectively, and strain response graph also being relatively more smoothly (sinusoidal).

Table 1. Some examples of maximum strain values at R10, R25 and R40 mm

Crank Radius, R (mm)	Crank Speed, ω (rad/s)	Maximum Strain, ε (mm/mm)
R=10 mm	37.8	2.0184×10^{-5}
	84.3	2.5358×10^{-5}
	180.12	6.3560×10^{-5}
R=25 mm	39.27	1.3241×10^{-5}
	83.75	2.5664×10^{-5}
	149.54	1.0799×10^{-4}
R=40 mm	36.44	1.1921×10^{-5}
	67.54	3.9590×10^{-5}
	130.9	1.3435×10^{-4}

The maximum strain value on each crank radius then plotted against variation of crank rotation speed. From the plot results, it shows the fluctuations in the value of the strain in each variation of the radius and rotational speed crank. This phenomenon is called a sub-harmonic resonance. The examples of sub-harmonic resonances frequency at R10, 25 and 40 mm.

Furthermore, when viewed from the frequency domain also showed that when the crank rotational speed is relatively low for each variation of the crank radius then obtained other peaks (frequencies) apart than the system frequency [18] (caused by crank speed). At relatively high rotational speed, the peak frequency from the system is more dominant and the frequency curve is “smoother”.

4 Conclusions

Experimental investigation of vibration response of flexible coupler in a four bar mechanism show its dynamic behaviours due to crank length and crank speed. The shape of strain versus time graph is obtained from each variation of condition of parameter test. In general, the crank rotational speed is relatively low (<1000 rpm) for each variation of the crank radius (R10 to R40 mm), the strain versus time curves tend to experience periodic in every 2 to 3 times of periods. When the crank speed is relatively high (> 1000 rpm), the shape of strain versus time curve tends periodically at any one period. From the comparative analysis with slider-crank mechanism [5], also obtained the same phenomenon.

References

1. J. A. Richards, *Analysis of Periodically Time-Varying Systems*. Berlin, Heidelberg, New York: Springer-Verlag, (1983).
2. W. R. Chen, *Acta Mech.*, **216**, no. 1–4, pp. 207–223, (2011).
3. Z. Jinfu, Q. Weiyang, and H. Xinsuo, "Stability Analysis Of Elastic Vibration Of Connecting Rod Of A Slider-Crank Mechanism," vol. **2**, no. 1, pp. 31–40, (2006).
4. S. V. Levyakov and V. V. Kuznetsov, *Acta Mech.*, **211** 1–2, 73–87, (2010)
5. K. Farhang, "Analytical and Experimental Parametric Vibration Stability Studies in Elastic Mechanisms," Purdue University, W. Lafeyette, IN, USA, (1989).
6. A. I. Mahyuddin, "Analytical and Experimental Investigations of Dynamic Response and Parametric Instability in Flexibel Cam-Follower System," Purdue University, W. Lafeyette, IN, USA, (1993).
7. A. L. Schwab, "Dynamics of Flexible Multibody Systems; Small Vibrations Superimposed on a General Rigid Body Motion," Delft University of Technology, Delft, The Netherlands,)2002).
8. C. Chen, J.S., Chian, *ASME, J. Vib. Acoust.*, **123**, pp. 318–323, (2001).
9. S. M. Hashemi-Dehkordi, A. R. Abu-Bakar, and M. Mailah, *Adv. Mech. Eng.*, (2014).
10. Y. Bian, Z. Gao, and M. Fan, "A Vibration Control Method for the Flexible Arm Based on Energy Migration," vol. 2015, (2015).
11. A. Tusnin, *Procedia Eng.*, **150**,1673–1679, (2016).
12. J. P. Fuertes, O. Murillo, J. León, C. Luis, D. Salcedo, I. Puertas, and R. Luri, *Procedia Eng.*, **132** 313–318, (2015).
13. C. E. Majorana and B. Pomaro, *Eng. Comput.*, **28,** 2,. 114–129, (2011).
14. V. D. Potapov, *Acta Mech.*, **212,** 1–2. 123–129, (2010).
15. A. Talimian and G. M. V??r??s, *Period. Polytech. Civ. Eng.*, **57**, 2, 123–128, (2013).
16. A. Yildiz, O. Kopmaz, and S. T. Cetin, *J. Mech. Sci. Technol.*, **29** 3, 1001–1006, (2015).
17. D. Q. Cao and R. W. Tucker, *Int. J. Solids Struct.*, **45** 2, 460–477, (2008).
18. A. Cabboi, T. Putelat, and J. Woodhouse, *J. Mech. Phys. Solids*, **92** 210–236, (2016).

The in-plane electrical conductivity: the rotation parameters effect on producing graphite/epoxy composites

Hendra Suherman[1*], *Andro* Hamdani[1], *Edi* septe[1], *Yovial*[1], *Irmayani*[2]

[1]Department of Mechanical Engineering, Universitas Bung Hatta, 25143 Padang, Indonesia
[2]Department of Industrial Engineering, Universitas Ekasakti, 25133 Padang, Indonesia

Abstract. The rotation parameters used to produce graphite/ epoxy composites material is greatly determine the properties of in-plane electrical conductivity. However, the appropriate rotation parameters are different for each composition, conductive materials, and matrix used. In this study, variations of rotation parameters (250 rpm, 350 rpm and 450 rpm) and rotation time (5 min, 10 min and 15 min) with constant molding parameter such as molding temperature (150 ^{0}C), molding time (120 minutes) and molding pressure (50 kg / cm^2) have been used to produce graphite/epoxy composites. The results showed that the highest in-plane electrical conductivity 65 S/cm on the resulting graphite/epoxy composites.

1 Introduction

Rotation parameters, such as rotation time, rotation speed, and temperatures during rotation are the significant parameters affecting the quality of the resulting composite material [1-5]. The dispersed or coagulated filler or conductive material in the matrix were greatly depends on the mixing parameters used in producing the composite material [1, 2]. Mixing matrix and filler can be done using different method and parameter variation. In situ polymerization and melt mixing compounding have been used to mix matrix as a binder and filler as conductive material [6-8]. Hu et al. [1] found that short rotation processes using modest shear forces increased the formation of conductive networks within the matrix. Meanwhile, high-speed stirrer can produce a homogeneous dispersion of filler, resulting in higher in-plane electrical conductivity [9]. Suherman et al. [10] using carbon nanotubes (CNTs) as a second filler in producing CNTs/graphite/epoxy composite using Taguchi method. They found that the electrical conductivity was significantly influenced by the rotation parameters in generating conductive networks. Zakaria et al. [11] conducted a study on the effect of rotation parameters on the electrical conductivity of carbon black/graphite/epoxy nanocomposites using Taguchi method. The results showed that the significant agitation parameters affect the electrical conductivity produced. These previous researches use multi filler in investigating the effect of rotation parameters on the electrical conductivity. This study use 80 wt.% single filler of high conductive filler to investigete the

* Corresponding author: hendras@bunghatta.ac.id

effect of rotation parameters in producing graphite/epoxy composite material to the in-plane electrical conductivity.

2 Experimental procedure

2.1 Materials

Graphite is obtained from Asbury Carbons, Inc. U.S. Graphite type 635 fine flake has particle size of 150 μm is used in this study. The 635 thin epoxy resin, with a viscosity of 6 poise, and fast epoxy hardener was obtained from US Composites.

2.2 Fabrication of graphite/epoxy composites.

The composition of graphite and epoxy used to produce graphite/epoxy composite is 80/20 by weight percent (wt.%). In the early stage, epoxy and hardener are mixed in a 4: 1 composition based on weight percent (wt.%), as recommended by the manufacturer. The mixture was stirred using a RM 20-KIKA-WERK mechanical stirrer at 200 rpm for 5 minutes. On the next step, 80 wt.% of graphite as a conductive filler is stirred with various rotation speeds (250, 350, 450 rpm) and various rotation times (5, 10, 15 minutes). Furthermore, the mixture was poured into an aluminum mold at constant molding parameters of molding pressure 50 kg/cm^2 ; molding temperature 150^{0}C; and molding time 120 minutes.

2.3 Characterization

ASTM C 61 standard is used to measure the in-plane electrical conductivity of graphite/epoxy composites.

3 Results and discussion

3.1 Effect of rotation speed on the in-plane electrical conductivity

The effect of rotation speed on the in-plane electrical conductivity of the graphite/ epoxy composites material is shown in Figure 1. Figure 1 shows three variations of rotation speed (250 rpm, 350 rpm and 450 rpm) at constant rotation time of 5 min.

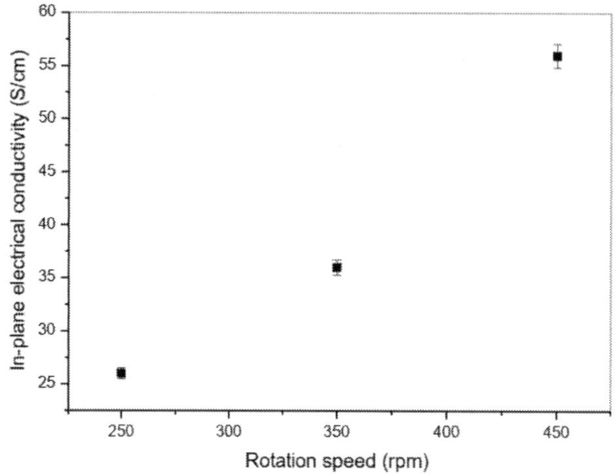

Fig. 1. Effect of rotation speed on the in-plane electrical conductivity

The in-plane electrical conductivity of graphite/epoxy composite at rotation speeds of 250 rpm, 350 rpm and 450 rpm sequentially are 26 S/cm, 36 S/cm and 56 S/cm. The in-plane electrical conductivity increases by the increasing of rotation speed used. The increasing electrical conductivity due to higher rotation speed (ie 450 rpm) is capable to form a better conductive network. This resulted the highest rotation speed of 450 rpm having the highest in-plane electrical conductivity (56 S/cm) compared with 250 rpm rotation (26 S/cm) and 350 rpm (36 S/cm) [1-3, 5].

3.2 Effect of rotation time on the in-plane electrical conductivity

Figure 2. shows the effect of rotation time on the in-plane electrical conductivity of graphite/ epoxy composites. Overall, the in-plane electrical conductivity of graphite/ epoxy composites increase by the increasing of rotation time from 5 minutes to 15 minutes. This proves that rotation time significantly influences in forming conductive network within polymer matrix, resulting in higher in-plane electrical conductivity [1, 12]

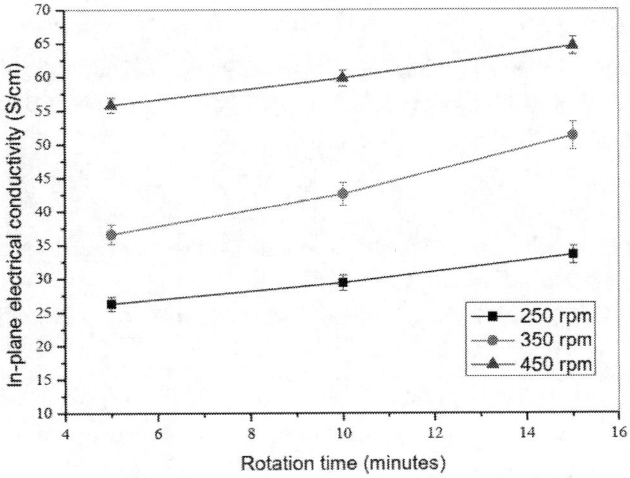

Fig. 2. Effect of rotation time on the in-plane electrical conductivity

At rotation speed of 250 rpm, in-plane electrical conductivity of 5 minutes (26 S/cm) to 15 minutes rotation time (33 S/cm) increased by 26.92%. At a rotation speed of 350 rpm, the increase in in-plane electrical conductivity from 5 min (36 S/cm) to 15 min rotation time (51 S/cm) was 41.66%, while at 450 rpm rotation speed increased in-plane electrical conductivity from 5 min (56 S / cm) to 15 minutes (65 S / cm) was 16.07%. Thus, the largest increase in in-plane electrical conductivity with the increase of rotation time from 5 to 15 minute occurs at 350 rpm rotation speed. However, the best in-plane electrical conductivity of graphite/epoxy composites is at a rotation speed of 450 rpm, and rotation time of 15 min. The increase of rotation time is capable of generating more electrical conductivity network in the polymer matrix thus increasing the in-plane electrical conductivity to the resulting graphite/epoxy composites [12, 13].

4 Conclusion

Graphite/epoxy composites with simple methods have been produced. Some conclusions are:

1. The increasing of rotation speed (from 250 rpm to 450 rpm) and rotation time (5 minutes to 15 minutes) increase the in-plane electrical conductivity of graphite/epoxy composites produced.
2. The highest in-plane electrical conductivity (65 S/cm) is obtaine at 15 minutes rotation time and 450 rpm rotation speed.

The author wishes to thank the Directorate General of Higher Education through the applied research of excellent university, so this research can run well based on the time set through contract number: 023-34/LPPM-Penelitian/Hatta/II-2018.

References

1. N. Hu, Z. Masuda, G. Yamamoto, H. Fukunaga, T. Hashida, J. Qiu, Compos. Part A-Appl. S, **39**, 893 (2008)

2. P.C. Ma, M.Y. Liu, H. Zhang, S.Q. Wang, R. Wang, K.Wang , Y.K. Wong, B.Z. Tang, S.H. Hong, K.W. Paik, J.K. Kim, ACS Appl.Mater.Interfaces 1, 1090 (2009)
3. H. Suherman, A.B. Sulong, J. Sahari, Int. J. Mech. Mater. Eng. **5,** 74 (2010)
4. H. Suherman, A.B. Sulong, J. Sahari. Adv. Mater. Res. **264-265,** 559 (2011)
5. H. Suherman, A.B. Sulong, M.Y. Zakaria, N.R. Rajendra Royan, J. Sahari, Songklanakarin J. Sci. Technol. **40**, 105 (2018)
6. H. Deng, T. Skipa, R. Zhang, D. Lellinger, E. Bilotti, I. Alig, T. Peijs, Polymer, **50**, 3747 (2009)
7. M. Murariu, A.L. Dechief, L. Bonnaud, Y. Paint, A. Gallos, G. Fontaine, S. Bourbigot, P. Dubois, Polym. Degrad. Stabil, **95**, 889 (2010)
8. Y. Pan, L. Li, S.H. Chan, J. Zhao, Compos. Part A-Appl. S, **41**, 419 (2010)
9. X. Xie, Y. Mai, X. Zhou, Mat. Sci. Eng. R, **49**, 89 (2005)
10. H. Suherman, J. Sahari, A.B. Sulong, Advan. Sci. Lett, **19,** 334 (2013)
11. M.Y. Zakaria, H. Suherman, J. Sahari, A.B. Sulong, Appl. Mech. Mater, **393**, 68 (2013)
12. R. Taherian, A.N. Golikand, M.J.Hadianfard, Mater. Design **32**, 3883 (2011)
13. N.R. Rajendran Royan, A.B. Sulong, J. Sahari, H. Suherman, J. Nanomater, **2013** Article ID 717459 (2013)

Design and Production of Rotary Type Machine for Chop Up Organic Waste Plantation

Wenny Marthiana[1*], Duskiardi[1], *Rizky* Arman[1], Yovial Mahyoedin[1], *Dedi* Wardiyanto [2].

[1] Department of Mechanical Engineering, Universitas Bung Hatta
[2] Department of Mechanical Engineering, Universitas Muhammadiyah Sumatera barat

Abstract. Agricultural organic waste chopping machine are tools that can be used to chop leaves, twigs and could be used to help farmers in processing of the agricultural disposal products such as stalks and corn stalks, cassava stems, rice stalks in order to accelerate farmers in processing of those disposal in order that the plantation land can be prepared immediately for subsequent cultivation, as well as the chopped products can be reused for various purposes such as animal feeds and composts fertilizer. The research of this engineering method is carried out with the following stages: (1) study of the characteristics of several organic wastes to be enumerated, (2) analysis of the Chopping machine elements, consists of Chopping Blades, the house of the chopper blades, hopper, frame and transmission system. (3) A Prototype chop up mechanism manufacturing. And (4) Functional testing machine. The Copping Machine design resume as follows: Production capacity of enumeration 220 kg / hour. The cutting mechanism uses 2 static blades and 6 rotary blades. Transmission system using Belt-Pulley mechanism. The 24 Hp Diesel engine is used.

1 Introduction

Waste from organic agriculture such as stems, leaves, generally still reusable, for example for animal feed, and fertilizer, unfortunately agricultural waste is not all being utilized by farmers, the causes are: a) generally farmers burn food crops waste as soon as it will be done soil processing, b) crop waste is kamba so it makes difficulty for farmers to transport in large quantities to be given to livestock, and generally agricultural land is far enough from farmers' settlements so that they require transportation costs, c) unavailability of food crop storage, and farmers are unwilling to store / accumulate waste around the house / under the house for fear of fire hazard, d) breeders assume that the availability of forage in the yard, garden, rice field is still sufficient as animal feed [1] . Most of the corn waste can be used as feed ingredients. With a touch of simple technology, the waste can be converted into nutritious feed and energy source for livestock [2]. Beside green feed agricultural waste can also be used as animal feed. Agricultural waste is a potential local feed for supporting livestock development especially in agriculture-based areas [3].

* Corresponding author: wenny_ma@yahoo.com

The bad management of agricultural waste can lead to a decrease in agricultural productivity in the fulfillment of food needs; therefore, appropriate handling the rest of this farm is very important and should be of concern [4]. To Burn the rest of the farm is common practice in waste management and as a quick way, as consequences that method causing smoke pollution [5], therefore, the management of agricultural waste should not be burned but in other ways such as by reducing the volume to be easily carried or processed into other uses, the process is by chopping. Based on the research, the absorption of Carbon content in the soil was found to be better when the soil was fed with corn stalks rather than the burning of corn stems. [6] In order to cultivate large quantities of agricultural waste and the limited time and manpower to manage agricultural waste efficient and quick, so the mechanization application is necessarily

Various tools for treating plantation waste have been widely available, such as the application of hammer mill model crusher [7]. Reel type elephant grass chopper machine [8]. Bionic Saw Blade for Corn Stalk Cutting mechanism, this machine type apply a serrated Insisivus design [9]. The schered blade mechanism for cutting coconut midrib [10] and many more. Since the agricultural types and wastes are various, whether influenced by the local conditions or influenced by the price of agricultural commodities, it becomes an advantage if the process tool can be used for several type of raw materials. The difference of the tool in this research is the design of the enumerator machine that can be used for various organic materials such as corn cobs, corn stalks, lumps of manure dried. The result of such shrimp can be used for various purposes, such as cattle feed mixture, corncob mixed flakes, corn stalk can be used as compost, dried livestock debris can be used directly as ready-made manure. Several studies have shown that corn stover results can be used as a source of renewable energy, and is an abundant gold mine for farmers.

Consideration of product design criteria for agricultural waste include: the physical properties of the material to be chopped that is shear stress of the cutting materials and the mechanical properties of the enumerator machine, mechanical properties of the machine are Machine Capacity, component dimensions and cutting power requirements.

2 Methodology

Research Phase:

Research phase as described in Figure 1 below

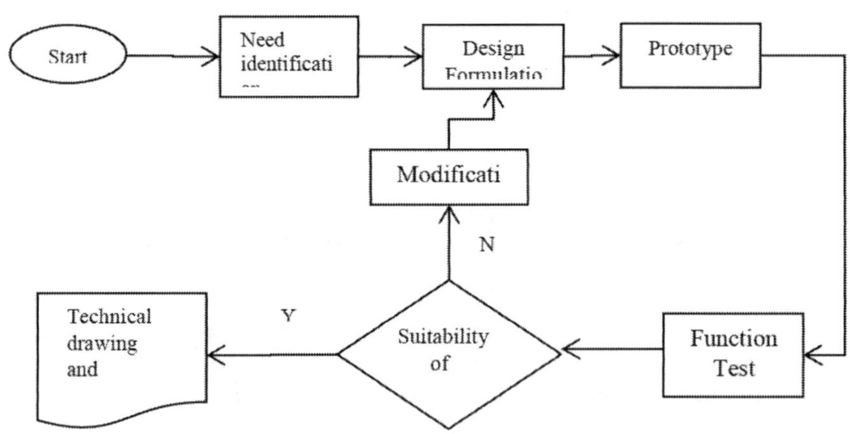

Fig 1. The Research Flow

A. Identification of needs. Based on the problems faced by the farmers of crops in general. That is to clear the land fields from the organic residues of the harvest quickly and the processed of residual organic products can be reused as organic fertilizer.
B. Formulation of Design. This stage is the crystallization of ideas. In this phase, the formulation of the design specified product specifications desired as the identification needs. Then continue with machine components design. The outputs of this stage are Machine Technical drawings and components specification.
C. Prototype construction. Based on engineering drawings, then the machine prototype was created. The Machine prototype was processed in production process Laboratory of Universitas Bung Hatta Padang
D. Functional testing. This test is performed to see if all components can be unified and whether the machine can function as design accordingly. Where the prototype is appropriate to the design, next the engineering drawings and technical specifications are defined. While not, then Design modification is needed. .
E. Design Modification. This phase is done when there is a mismatch between functional design and machine operation at run time.
F. Technical Drawing and Specification phase of Preparation of engineering drawings and product specifications to be used for further production purposes.

3 Results and Discussion

3.1 Unit Design Enumerator

cutting knifes being used are a Plate type knife, which is composed of 2 stationary blades and 6 rotary blades. This knife used is taken from paper knife of paper cutting machine that is no longer used. Knife material is S50C, placed with a slope of 5^0 from the cutting mechanism shaft. The cutting slope is expected that cutting process will happen as the shearing process. The Blades position as shown in Figure 2.

3.2 Frame Design.

The Construction Frame is designed to withstand cutting mechanism and power engine as well. The Frame construction dimension is 740 mm in long, 660 mm in wide, 834 mm in high. The construction material are the iron elbow and iron UNP. [6-8].

3.3 The Hopper design

Hopper, use as a feeding unit is part of a chopping machine that serves material cutting into the blade housing. The cutting material is fed manually through the hopper from the top side. The hopper dimensions are Length 468 mm, Width 250 mm, Height 235 mm

3.4 Filter Design

Screening is used to obtain the uniform and smallest possible results, since the organic material of the enumeration will be used as raw material for organic fertilizer then. To satisfy the cutting material can be used in the manufacture of fertilizer in accordance with the desire to quickly decay, the dimensions of the enumeration results are kept to a minimum Fig. 3

The Mechanism design can be seen as follow

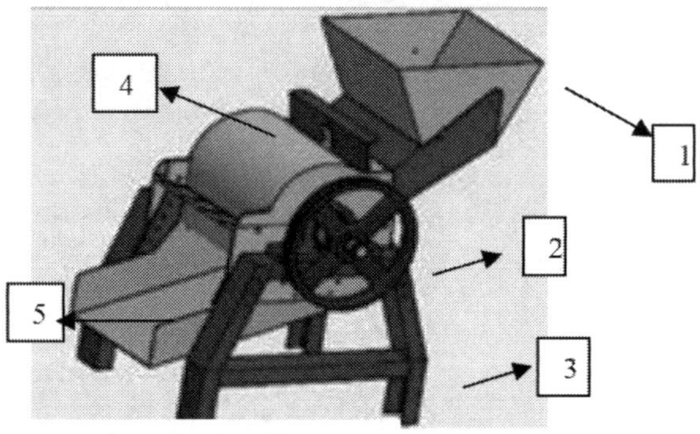

Fig 2. Chopping machine

Fig 2 Information.

1. Hopper
2. Pulley
3. Frame
4. Casing Blade
5. Chute

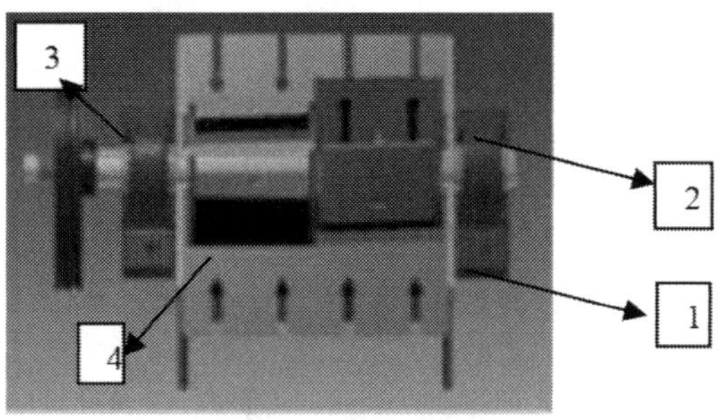

Fig 3. Chopping Blade

Fig 3 information

1. Stationary Blade
2. Rotary Blade
3. Pulley
4. Bearing

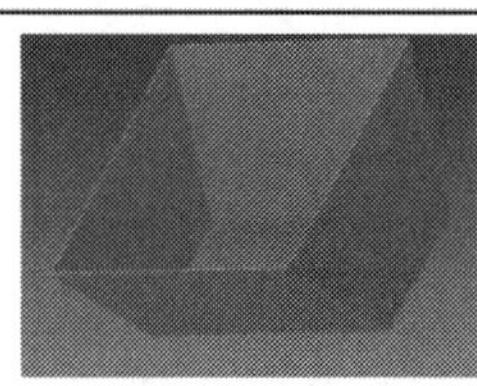

Fig 4. Hopper

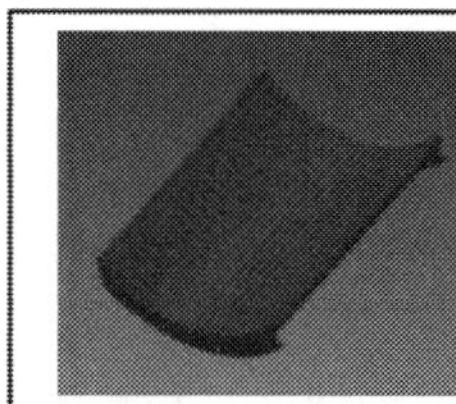

Fig 5. Screening

3.5 Chopping machine Work mechanism

The working mechanism is describe as follows :

1. When the main drive motor (diesel motor) is turned on, the pulley of the primarily driving motor transmits the spin to the pulley of the enumerator mechanism through the v-belt belt.
2. Pulley then rotates the dynamic knife shaft.
3. While the organic material is inserted through the hopper, the chopping knife will cut the material when the rotary blade meets with the stationary blade.
4. Corn stover or any other material had been cooped then will be filtered by a filter plate or screen in the bottom of the machine. Screening size is a hole of 5mm in diameter and 10 mm long
5. Flakes that are already in 1mm or smaller will pass through the filter and will come out through the chute.

4 Conclusion

From design process, the chop up mechanism specification can describe as below.
Machine Specification

1. The organic machine enumerator specifications are as follows:
   Dimensions of chopper machine: Length 1560 mm, Width 870 mm, Height 1350 mm.
2. Machine Weight 200 kg
3. Production capacity of enumeration 220 kg / hour, for dry organic matter (moisture content less than 25%).
4. The cutting mechanism uses 2 static blades and 6 rotary knives with S50C material
5. Transmission system using 2 units Belts -Pulley mechanism. With V-Belt type B-105.
6. Pulley dimension : 5 inch and 10 inch, Material cast iron Pulley.
7. The Driving force Requirement to rotate the knife and chop material is 20 hp.
8. Power drive is Dong Feng diesel Engine, 24 hp, with a maximum of 2.200 rpm rotation.

References

1. Liana and Febriana. Utilization of Agricultural Waste as Ruminant Feed on Community Farmers in Kec. West Rengat Kab. Indragiri Hulu. Jurnal Peternakan Vol 5 No 1, (28-37), (2011)
2. E. T. Yuniarsih and M. Basir N. Balai Pengkajian Teknologi Pertanian Sulawesi Selatan. Utilization of Maize Waste as an Eating Feed in South Sulawesi. Seminar Nasional Serealia, (2013).
3. Anita Sari, Liman, and Muhtarudin. Supporting of Agricultural By Product as Ruminant Feed in District Pringsewu Regency. Jurnal Ilmiah Peternakan Terpadu Vol. 4(2): 100-107, Mei, (2016).
4. Vaclav Smil. Crop Residues: Agriculture's Largest Harvest: Crop residues incorporate more than half of the world's agricultural phytomass. *BioScience*, Volume 49, Issue 4, 1 April, Pages 299–308, https://doi.org/10.2307/1313613. (1999)
5. Adeoye, Peter Aderemi, Adebayo, Segun Emmanuel And Musa, John Jiya. Agricultural Post-Harvest Waste Generation And Management For Selected Crops In Minna, Niger State, North Central Nigeria. Journal Sciences and Environmental sanitation Volume 6, Number 4: 427-435, December, (2011)
6. Francisco J. Calderón, Joseph Benjamin, Merle F. Vigi . A Comparison of Corn (Zeamays L.) Residue and Its Bio-char on Soil C and Plant Growth. PLOS ONE | DOI:10.1371/journal.pone.0121006 April 2, (2015)
7. Rifki Zulkarnain, Sugeng Slamet and Taufiq Hidayat, Design Of Hammer Mill Machine Destroying Corn Bongcol With 100kg / Hour Capacity As Animal Feed. Prosiding SNATIF Ke -1 Tahun 2014 ISBN: 978-602-1180-04-4. (2014)
8. Wahyu K Sugandi, Asep Yusuf, Muhammad Saukat.; Design And Performance Test Machine Engine Tube Reel Eye Rey. Design and Test Performance Of Cutter Machines Reel Type for Elephant Grass. Jurnal Teknotan Vol. 10 No. 1, Agustus. P - ISSN :1978-1067; E - ISSN: 2528-6285. (2016).
9. Honglei Jia, Changying Li, Zhihong Zhang, Gang Wang. J of Bionic Eng 10- 497–505. (2013)
10. I.M.Sanjay Kumar , DR. T.R. Hemanth Kumar Design and Development Of Agricultural Waste Shredder Machine. IJISET - International Journal of Innovative Science, Engineering & Technology, Vol. 2 Issue 10, October (2015)

Strain-Stress Formation and Deformation of Molded Materials by Drying-Induced Shrinkage

Yoshinori Itaya[1,*]

[1]Environmental and Renewable Energy Systems Division, Graduate School of Engineering, Gifu University, 1-1, Yanagido, Gifu, 501-1193, Japan

Abstract. Drying-induced strain-stress and deformation was modeled for molded materials such as ceramics and foods. The transient three-dimensional problem of strain-stress and heat and moisture transfer was solved simultaneously by the finite element method. The behaviors of internal strain-stress formation and deformation are compared among three modes of hot air heating, intermittent heating and internal heating for a ceramic slab. The concepts of the mass transfer potential and a linear viscoelasticity were introduced to consider the different sorption isotherm for layered foods consisting of two ingredients and creep phenomenon. The tensile and compressive stresses fluctuate and fall remarkably during lower heating period when the slab is heated intermittently. In the internal heating mode, drying proceeds fastest as well as stress formation is maintained at the lowest level in the three modes. This effectiveness of the internal heating is investigated experimentally by employing the microwave heating as well. The combination of different foods influences significantly not only the drying characteristic but also the internal strain-stress generation. The drying behavior could be analyzed with a high accuracy even if only heat and moisture transfer were solved without strain-stress analysis. This fact results in that heat and mass transfer and strain-stress are not always necessary to be analyzed simultaneously for saving the CPU time.

1 Introduction

Drying-induced failures and deformation of molded materials are substantially undesirable problems for the quality control in the production process. If drying of molded materials is hastened, the surface does not only dry up immediately and the drying rate falls markedly maintaining a wet state in the inside, but also medium is subjected to strong internal stress and the product results in generation of failures in the worst case. In order to maintain the minimum stress formation during drying, the internal migration rate of moisture should be improved or an appropriate operation of the drying should be carried out so to reduce the gradient of the moisture content profile. Theoretical studies have been carried out for solving numerically the drying problems including shrinkage of clay in the past works [1-3]. However, there have been rather few researches from the viewpoint of failure and effective drying processes.

*Corresponding author: yitaya@gifu-u.ac.jp

The present author performed R&D on modeling of drying-induced strain-stress [4-6] and some parametric analyses of the strain-stress in the drying process are performed to investigate the influence of heating modes on the behaviors of drying-induced strain-stress and deformation as well as the drying characteristic [7]. The transient three-dimensional problem of strain-stress and heat and moisture transfer in a slab is solved simultaneously by the finite element method. The modes of convective, intermittent and internal heating are introduced by modeling them in the normalized parameters. The experiment is also carried out to examine the effect of drying characteristic by microwave heating which is one of the promising internal heating modes [8, 9].

Drying of foods is an important process for preservation preventing deterioration and maintaining nutrition for a long term. There are several foods consisting of multi-ingredients whose sorption isotherm of moisture is different. The most foods generate also drying-induced shrinkage that causes strain and stress. In order to produce high quality molded foods in a drying process, accurately appropriate controls are necessary not so to form failures and unacceptable irregular deformation. However, there has been only a little works on modeling to analytically predict such a complicated drying behavior except for some works [10-14] Those model should combine simultaneously the drying-induced strain-stress and moisture migration taking account for different sorption isotherm or concept of chemical potential [15] between wet multi-solids to carry out accurate analyses. This author has developed also a generally applicable computerized method for predicting three dimensional heat and moisture transfer with viscoelastic strain-stress in macroscopically non-homogeneous food during drying and to simultaneously solve them. A triply layered food of a slab shape consisting of starch and a mixture of starch and sugar, whose properties are well known and the moisture sorption characteristic is different with each other, was chosen [16]. Furthermore, an attempt to reduce and save the memory and CPU time for computer simulation of the model was performed [17]. The sensitivity of the analytical results on the drying behavior was discussed with comparing with those obtained by solving independently heat-moisture transfer and strain-stress of the layered foods, whose body was mesh-generated in different number of finite elements.

This paper is to introduce the outline of a series of the works performed in the past by the present author and his co-authors on drying-induced strain-stress and deformation of modeled media

2 Parametric analyses on effect of different heating modes

2.1 Modeling

Parametric analyses of the strain-stress combined with heat and moisture transfer in the drying process are performed. The transient three-dimensional problem of strain-stress as well as heat and moisture transfer in a slab of clay is solved simultaneously by the finite element method. Introducing the similar assumptions to Itaya *et al.* [5]to simplify the model, the governing equations of the heat and mass transfer are expressed in dimensionless form as:

$$\frac{\partial T^*}{\partial \theta} = -div(\mathbf{q}^*) + \dot{q}^*, \frac{fW^*}{f\theta} = -div(\mathbf{J}^*) \tag{1}$$

where

$$\mathbf{q}^* = -grad(T^*)\ ,\ \mathbf{J}^* = -\frac{1}{Le} grad(W^*) \tag{2}$$

The initial and boundary conditions are given by:

$$\theta = 0 \; ; T^* = 0, \; W^* = 1 \tag{3}$$

$$\theta > 0 \; ; Bi(1 - T^*) + Bi_r(1 - T^*) - BiN_h(H_e - H_a) = \mathbf{q}^* \cdot \mathbf{n} \tag{4}$$

$$BiLeN_m(H_e - H_a) = \mathbf{J}^* \cdot \mathbf{n} \tag{5}$$

In the equations, the dimensionless parameters are defined by:

$$\theta = Fo = t\alpha / l_0 \;,\; T^* = (T - T_0)/(T_a - T_0) \;,\; W^* = W / W_0 \;,\; X = x / l_0 \;,\; Y = y / l_0 \;,$$
$$Z = z / l_0 \;,\; Bi = hl_0 / k \;,\; Le = \alpha / D_W \;,\; Bi_r = \sigma_r \varphi l_0 (T_a^2 + T_s^2)(T_a + T_s) / k \;,$$
$$N_h = L_a / c_{pa}(T_a - T_s),\; N_m = c\rho V / c_{pa}\rho_0 V_0 W_0,\; \dot{q}^* = l_0^2 \dot{q} / k(T_a - T_0) \tag{6}$$

where t is time, α and D_W are the apparent thermal and overall diffusion coefficients, T_a and T_s are temperatures of ambient and surface, respectively, W is moisture content, H_a and H_e are ambient humidity and equilibrium humidity at the surface temperature, c and c_{pa} are specific heat of clay and air, L_a is latent heat, V is the volume, σ_r and ϕ are Stefan-Boltzmann constant and blackness, $\dot{q}$ is the heat generation rate inside the body and the subscript 0 represents the initial condition. Introducing those parameters, we can discuss the effect of more general parameters on the drying behavior.

Similarly to the works by Itaya et al.[4, 5], approximating the free strain, that is the strain taking place under no internal stress defined by a function of only the local moisture content, the strain and stress are analyzed by a set of governing equations represented for viscoelastic media.

Constitutive equation:

$$\sigma_{ij} = \int_{t=0}^{t=\tau} G_{ijkl}\left(\xi(t) - \xi'(\tau)\right) \frac{\partial \varepsilon^*_{kl}\left(\xi(\tau)\right)}{\partial \tau} d\tau \tag{7}$$

where

$$G_{ijkl}(t) = \frac{1}{3}\left[G_1(t) - G_2(t)\right]\delta_{ij}\delta_{kl} + \frac{1}{2}G_1(t)\left[\delta_{ik}\delta_{jl} + \delta_{il}\delta_{jk}\right] \tag{8}$$

$$G_1(t) = 2G(t) \;,\; G_2(t) = 3K(t) \tag{9}$$

$$\varepsilon^*_{ij}(t) = \varepsilon^*_{ij}(t) - \varepsilon^s_{ij}(t) \tag{10}$$

Equilibrium equation:

$$\sigma_{ij,j} + F_i = 0 \tag{11}$$

Strain-displacement relation:

$$\varepsilon_{ij} = \frac{1}{2}\left(U_{i,j} + U_{j,i}\right) \tag{12}$$

where ξ and ξ' are the reduced times and equal t and τ, respectively. σ_{ij} and ε_{ij} are ij-th elements of the stress and strain tensor σ and ε, $\sigma_{ij,j}$ and $U_{i,j}$ denote differentiation of σ_{ij} and U_i with respect to j, G and K are shear and bulk moduli, F_i is body force, ε^*_{kl} and ε^s_{kl} are strain caused by the stress and shrinkage and δ is delta function. These equations are solved by the finite element method. The detail formulation appears in the papers by Itaya et al.[4, 5]

2.2 Analytical results and discussion

Intermittent heating is modeled so that a slab is heated alternatively with a high and low Biot number in every $\Delta\theta$ interval time. In the internal heating mode, the heat generation rate is given so that the sum of the convective, radiative and internal heating rates at $\theta = 0$ are equal to the rate in the hot air heating mode. Then the dimensionless heat generation rate is derived as

$$\dot{q}^* = 2\left(Bi + Bi_r\right)\left(\frac{1}{x_0}+\frac{1}{y_0}+\frac{1}{z_0}\right)l_0\left(\frac{T_a^c - T_a}{T_a^c - T_0}\right) \tag{13}$$

where T_a^c denotes air temperature in the hot air heating mode. Referring Hasatani et al. [3], and Itaya et al. [4],Bi,$1/Le$ and Bi_r are set a constant at 0.11, 0.2 and 0.15 respectively. In the intermittent heating mode, Bi is reduced stepwise to one-tenth at the low heating period. The internal heating is assumed to be operated under the temperature ratio, $(T_a^c - T_a)/(T_a^c - T_0)$= 5, in Eq.(13).

It is quite difficult to normalize the strain-stress equations, the saturated humidity and the shrinkage factor, relating to the heat and mass transfer since they are nonlinear with temperature and moisture content. Thus the same heat and moisture transfer properties as the past work [3] are employed for the conversion to dimensional parameters from the dimensionless ones. The initial temperatures of 293 K and the initial moisture content 0.249 kg-water/kg-dry solid are fixed for all runs. The air temperature is 343 K for the hot air and intermittent heating modes and 303 K for the internal heating mode. For the mechanical properties and the shrinkage factor, values given by the reference [3] are used, i.e. the bulk and shear moduli, $K(t)$ and $G(t)$ are determined from the following expressions.

$$K(t) = E(t) / 3(1-2\nu)\,,\quad G(t) = E(t) / 2(1+\nu) \tag{14}$$

$$E(t) = 7.0\text{x}10^7\exp(-\,t\,/\,1.7\text{x}10^4)\;\;[\,\text{Pa}\,]\,,\quad \nu = 0.35 \tag{15}$$

The drying conditions are assumed to be symmetric with respect to X-axis. Hence the simulation is carried out for a quarter part of the slab. The initial size of the calculation domain is 1x1x0.4 (X_0 x Y_0 x Z_0) and the equivalence length l_0 is 0.03 m.

Figure **1** shows theoretical results on the profiles of moisture content and temperature in a several heating modes at a dimensionless time θ = 0.5. The profiles are drawn by contour curves on a half of the bisected face. As the air temperature only in the run of the internal heating mode is different from the other modes, the normalized temperature can not be compared at the same level with each other. Thus the temperature in the figure is modified so to adjust the same dimensionless temperature as one in the other heating modes. The dotted line is an original configuration of the slab before drying starts. In the intermittent heating mode, hot air heating by high and low Biot number, 0.2 and 0.02 is supposed to be repeated alternatively in every time interval of 0.1. While the gradient of the moisture content distribution seems to be larger than the other modes at θ= 0.5 in the high Biot number period, the gradient is significantly relaxed at the end of the low heating period as showing the profile at θ= 0.4 in Figure **2**. The temperature profile achieves almost uniform within the deviation of 0.01 at this time during continuous drying by the hot air heating mode. In the intermittent mode, the gradient of the temperature shows the opposite behavior to the moisture content if they are compared between the profiles in the low and high Biot number periods. This characteristic is explained by the effect of radiative heat transfer from the dryer wall, i.e. if the rate of convective heat transfer falls down, the contribution of radiative heat transfer appears significantly and the surface of a wet sample is heated so to rise toward a higher value than the wet-bulb temperature. In the internal heating mode, the temperature rises gradually inside the body and the gradient becomes significant as well since the heat sink region due to evaporation and/or convective heat loss lies on the surface exposed to the ambient.

The transient behavior of average moisture content and temperature in the slab, which is determined theoretically in a several modes, are shown in Figure **3**. The moisture content and temperature in the intermittent heating mode fluctuates above and under the results in the continuous convective heating mode, i.e. while the drying rate increases in the high heating period, temperature falls as described in the above section. Both the drying rate and

 https://doi.org/10.1051/matecconf/201824801011

temperature rise significantly in the internal heating mode compared with the other modes although the heating rate given into the body is assumed to be equal to the other modes at the beginning of drying. The reason can be explained by the fact that the net heating rate falls gradually due to reduction of the temperature difference between hot air and the surface with elevation of the surface temperature in the convective heating while the internal heating is adopted with a constant rate through the drying period.

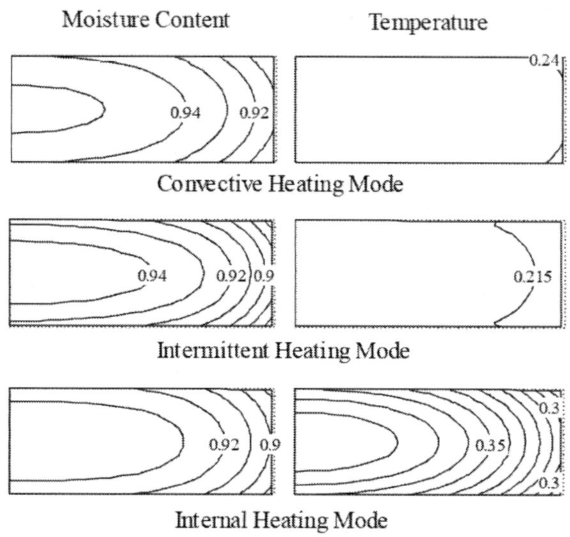

Fig. 1. Profiles of moisture content and temperature in different heating modes at $\theta = 0.5$.

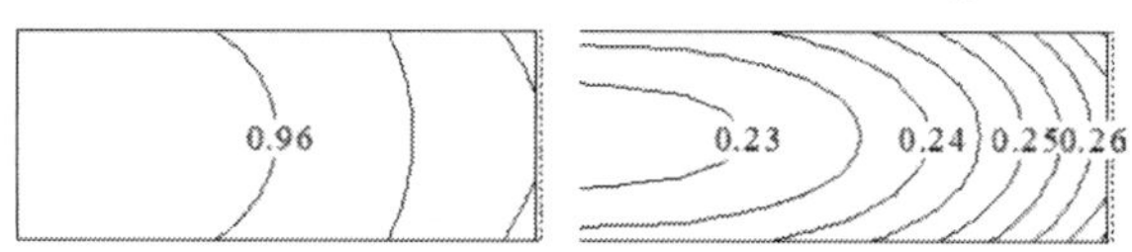

Fig. 2. Profiles of moisture content and temperature in the intermittent heating at $\theta = 0.4$.

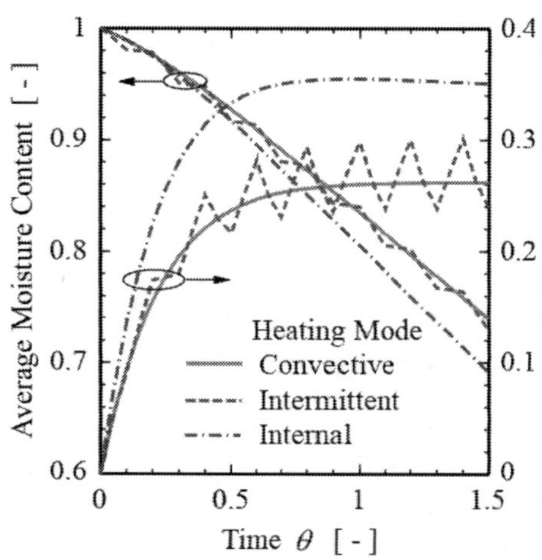

Fig. 3. Transient moisture content and temperature.

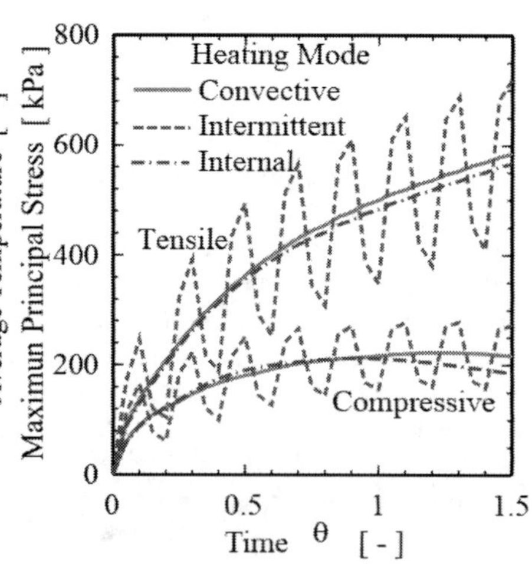

Fig. 4. Tensile stress against moisture content.

The time behavior of the maximum tensile and compressive stresses is compared among the heating modes in Figure **4**. Both the stresses oscillate and reduce appreciably during the low Biot number period when the slab is heated intermittently. These behaviors can be predicted also from the average moisture content and temperature. In the internal heating mode, the maximum stresses are a slightly smaller than the convective heating particularly in the later period of drying. If tensile stress, which can be related with the cause of crack generation, is plotted against the moisture content, the effect of the internal heating on reduction of the stress is made clearer as drawn in Figure **5**. In general, as hot air drying proceeds rapidly at the corners of the slab where the temperature maintains the highest, the maximum stress appears around these portions. However, the drying rate is somewhat controlled there in the case of internal heating because the temperature becomes the lowest at the corners. This fact is considered as a reason of lower stress formation by internal heating than convective heating. Indeed the present analytical model takes account for no influence of temperature on the apparatus diffusivity of moisture in the slab. However, as vapor pressure increases exponentially with rising temperature, the moisture movement in the body will be enhanced by the internal pressure diffusion of vapor. This effect on the drying performance must be studied by further advanced model and experiment in the future. Intermittent heating contributes also a remarkable fall of the stress in the low heating period, but the stress results in loaded much strongly again in the high heating period. Hence it is implied that the internal heating mode may be the most effective one to drying of the slab if the drying-induced stress is compared among those heating modes under a limited simulation in the present work.

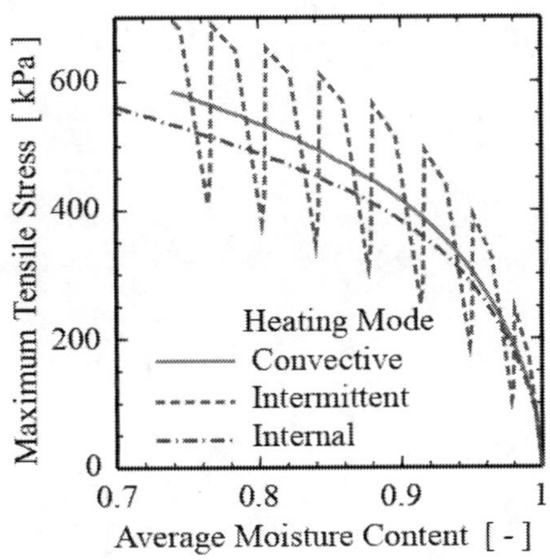

Fig. 5. Tensile stress against moisture content.

3 Drying experiments by microwave heating

Drying experiments of ceramics molded into a slab are conducted by applying microwave to examine the effect of the internal heating on drying characteristic and crack generation behavior. Figure **6** shows an outline of the applicator in the experimental setup. The applicator is made of a steel duct with 100 x 100 mm in cross section and 500 mm in length. The test section is located in the center of the duct, and a plate of ϕ50 mm of diameter for a sample holder is connected to an electronic balance to measure the time behavior of sample weight during drying. A window with a door of steel plate is placed at the side of the test section for putting the sample in and out. Microwave generated in a magnetron (Tokyo Electronics Inc., IMG-2501S) was transmitted to the sample through the waveguide joint on the test section from the above. The frequency is 2,450 MHz and the power varies from 100 to 1,000 W. An isolator, a stab-tuner and a power monitor are installed for tuning microwave between the magnetron and the applicator. Convective heating can take place simultaneously by flowing hot air in the duct of the applicator. Steel wire nets are set at both the inlet and outlet of the test section in the duct to seal the microwave from leaking out. The open face on the duct joint with the waveguide is covered by a glass plate to prevent the turbulence of hot air flowing in the duct and diffusion of steam evaporated into the waveguide.

Drying by convective heating was performed without irradiation of microwave using the same dryer as in Fig. 6. The drying behavior by microwave heating and convective heating are compared with the data by radiation heating in an oven type dryer (Yamato, Drying Oven DX31).The samples were prepared by molding kaolin into a slab as following: the dry powder of kaolin 50 g mixed with 20 g water was pressed in the mold made of aluminum with the pressure 200 kPa. The slab size was 60x60x10 mm. The kaolin is used industrially for tile production with the median particle size, 0.75 μm. The sample was put horizontally keeping out a clearance with small spacer on the holder in the test section so that a surface of the slab faced to incident microwave and the both sides were subjected to parallel flow of air. The drying rate was determined from the time behavior of the sample weight measured by electronic balance. The transient temperatures were measured by luminescence types of optical fiber thermometer (Anritsu Instrument, FX-8500) placed at the center core and the surface of the sample.

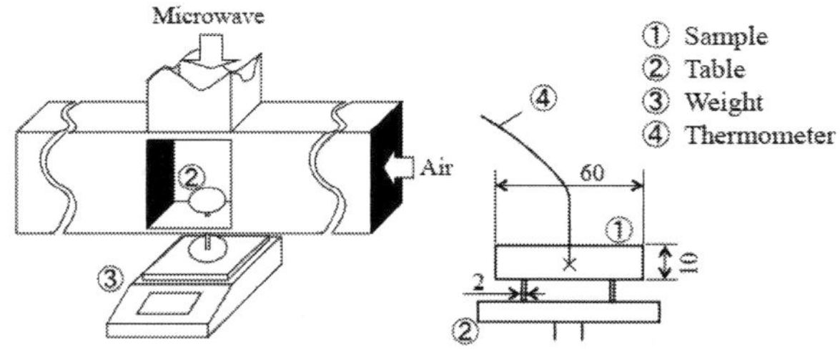

Fig. 6. Outline of applicator for microwave drying experiment.

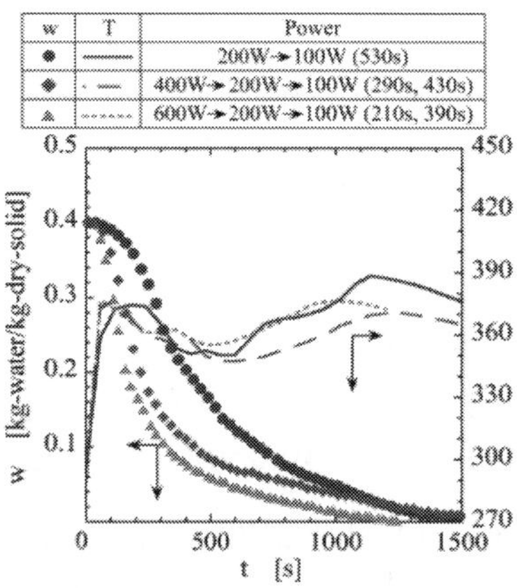

Fig. 7. Transient behavior of kaolin slab by microwave heating controlled dynamically.

It was implied that a rise in the internal temperature up to the boiling temperature of water causes crack by the expansion throughthe experiments. In order to complete microwave drying of molded samples without crack generation in as short time as possible, a control of the internal temperature must be a key factor. Here, microwave power was controlled dynamically in the progress of drying. The time behaviors of the moisture content and the internal temperature were plotted in Figure **7** when the experiments were performed in three dynamic patterns of the power: the first one is 200 W for the first 530 seconds and sequentially maintaining 100 W until the drying completion, the second one 400 W for the first 290 seconds, 200 W for the next 430 seconds and 100 W, and the last one 600 W for the first 210 seconds, 200 W for the next 390 seconds and 100 W. The internal temperature was maintained under the boiling point except in the last period of drying for any dynamic control pattern. Crack generation was not observed on the sample through the drying period as well. The drying time was reduced remarkably until approximate 1,500 seconds and less while it was the order of 10^4 seconds in the oven dryer and over 4,000 seconds in hot air drying. Particularly the third pattern was only 1,250 seconds. From these facts, it is found that dynamic microwave heating contributes greatly to drying enhancement of molded ceramics.

4 Drying of triply layered nonhomogeneous foods

4.1 Modeling

Governing equations for the transient heat and moisture transfer in food are expressed by modifying the conservation equations [15] on the phase change of moisture as follows, respectively:

$$c_p \rho_b \frac{\partial T}{\partial t} = -\mathrm{div}(\mathbf{J_h}) + \sum_j \Delta H_r R_j + \Delta H_v \varepsilon \frac{\partial C}{\partial t} \tag{16}$$

$$\frac{\partial C}{\partial t} = -\mathrm{div}(\mathbf{J_m}) + \sum_j R_j + \varepsilon \frac{\partial C}{\partial t} \tag{17}$$

where C is moisture concentration given by the unit kg-moisture/m^3-wet solid, ΔH is enthalpy, R reaction rate, T temperature, t time, ε is the parameter on the rate of internal moisture vaporization or condensation. Assuming the fluxes due to heat conduction and moisture transfer for heat transfer and due to the gradients of moisture concentration, temperature and pressure for moisture transfer, the total fluxes, $\mathbf{J_h}$ and $\mathbf{J_m}$ in Eqs.(16) and (17) are expressed as

$$\mathbf{J_h} = -\mathbf{k_t} \cdot \mathrm{grad}(T) - (-1)^m \mathbf{k_c} \cdot \mathrm{grad}(C) - (-1)^n \mathbf{k_p} \cdot \mathrm{grad}(P) \tag{18}$$

$$\mathbf{J_m} = -\mathbf{D_w} \cdot \mathrm{grad}(C) - \mathbf{D_t} \cdot \mathrm{grad}(T) - \mathbf{D_p} \cdot \mathrm{grad}(P) \tag{19}$$

where $\mathbf{k_t}$, $\mathbf{k_c}$ and $\mathbf{k_p}$ are the thermal conductivity, the Dufour thermal conductivity and the filtrational thermal conductivity, respectively and $\mathbf{D_w}$, $\mathbf{D_t}$ and $\mathbf{D_p}$ are the mass diffusivity, the Soret mass diffusivity and the pressure mass diffusivity, respectively. To apply the governing equations to the moisture transfer in non-homogeneous food, whose components have generally different moisture sorption characteristics with each other, the concept of the mass transfer potential ϕ [15] defined by following equation was introduced

$$dW = \left(\frac{\partial W}{\partial \phi}\right)_T d\phi + \left(\frac{\partial W}{\partial T}\right)_\phi dT \tag{20}$$

where W represents the moisture content and is related easily to the water concentration. Substituting Eq. (20) into Eqs. (16) and (17), the equations are rewritten as:

$$\left\{ c_p \rho_b - \Delta H_v \varepsilon \left(\rho_s + \frac{\partial \rho_s}{\partial W} W \right) C_T \right\} \frac{\partial T}{\partial t} - \Delta H_v \varepsilon \left(\rho_s + \frac{\partial \rho_s}{\partial W} W \right) C_m \frac{\partial \phi}{\partial t} = -\mathrm{div}(\mathbf{J_h}) + \sum_j \Delta H_r R_j \tag{21}$$

$$\left(\rho_s + \frac{\partial \rho_s}{\partial W} W \right) \left(C_m \frac{\partial \phi}{\partial t} + C_T \frac{\partial T}{\partial t} \right) = -\mathrm{div}(\mathbf{J_m}) + \sum_j R_j \tag{22}$$

$$C_T = \left(\frac{fW}{fT}\right)_\phi, C_m = \left(\frac{fW}{f\phi}\right)_T \quad (23)$$

The mass transfer potential can be substituted by the potential corresponding to Gibbs free energy in a wet materials. Then the mass transfer potential is determined by (Luikov, 1975)

$$\phi = \mu_0(T) + RT\ln(a_w) \quad (24)$$

where a_w is an activity and is related to W by GAB equation form (Lomauro et al., 1985) or an isotherm curve of food.

$$W = \frac{w_m C K_a a_w}{(1-K_a a_w)(1-K_a a_w + C K_a a_w)} \quad (25)$$

where K_a is a constant. μ_0 is the chemical potential in the state of pure saturated water.

$$\mu_0 = \Delta G^0 = \Delta H^0 - T\Delta S^0 \quad (26)$$

ΔG^0 and ΔS^0 are the Gibbs free energy and the entropy, respectively.

A linear viscoelasticity, that is suggested as a reliable behavior of strain-stress for several foods, was assumed. The governing equations for a viscoelastic boundary value problem are represented as the same formulas in Eqs. (7) to (12). Here the reduced time, ξ and ξ' in Eq. (7) are defined to modify temperature and moisture content dependency on stress by extending or reducing the effective time as

$$\xi = \frac{t}{a_T(T)a_M(W)}, \quad \xi' = \frac{\tau}{a_T(T)a_M(W)} \quad (26)$$

where a_T and a_M are temperature and moisture shift factors, respectively.

Additionally the equilibrium equation and the strain displacement relation are necessary, but they are the same as conventional ones [16].

4.2 Sample preparation

The hydrates of two different components were used as a representatively nonhomogeneous food sample: 1) Hylon VII starch (63.2% amylose starch provided by the National Starch and Chemicals Co.), 2) mixture of Hylon VII (75%) and sucrose (25%). The former one abbreviated by H while the latter one by S later. Each component was layered as seen in Figure **8**. The moisture sorption characteristics of two components are shown in Figure **9**.

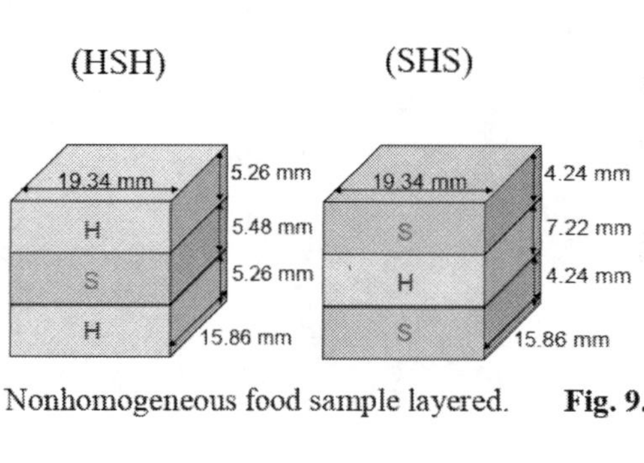

Fig. 8. Nonhomogeneous food sample layered.

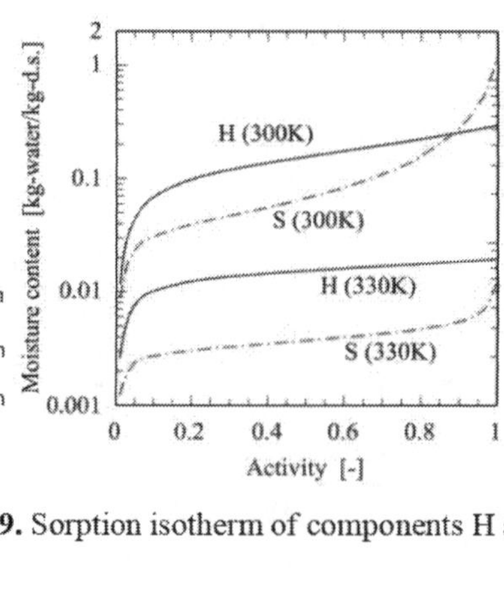

Fig. 9. Sorption isotherm of components H and S.

4.3 Drying behavior and internal stress formation

Figure 10 and 11 compare the experimental data on the temperature and moisture content of the triply layered sample subjected to drying with theoretical results obtained from the present analytical method. In the figures, W_1, W_2 and W_{av} are average moisture content in the outside, the inside and the whole layers, respectively. The contents are normalized by the initial moisture content. The physical properties of the sample required for the simulation are referred from the literature [14]. A large difference of time behavior of average moisture content was observed between HSH and SHS due to the difference of sorption isotherm between H and S. The moisture content in the outside layers for SHS is remarkably reduced with the drying time and is lower than that in the inside layer through whole drying time. The moisture content in the outside layers for HSH is much close to that in the inside layer. "Calc. 1" and "Calc. 2" correspond to the results solved numerically by using meshes divided into 150 and 1200 elements, respectively. The numerical result solved with 1200 elements agrees better with the experimental results than that with 150 elements for both the HSH and SHS, and a little deviation was observed between Calc. 1 and Calc. 2.

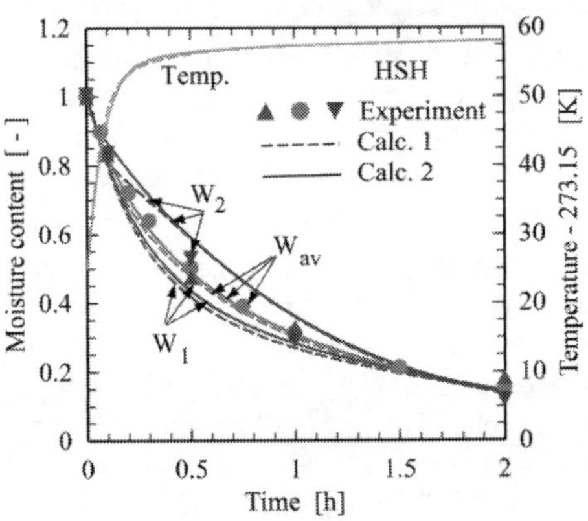

Fig. 10. Transient moisture content of each layer and temperature in the core center (HSH).

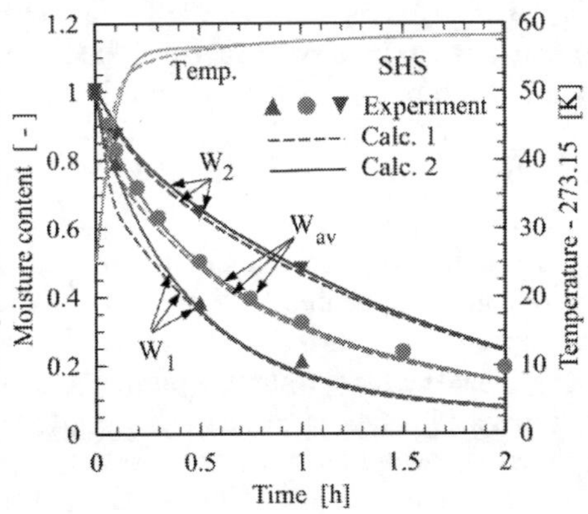

Fig. 11. Transient moisture content of each layer and temperature in the core center (SHS).

Figure 12 shows the maximum principal stress analyzed by the theoretical model against the drying time. The maximum magnitude of tensile and compressive stresses generated inside the body is plotted to compare the analytical accuracy between Calc. 1 and Calc. 2. It is considered that a crack may be formed by tensional stress of a critical magnitude. In the beginning of drying of 0.1 to 0.2 hour, a peak of tensile and compressive stresses was observed in the sample. The location was identified in the area of exposed surface for tension and the zone close to the interface of two components for compression from the stress distribution.

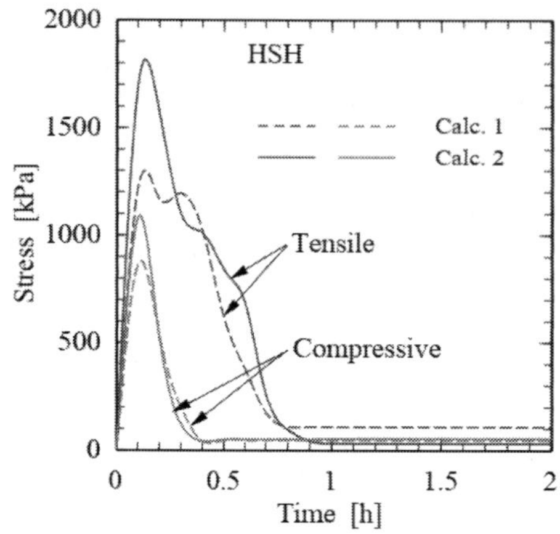

Fig. 12. Transient maximum tensile and compressive stresses in HSH.

Figure 13 shows examples of moisture content and temperature profilespredicted analytically at 1.0 hour of drying for HSH. The profiles compare the solution obtained from the combined analysis of heat and moisture transfer and strain-stress with strain-stress free analysis. The contours indicate the distribution on one-fourth of the bisected face of the samples because of its symmetric shape. Thus the point of $x = y = 0$ corresponds to a core center of the sample. The horizontal line in the body is an interface of two components. The temperature distributions are fairly uniform at this short drying time. The discontinuity of moisture content appeared on the interface. The moisture content in the inside layer close to the interface is lower than that of the outside layer for the sample of HSH. The reason is due to the fact that the mass transfer potential of S is greater than H at this moisture levels around the interface. This fact means that the moisture content of H becomes higher than S when the mass transfer potential or activity is same.On the other hand, both the analyses of heat-moisture transfer with and without strain stress give the almost same profiles of moisture content and temperature. From these analyses, the profiles of moisture content and temperature predicted will be less sensitive to strain-stress formation as far as weak drying-induced shrinkage takes place in a similar level to the present samples at least (<1%). This fact suggests that the drying behaviors can be analyzed accurately with short CPU time unless strain-stress equations are solved simultaneously.

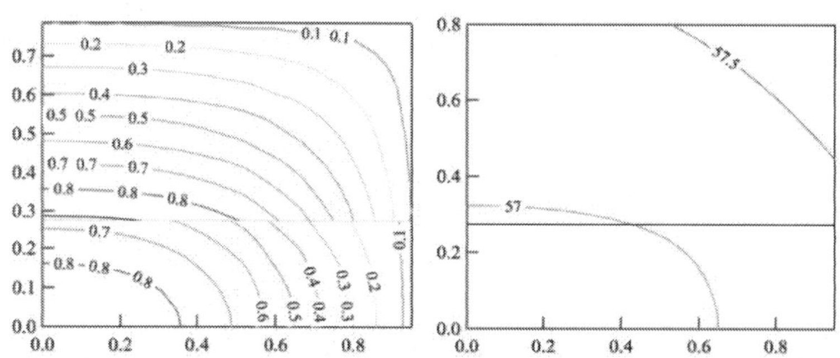

(a) Moisture content (strain-stress) (b) Temperature (strain-stress)

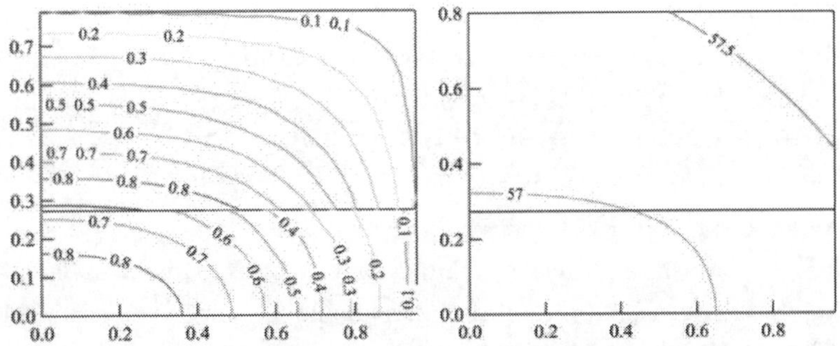

(c) Moisture content (strain-stress free) (d) Temperature (strain-stress free)

Fig. 13. Profiles of moisture content and temperature in HSH at drying time 1.0 h (1200 elements).

5 Conclusions

The outline of a series of the works was introduced on drying-induced strain-stress and deformation of modeled media performed by the present author and his co-authors. They are summarized as:

1. The transient three-dimensional problem of strain-stress and heat and moisture transfer was solved simultaneously by the finite element method.
2. The behaviors of internal strain-stress formation and deformation are compared among three modes of hot air heating, intermittent heating and internal heating for a ceramic slab.
3. The internal heating modemaintains stress formation at the lowest level and can enhance drying rate in the three modes.This effectiveness of the internal heating is examined by employing the microwave heating. Microwave heating enhanced significantly drying rate without crack formation compared with convective heating.
4. Drying of three layered foods consisting of two ingredients wasmodeled introducing the concepts of the mass transfer potential and a linear viscoelasticity to consider the different sorption isotherm and creep phenomenon.
5. The combination of different foods influences significantly not only the drying characteristic but also the internal strain-stress generation.
6. The drying behavior could be analyzed with a high accuracy even if only heat and moisture transfer were solved without strain-stress analysis.

References

1. I. Shishido, T. Maruyama, M. Funaki, S. Ohtani, Kagaku KogakuRonbunshu, **13**(1), 78 (1987)
2. I. Shishido, T. Muramatsu, S. Ohtani, Kagaku KogakuRonbunshu, **14**(1), 87 (1988)
3. M.Hasatani, Y. Itaya, K. Hayakawa, Drying Technology, **10**(4), 1013 (1992)
4. Y.Itaya, S. Mabuchi, M. Hasatani, Drying Technology, **13**(3), 801 (1995)
5. Y. Itaya, S. Taniguchi, M. Hasatani, Drying Technology, **15**(1), 1 (1997)
6. Z. X. Gong, A. S. Mujumdar, Y. Itaya, S. Mori, M. Hasatani, Drying Technology, **16**(6), 1119 (1998)
7. Y. Itaya, K. Okouchi, S. Mori, Drying Technology, **19**(7), 1491 (2001)
8. Y. Itaya, S. Uchiyama, S. Hatano, S. Mori, Drying Technology, **23**(6), 1243 (2005)
9. Y. Itaya, S. Uchiyama, S. Mori, Transport in Porous Media, **66**(1-2), 29 (2007)

10. R. J. Gustafson, D. R. Thompson, S. Sokhansany, Trans. ASAE, **22**, 955 (1979)
11. R.N. Misra, J.H. Young, Trans. ASAE, **24**(3), 751 (1981)
12. K. Haghighi, L.J. Segerlind, Trans. ASAE, **31**(3), 930 (1988)
13. K. Haghighi, L.J. Segerlind, Trans. ASAE, **31**(3), 938 (1988)
14. T.Tsukada, N. Sakai, K. Hayakawa, J. Food Sci., **56**(5), 1438 (1991)
15. A.V.Luikov, Int. J. Heat Mass Trans., **18**(1), 1 (1975)
16. Y. Itaya, T. Kobayashi, K. Hayakawa, Int. J. Heat and Mass Transfer, **38**(7), 1173 (1995)
17. Y. Itaya, N. Kobayashi, Proc. 17th International Drying Symposium (IDS 2010), Magdeburg, Germany, 3-6 October (2010)

Extrusion Process of Polypropylene Composites Reinforced Milled Carbon Fibre for Conductive Polymer Composite Application

Nabilah Afiqah Mohd Radzuan[1,*], *Abu* Bakar Sulong[1,2] , and *Mahendra* Rao Somalu[1]

[1]Fuel Cell Institute, Universiti Kebangsaan Malaysia, 43600, UKM Bangi, Selangor, Malaysia
[2]Centre for Materials Engineering and Smart Manufacturing, Mechanical Engineering Programme, Faculty of Engineering and Built Enviroment, Universiti Kebangsaan Malaysia, 43600, UKM Bangi, Selangor, Malaysia

Abstract. A polypropylene (PP) reinforced milled carbon fibre (MCF) was developed to produce high conductive polymer composites. Theoretically, by altering the filler orientation, the electrical conductivity and mechanical properties can be controlled. However, the orientation techniques which influence the MCF fibre are difficult to performed. Therefore, this study focused on controlling the filler orientation through the extrusion process. Hence, the extrusion temperature of 230°C and rotational speed of 50 rpm at 70 wt.% of MCF and 30 wt.% of PP were used. The electrical conductivity in perpendicular to the extrusion direction was higher at 2.0 S/cm as compared to 0.66 S/cm in the parallel direction. Besides, the extruded composite of rod and sheet were studied in which rod dies offers higher electrical conductivity of 3.0 S/cm and better mechanical properties of 1225 MPa than the sheet dies. Alteration in filler orientation aid in enhancing the electrical conductivity as minimum fillers breakage occurred due to the low shear rate of 2.2 s^{-1} which indirectly induces the filler to the desired orientation. Therefore, the extrusion process able to improve the electrical conductivity and mechanical properties of composite materials, as the filler oriented perpendicular to extrusion direction.

1 Introduction

Conductive polymer composites (CPCs) have been extensively studied for various field of application due to its remarkable features and performance [1]. In order to produce a highly conductive material, the conductive phase above the critical value is needed to ensure composite material stay electrically conductive [2]. However, the electrical conductivity of CPCs material depends on many factors including filler volume faction, filler shape and size, filler orientation, shear rate and the morphology of the matrix. The optimization of the processing parameter is a method which influence the electrical conductivity and mechanical properties of the CPC. Recently, compression molding and injection molding are the two common technique used to fabricate CPC materials [3]. Although this method has been used

* Corresponding author:afiqah@ukm.edu.my

extensively, studied shows that pre-mixing process are able to enhance the electrical and mechanical performance of the CPCs [2]. This pre-mixing process is crucial for achieving a proper filler dispersion in polymer matrix, modest shear forces and increase the ability to form a homogeneous matrix in combination with polymer resin [4, 5]. Extrusion process using twin screw extruder has been used widely in various industries including polymer processing and food processing due to its ability to run multiple processes that can be conducted continuously [6]. Parameters including shear rate, temperature and filler aspect ratio need to be considered in order to optimize the extrusion process. In this study, to ensure a highly electrically conductive polymer composite optimum die configurations were studied and analyzed.

2 Methodology

2.1 Materials and Fabrication Process

Milled carbon fibre (MCF), Grade CFP-7-50 was supplied by ShenZhe Yataida High-Tech. Co. Ltd.with a diameter of 9 μm, average length of 300 μm, aspect ratio of 43 and a density of 1.75 g/cm^3. Meanwhile, the polypropylene (PP) selected as a binder in this study was supplied by Goonvean Fibres Ltd. due to its extraordinary properties; mechanical strength, low density, electrical resistance, and greater permeability by gases [1]. This PP polymer has a maximum average size of 90 μm, density of 910 kg/m^3 and melting index of 10g/10 min at 160 ^{0}C. The PP was supplied in powder form with grade number Titan-600. Initially, the PP powder and the MCF, were pre-mixed based at 30 wt.% of PP and 70 wt.% of MCF. The pre-mixed process was conducted using a mechanical mixer (Model RM 20-KIKA-WERK) at room temperature before being melt extruded using a twin-screw extruder, (Thermo Haake TSE Twin-Screw Extruder). The extruder was set at a temperature of 230 °C with a rotational speed at 50 rpm for 30 minutes using a different dies shapes [2]. The fibre orientations were analysized using the field emission scanning electron microscope (FESEM) (Zeiss Supra-55VP, Germany). Meanwhile, the electrical conductivity was measured for both in-plane conductivity using Jandel four-point probe and an RM3 test unit [4-6].

2.2 Simulation of pre-mixing process

This study focused on three main dies configuration which are shapes and dimensions. Hence, series of simulation has been conducted to verify the pre-mixed process and its ability to orientate the fibre. The simulations were performed using the MoldFlow simulations software in which the input data were fist sketched using the AutoCAD Inventor. The dimensions of 3 mm and 5 mm opening diameter with rod and sheet shapes were considered in this study.

3 Results and Discussions

The simulations conducted were shows in Figure 1, in which the different fibre orientations within the polymer matrix were exposed. Previous studies on fibre orientations reported that the fibre orientation were depends on the dies opening and the shapes used [9]. As shows in Figure 1(c), rod die demonstrates the highest fibre orientation at parallel to the extrusion directions compared to the sheet dies at various die opening. These exhibited that the fibre tends to orientate in converging die compared to diverging die and results in fibre to orientate parallel to extrusion directions [10]. Differ results shows in sheet dies in Figure 1 (a) and

 https://doi.org/10.1051/matecconf/201824801012

Figure 1 (b) in which the fibre tends to orientate randomly within the process. This phenomenon was reported in previous studies in which the diverging dies will developed a fibre deformations and results in random fibre orientations [11]. Despite that, the die opening thickness also contribute in fibre orientations. As the die opening thickness reduces from 5 mm to 3 mm, the fibre tends to become compact and compress. These further results in fibre to orientate parallel to extrusion directions. Moreover, Figure 1 also indicated the orientation tensor of each fibre at different die configurations. As shown in Figure 1, rod die exhibited highest orientation tensor at 0.9337 compared to other 5 mm and 3 mm sheet dies at 0.9243 and 0.9254, respectively. Hence, these simulations demonstrate that the experimental performed will encounter these similar results.

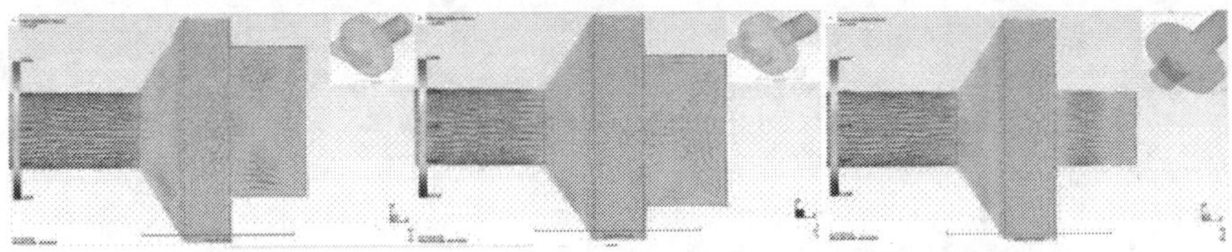

Fig. 1. Fibre orientation simulations of (a) 3mm sheet die, (b) 5 mm sheet die and (c) 5 mm rod die.

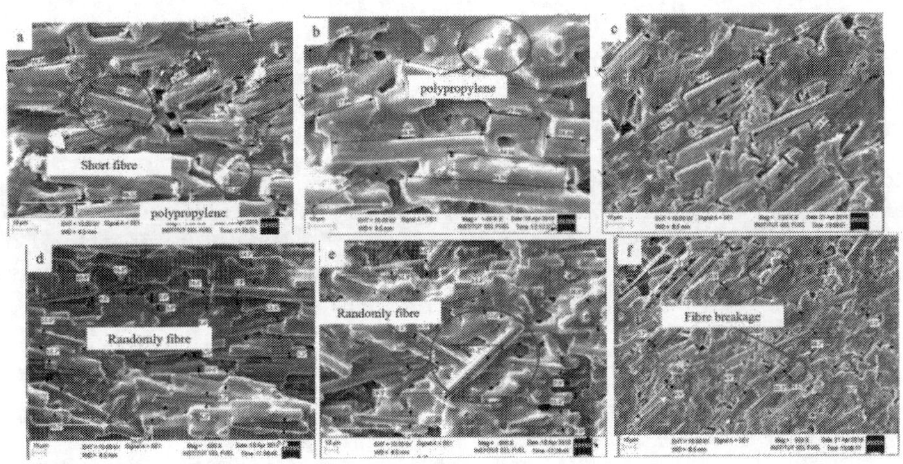

Fig. 2. Microstructural analysis at 1.0K X observations for (a) 3mm sheet die, (b) 5 mm sheet die and (c) 5 mm rod die, while at 500 X observations for (d) 3mm sheet die, (e) 5mm sheet die and (f) 5 mm rod die

Based on the simulation conducted, Figure 2 (a) and Figure 2 (b) shows the FESEM images in which the fibre tends to randomly orientate when using the sheet dies. The randomly orientated fibre was caused by the high shear stress which transfer into the fibre as the die opening were decreased. Hence, these results on fibre deformations especially at the die surfaces which result the fibres to randomly orientated [10–12]. Recent studies reported that the fibre which travel using the slit die will results in more randomly orientated fibre [12]. Despite that, Figure 2 (a) shows the higher fibre brakeage experience compared to Figure 2 (b). These are because the high shear rate was developed as the die opening were decreased. These were shown by Figure 2(b) of the sheet die with 5 mm die opening which experience the fibre length of 42 μm. Meanwhile, the 3mm sheet die experience extreme fibre breakage at 28 μm of fibre length. In contrast, Figure 2 (c) indicated the fibre length of 55 μm, which are the longest among all the dies used. These occurred due to the converging die which enable the fibre to orientate uniformly within the polymer matrix [10, 12]. Figure 2 (d) and Figure 2 (e) shows the sheet dies at 3 mm and 5 mm die opening in which the orientation angles are far below the 90°. Studies reported that the reduced in orientation angles indicated that the fires tend to orientate randomly due to high viscosity, low shear rate, and longer cooling time [13]. Table 1 shows the electrical conductivity of the composite materials at different die configurations. The findings are significant with the microstructural and simulation analysis. The rod die exhibited the highest electrical conductivity with 3.66 S/cm, while 3 mm sheet die and 5 mm sheet die offered 0.56 S/cm and 1.26 S/cm, respectively. The excellent in performance occurred on the rod dies was due to the fact that the fibre able to orientate uniformly within the polymer matrix. Recent study suggested that this phenomena was due to the electron tunnelling as the fibre developed an excellent conducing pathways [15].

Table 1. Electrical conductivity at different die configurations

Die	Electrical conductivity
3 mm sheet	0.56 S/cm
5 mm sheet	1.26 S/cm
5 mm rod	3.66 S/cm

5 Conclusions

The experimental conducted shows an excellent correlation when compared to the simulation developed. Studies exhibited that the extruded MCF/PP composite using a 5 mm rod die geometry results in better electrical conductivity due to its better fibre orientation in the converging dies. The electrical conductivity produced using the rod die is 3.66 S/cm shows that the converging dies able to orientate the fibre parallel to the extrusion direction.

The authors gratefully thank the Universiti Kebangsaan Malaysia for the financial support to complete this study under grant number MI-2017-004.

References

1. R. Dweiri, J. Sahari, Journal of Power Sources, **171** (2): 424–432.
2. M. Y. Zakaria, A. B. Sulong, J. Sahari, H. Suherman, Composites Part B: Engineering, **83**: 75–80.
3. B.K. Kakati, D. Sathiyamoorthy, A. Verma, Int. Journal of Hydrogen Energy, **36**(22): 14851–14857.
4. R. Dweiri, J. Sahari, Comp. Sci. and Tech., **68**(7–8): 1679–1687.
5. N. Hu, Z. Masuda, G. Yamamoto, H. Fukunaga, T. Hashida, J. Qiu, Composites Part A: Applied Science and Manufacturing, **39**(5): 893–903.
6. Y. Nakayama, E. Takeda, T. Shigeishi, H. Tomiyama, T. Kajiwara, T. Chemical Engineering Science, **66**(1): 103–110.
7. A. B. Sulong, M. I. Ramli, S. L. Hau, J. Sahari, N. Muhamad, H. Suherman, Composites Part B: Engineering, **50**(0): 54–61.
8. S. P. Liu, S. S. Hwang, J. M. Yeh, C. C. Hung, International Communications in Heat and Mass Transfer, **38**(1): 37–43.

9. M. Kostic, L. Reifschneider, *Encyclopedia of Chemical Processing.*
10. P. J. Hine, N. Davidson, R. A. Duckett, I. M. Ward, Composites Science and Technology, **53**(2): 125–131.
11. G. Ausias, J. Jarrin, M. Vincent, Composites Science and Technology, **56**(7): 719–724.
12. T. Köpplmayr, I. Milosavljevic, M. Aigner, R. Hasslacher, B. Plank, D. Salaberger, J. Miethlinger, Polymer Testing, **32**(3), 535–544.
13. P. Pötschke, A. R. Bhattacharyya, A. Janke, European Polymer Journal, **40**(1): 137–148.
14. M. Sharma, I. M. Rao, J. Bijwe, Wear, **267**(5–8): 839–845.
15. J. Feller, I. Linossier, Y. Grohens, Materials Letters, **57**(1): 64–71.

Development of a Web-Based Convergent Hospital Billing System

Adhistya Erna Permanasari*, *Silmi* Fauziati, *Argi* Kartika Candri

Department of Electrical Engineering and Information Technology, Universitas Gadjah Mada, Jl. Grafika no 2, Yogyakarta, 55281, Indonesia

Abstract. Given the large volume of data generated in hospitals, the usage of hospital information system (HIS) that may ease the hospital's workload on managing patient bills is critical. Thus, this paper presents the development of a web based hospital billing system which allows the staffs to manage and store patient bills in such a way that it can integrate bills from various departments. In addition to that, the system has additional features to allow down payments, automatically generate accounting reports, as well as integrate some features of the registration and the pharmacy system. The system design followed the Waterfall approach and was done using the use case diagram, activity diagram, sequence diagram, and entity relationship diagram. The data processed was actual data from the hospital and dummy data for the pharmacy system. The system's functionality was tested using the black-box method and then evaluated by the hospital staffs using the SUS (System Usability Scale) method. From there, we obtained a score of 77.5 out of 100 from the SUS evaluation and 100% success rate from 102 features tested with the black-box testing method. Therefore, we can conclude that the new system works well and has good usability.

1 Introduction

As one of the most important social organizations that play a major role in providing health–treatment services to the citizens, hospitals always deal with the large volume of collected data, mainly transactions and bills. Hospitals are no exception to this. Thus, implementing Information Technology (IT) in the form of a Hospital Information System (HIS) is essential to manage patient bills in hospitals efficiently.

Vegoda defined Hospital Information System (HIS) as an information system which is integrated in such a way to improve patient care by providing information necessary to allow an increase in the user's knowledge and reduce in uncertainty, thus allowing rational decisions to be made [1]. In HIS, different software are integrated to grasp the whole data in certain hospital sections. A good HIS should be able to cover various departments and personnel of the hospital. Meanwhile, there are few ongoing trends in the development of hospital billing system to this date [2]. The trends include classifying patient bills based on the services given, integrating bills from the various department, integrating payment module as well as the accounting module.

* Corresponding author: adhistya@ugm.ac.id

Another popular trend in hospital billing system development is breaking down the bills to explain where the down payment went forth. In addition to that, Kavatra et al., also mentioned that one of the most important features of a billing system is to support down payment [3]. According to them, down payment should be compulsory and must be paid in advance to ease the hospital's burden when collecting the patient bills. Finally, Lin argued that hospital bills should be broken down into tiniest details and the one charged to the patient should have already been summarized from bills charged by various departments [4].

This paper presents a convergent hospital billing system, developed to help the staffs in calculating and printing patient bills. The term convergent means that we developed the system in such a way that it can integrate invoices from various services into one single invoice. In addition to that, we also integrated features to allow advance payment as well as automatically generated financial reports. Likewise, we also integrated some features of the registration and pharmacy system. We designed the system using waterfall software development method. Later, we asked the staffs of hospital 'X' where this study took place, to test the system usability using the System Usability Scale (SUS) method. Moreover, we also conducted a test on system functionality using the black-box method. The data processed in the conducted tests was actual patient data and dummy data for the pharmacy system.

The remainder of the paper is structured as follows. Section 2 describes methods. Section 3 presents system architecture. Section 4 reports results and analysis. Finally, Section 5 presents the conclusion of the study.

2 Methods

We developed the system on PCs using HTML, CSS, JS, and PHP programming language. The workflow processed in this system was adapted from the actual system flow running in Hospital 'X'. Meanwhile, we adapted the waterfall software development model in this research for its simplicity, flexibility and straightforward process. Those stages include requirement analysis and definition; system and software design; implementation and unit testing; and at last, integration and system testing [5].

2.1 Requirement analysis and definition

In this step, we gathered and analyzed requirements needed to build the system. The requirements were collected from various sources and a series of meetings with the hospital staffs. Throughout the gathering process, we sought to understand how the end-user hoped to use the system. Later, we successfully identified the actors and their use cases.

2.2 System and software design

After the requirements gathering was done, we designed some diagrams to understand the system workflow better. The diagrams are the use case diagram, activity diagrams, and sequence diagrams. The use case diagram was used to model the interactions between users and the tasks performed by the system. It is illustrated in Figure 1.

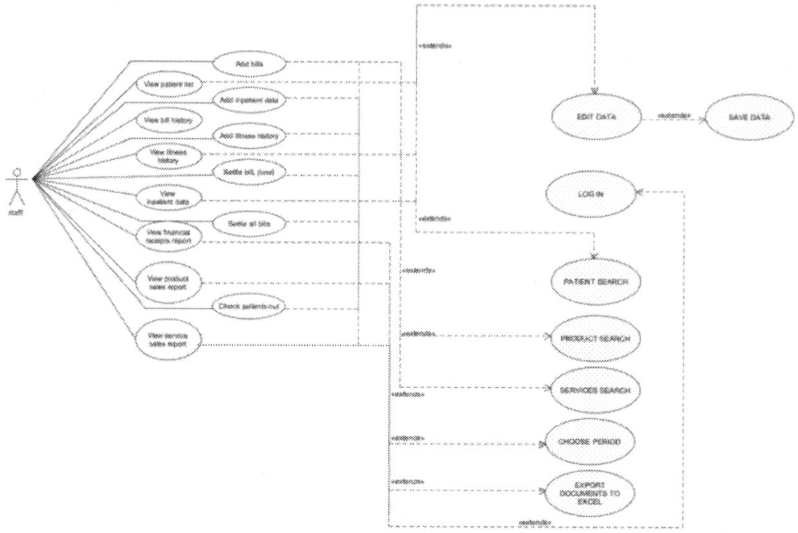

Fig. 1. Use case diagram of the system.

Here, we designed the front desk staff as the main actor, since they are the ones responsible for dealing with customers and document their bills. Later, the use case diagram was further described using interaction diagrams, which are the activity and sequence diagrams. After finished with the diagrams, we also designed low-fidelity prototypes to understand better the system workflow and how it should have worked.

2.3 Implementation and unit testing

In the last step, we implemented the features into one working system and tested it to ensure its reliability. The system's functionality was tested using the black-box method and then evaluated by the hospital staffs using the SUS (System Usability Scale) method.

3 System architecture

In this research, the system was implemented as a web application and accessible using a browser. The system architecture can be seen in Figure 2. Since it was still initial design, the system architecture is quite a simple one. The database was built in a local server and could be accessed through the HTTP protocol. The login rights assignment has yet to be implemented since there are still many changes to come, and thus implementing it at an early stage will probably disrupt the workings of the whole system.

Here, the data was stored in a local database, and the administrator could access the data with an internet connection. We developed the system by using the CodeIgniter framework for its easy configuration and flexibility of going into the very depth of development. In addition to that, CodeIgniter framework offers the easiest way to utilize modular programs and get the specific function through the framework.

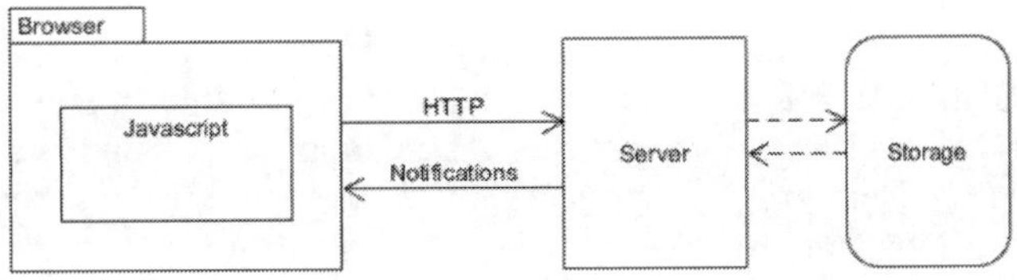

Fig. 2. System architecture.

4 Results

4.1 System implementation

By default, the user will be greeted with the home page when they started the system. The home page contains a handy collection of quick links to the features deemed most important to the user. In total, there are five options available on the menu, which are Home, Products and Services, Patient Directory, Financial Statements, and Master Data. Just as the name suggested, we could access the features to manage patients by clicking on the Patient Directory menu. We might also view the sales reports and financial statements by clicking on the Financial Statements menu. Figure 3 presents the homepage of the billing system.

Fig. 3. Billing system interface.

4.2 System evaluation

In this research, evaluation is done by using three methods: System Usability Scale (SUS) test, black box test, and implementation test at Hospital 'X'. SUS functions to test application usability, black box test is used to test whether the application is working properly or not, and the implementation test is intended to find out whether the application is ready to be implemented or not.

- Black box
  Black box testing is a test used to evaluate the fundamental aspects of the application is the application can work well or not without regard to the internal logic structure of the application. Testing is done by testing each function in the application. Based on the test results, the system functionality of the application can work as expected.

- System Usability Scale (SUS) test

We conducted Testing System Usability Scaling (SUS) with respondents from the Hospital 'X'. One of SUS advantage is that it can be used on small sample sizes with reliable results [6,7]. The respondents were three doctors. Selection of respondents tailored to the usefulness of the application is the application used to record patient medical record data in Hospital 'X'. Respondents are taken based on their direct association with the application and represent each related profession. The test results are shown in Table 1.

Table 1. SUS Evaluation Results.

	Old system			**New system**		
	U1	U2	U3	U1	U2	U3
Total	30	18	27	35	28	30
SUS Score (Total x 2.5)	75	45	67.5	87.5	70	75
Average Score	62.5			77.5		
Usability Value Improvement	$\frac{77.5-62.5}{62.5} x100\% = 24\%$					

SUS assessment is achieved by summing the scores for each question. Each point of the question contributes a value of 0 to 4. The value of odd number question (1, 3, 5, 7, and 9) is the position of the Likert value minus 1. As for the value of the even number (2, 4, 6, 8, and 10) is five minus Likert scale value. After the addition of each point value, the total value is multiplied by 2.5 to get the SUS value. The results of the system reusability test in Table 2 show that the user recognizes that the new system usability value is better than the old system (77.5> 62.5) and the new system usability value exceeds the acceptable usability limit value (77.5> 68). It can be seen that there is an increase in usability value of 24% compared to the previous system. Meanwhile, the functionality test results show a 100% success rate which means the system is running well.

5 Conclusions

This study presents a system of patient billing calculations and financial reports. The patient's bill is calculated based on the amount of product/drug bills and actions charged to the patient during the patient's treatment at the hospital. Financial statements are automatically generated based on the related input. Test results showed an increase of 24% compared to the old system. It indicates that the user well receives the proposed system.

References

1. P.R. Vegoda, *Int. J. Clin. Monit. Comput.*, **4**, 2 (1987).
2. P. Kumar and K. Kosalram, *Int. J. Inf. Eng. Electron. Bus.*, **1**, (2013).
3. S. Kavatra, A.N. Nagappa, P. Kumar, and R.K. Singla, *Indo Glob. J. Pharm. Sci.*, **3**, 3 (2013).
4. J. Lin, "Trends in Healthcare Payments and Billing", (2015).
5. I. Sommerville, "Software Engineering", 9th ed., Pearson Education, (2011).
6. J. Brooke, *Usability Eval. Ind.*, **189**, 194 (1995).
7. A. Bangor, P. Kortum, and J. Miller, *J. Usability Stud.*, **4**, 3 (2009).

Smart Clothline System Based on Internet of Thing (IoT)

**Zakiah* Mohd Yusoff[1,*], *Zuraida* Muhammad[2], *Amar* Faiz Zainal Abidin[3], *KA* Nur Dalila[1], *Noor* Fadzilah Razali[1], *Masmaria* Abdul Majid[1], *KK*.Hasan[1]

[1]Faculty of Electrical Engineering, Universiti Teknologi MARA, 81750 Masai, Johor,Malaysia
[2]Faculty of Electrical Enigneering, Universiti Teknologi MARA, Pulau Pinang,Malaysia
[3]Faculty of Electrical, Universiti Teknikal Malaysia Melaka, Hang Tuah Jaya, 76100 Durian Tunggal, Melaka, Malaysia

Abstract. Along with the growth of economic and infrastructure development, the level of invention or technologies is increasing steadily. These modern technology or invention is very useful to help people ease their daily life. Nowadays, there are many new inventions that have been developed due to the people demands. This is supported by the development of new technologies. This is because people nowadays are busy with their household management or have long working hours that makes they do not have enough time to do all home works such as pick up their clothes at their cloth line. To handle this problem, an Internet of Thing (IoT) Smart Cloth Line is one of the solutions. In this project, a smart cloth line is used to avoid the clothes from moistened with rainwater. When rain falls, the mechanical parts of cloth line will compress and expand when rains stop. The cloth line consists of Arduino UNO microcontroller, rain sensor, LDR, temperature and humidity sensor, servo motor and LCD display. During rainy day, the rain sensor will detect the present of water and servo motor will compress the mechanical parts of the cloth line. This is also an IoT based project so when anything happens when it is raining, it will automatically send notification to the user smartphone. The users also can get notification of temperature, humidity and intensity of light around their cloth line. Obviously, this project can help people in their daily life.

1 Introduction

Nowadays, housing is one of the important needs of society. In the modern world and better economic situation, the demands of having a safe and comfortable home is increasing rapidly for many people in urban areas since most of the people turning to technology in order to improving the daily life [1, 2]. The main purpose of smart home is to make sure homeowners can create comfort and safe when they inside or outside the house beside the home is built friendly to the environment, preventing social problems, how space it uses, or energy efficiency, maintaining privacy and recycling waste water[1, 2]. An intelligent smart home system has their own computer system which is able to monitor many aspects of daily

* Corresponding author: zakiah9018@johor.uitm.du.my

life. It's consist of home appliances, sensors, actuators, analyser, data processors [3]and can be applied either by wired or wireless[4].

The Internet of Things (IoT) is a device that connected to the network by any location within the globe that can be monitored and controlled remotely by user[5, 6]. The IoT that was first introduced by Kevin Ashton in 1999 is not something new since there are various implementation in many years ago in many places[2, 7, 8]. For example, [1] developed as smart home based on Raspberry PI and Android mobile device to manage smart home behaviour by easy to implement, flexible and scalable solution. The Development of Energy Saving Smart Home Prototype by Panna [9] developed an IoT smart home prototype using PIC18F458 microcontroller with the purposed to save electricity. Besides that, Piyare and Song built "Smart Home-Control and Monitoring System Using Smart Phone" to control and monitoring using micro-web server and smartphones. This work used an Arduino as a main controller to transmit data from the sensor to mobile user[10]. In IoT smart home development, there are many types of microcontroller that was implemented such as Arduino Uno, Arduino Mega, Raspberry PI, Global System for Mobile (GSM) and NodeMCU.[1, 2, 5, 10-12].

2 Methodologies

This project consists of 3 major parts: Hardware, Software and Simulation parts.

2.1 Hardware Design

The system hardware consists of three main modules which are sensing module, controller module and output module.

Figure 1 shows the overall block diagram for IoT Smart Cloth Line system. The sensing modules are temperature and humidity sensor, Light Dependent Resistor (LDR) and rain sensor. The rain sensor is used to detect the present of water. The water will complete the circuit board that has been printed onto the rain sensor. It will act like a variable resistor where the more water on the board, the less the resistance value. This work used a capacitive humidity sensor and the thermistor to measure the surrounding air. It sends the signal to the data pin in digital by changing the total capacitance. This capacitance value is proportional with the humidity value.

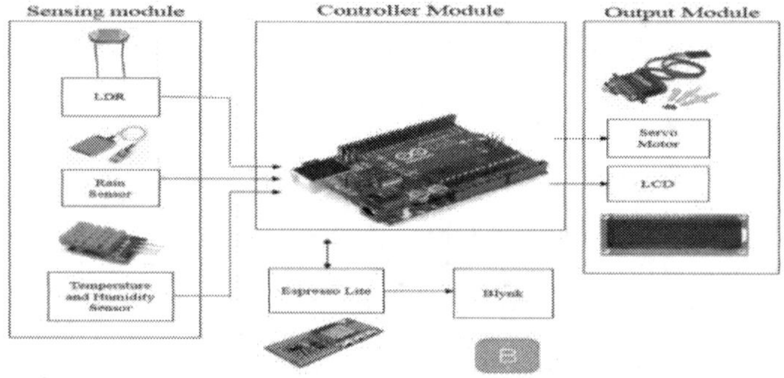

Fig 1: Block diagram of Smart Clothline System

The output module consists of DC motor and liquid crystal display (LCD). The DC motor can rotate until 180° in total. As the DC motor move, the potentiometer will move as well until shaft reach the desired position. The potentiometer is used to change the value of resistance according dc motor's rotation. Then, the control unit will calculate the movement either to the left or to the right. The LCD displays the actual reading of temperature and humidity at the surrounding area

The controller module includes of Arduino Uno and Espresso Lite that was added with the purposes to send the notification to the user via Blynk application. The Arduino UNO act as a microcontroller where it can be programmed by using Arduino IDE software. Arduino UNO has 14 digital input/output pins, USB connection, a power jack, and an ISCP header and reset button. It can be implemented by attach it to the computer with USB cable.

2.2 Software Design

Figure 2 shows the schematic diagram that shows the connections between all the components using Proteus software.

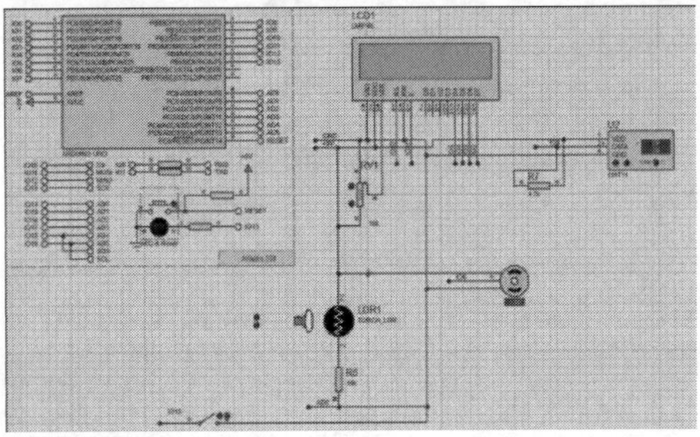

Fig 2: Overall schematic diagram

2.3 Simulation Design

The flowchart for the overall system is shown in Figure 3. In default condition, the Smart Cloth Line is originally expanding. Then, the rain sensor will detect the present of water and the LDR used to detect the present of light. Both sensors, LDR and rain sensor used to detect the respective elements. When the sensors detect these elements, it will send input data to the microcontroller, and then the Cloth Line will be compressed. Temperature and humidity sensor is added to add features in this work whereby it will detect the temperature and humidity from the surrounding. The microcontroller, Arduino Uno will coordinate the system by calculating the input from all sensors either to compress or expand the Smart Cloth Line system. Then, the system will send the notification to the user through mobile phone using Blynk application.

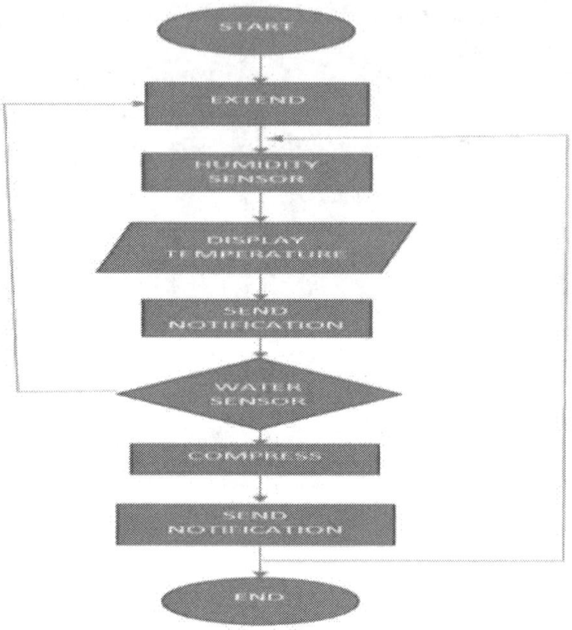

Fig 3: Flowchart for Smart Cloth line system

Arduino programming for the Smart Cloth Line system was done by using C++ programming language. This language consists of two main function which are 1) setup function which is used to initialize variables, initialize the input and output pin modes and other libraries needed in the program. Number 2) is loop function in which function loop is executed in the main program in the repeatedly manner. The example of coding is shown below:

```
void rainsensorLDR()
{
 int LDRstatus = analogRead(LDRpin);
 bIsRaining = !(digitalRead(nRainDigitalIn));
 if(bIsRaining==true || LDRstatus<=300){
  Blynk.notify("CLOTH LINE IS COMPRESS");
  Servoc.write(0);
 }
 else if(bIsRaining==false || LDRstatus>300){
  Servoc.write(45);
 }
 }
```

The *void* is used to read input from LDR and rain sensor. If the condition of rain sensor is true and LDR is below or equal to 300, Arduino will send notification to Blynk app according to the message and the servo is set to 0°. If the other condition is met the servo is set to 45°.

3 Results and Discussions

The prototype of Smart Cloth line system is shown in Figure 4.

Fig 4: Prototype of Smart Cloth Line System

At first the temperature and humidity sensor detect the surrounding temperature and humidity, then it will be display on the LCD display. If the rain sensor did not detect any presence of water, the servo motor in position 0° which mean the cloth line is expand as shown in Figure 5. Next is when the rain sensor detects the present of water, so the servo motor moves at angle 45° which makes the cloth line compress as shown in Figure 6. Then, LCD shows the actual reading of temperature and humidity from the surrounding as shown in Figure 7.

Fig 5: Servo motor at 0°

Fig 6: Servo motor at 45°

Fig 7: LCD Display

Then, Blynk application will send the notification to the user about the current temperature and humidity as shown in Figure 8.

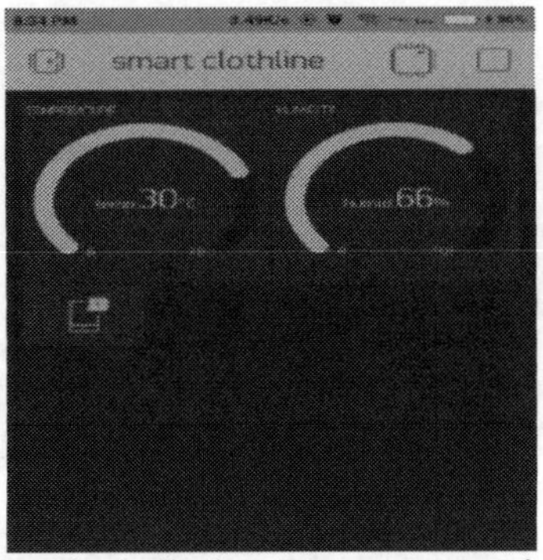

Fig 8a: Blynk displays temperature and humidity

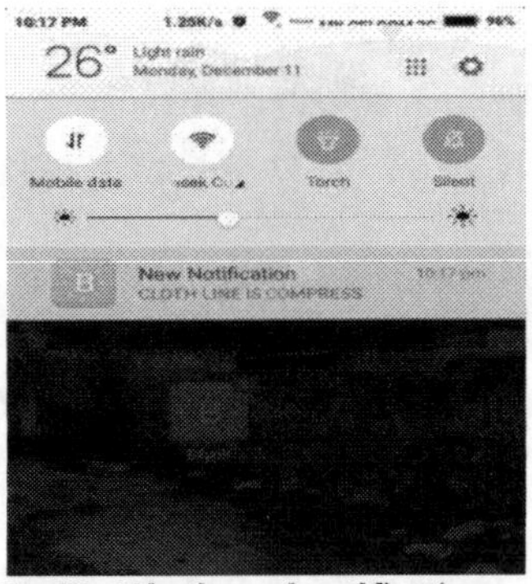

Fig 8b: Blynk send notification

4 Conclusions

This project is proposed as a method to improve the lifestyle of many people as the user can monitor and control the connected device using smartphone. The system which is based on Internet of Thing (IoT) detects the presence of water and light during the rainy day. The system will send the notification to the user through Blynk application. The result shows that the system has been successfully developed and has great potential for future commercialization since many people out there were too busy with their daily life. Thus, this kind of invention can help people to ease their daily life.

We would like to express our gratitude to UiTM Kampus Pasir Gudang and ARAS under project code: 600-IRMI/DANA 5/3/ARAS (0135/2016 for endless support and providing me an opportunity to this project, Smart Cloth Line System. Also thanks to friends for giving me an ideas and suggestions to embark this project. The author would like to express her thanks to the sponsor.

References

1. B. Davidović and A. Labus, Electronics and Energetics vol. **29**, pp. 451-460, (2016).
2. A. Yasirli and S. Mukhammad Andri. IOP Conference Series: Materials Science and Engineering, vol. 325, p. 012021, (2018).
3. H. Ghayvat, et al., "WSN- and IOT-Based Smart Homes and TheirExtension to Smart Buildings," Sensor, vol. 15, pp. 10350-10379, (2015).
4. W. S. Lee and S. H. Hong, "Implementation of a KNX-ZigBee gateway for home automation," presented at the Proceedings of the IEEE 13th International Symposium Consumer Electronics ISCE, (2009).
5. F. Tariq, et al., "Implementation of Smart Homes and Industrial Automation System with Secure Communication over GSM," Universal Journal of Electrical and Electronic Engineering, vol. 3, pp. 125-131, (2015).
6. PanthS and J. M., "Home Automation System (HAS) using Android for Mobile Phone," International Journal of Electronics and Computer Science Engineering (IJECSE), vol. 3, pp. 1-11, (2013).
7. "That 'Internet of Things' thing," RFID J., p. 4986, (2010).
8. "A Review on security in smart home development," Int. J. Adv. Sci. Technol., vol. 15, p. 13, (2010).
9. P. R, et al., "Development of Energy saving smart home prototype," Int. J. Smart Home, vol. 7, pp. 47-77, (2013).
10. P. R and S. R. L, "Smart Home-control and monitoring system using smart phone," in Proc. 1st Int. Conf. Converg. it's Appl. p. 83, (2013).
11. H. S. Bedi, et al., "Smart Trolley using Smart Phone and Arduino," Journal of Electrical & Electronic Systems, vol. 6, (2017).
12. M. Hidayat, et al., "Implementation of Wireless Sensor Network to reduce cashier queue," in 2014 Makassar International Conference on Electrical Engineering and Informatics (MICEEI), pp. 78-82. (2014)

Grid Study of Mini Hydro Power Plant (MHPP) of Palangai Hulu 2x4,9 MW South Pesisir

Hidayat[1*2], Cahayahati[1], Arnita[1], *Saiful* Jamaan[2]

[1]Electrical Engineering Department, Bung Hatta University, Padang, Indonesia
[2]PT. Multi Sukses Energi, Padang, Indonesia

Abstract. This paper presents a gridd study of Mini Hydro Power Generation (MHPP) of Palangai Hulu, 2 x 4.9 MW. It is owned by PT. Dempo Sumber Energi (DSE) located at Palangai of South Pesisir Regency. The objective is to fulfill the legal aspects, technical aspects and economic aspects to develop of MHPP. The legal aspect is one of the requirements that must be fulfilled prior to the signing of Power Purchase Agreement (PPA) between DSE and State Electricity Company that called PLN. Technical aspect is to analyze the operational impact of the distribution system after MHPP connecting to the PLN distribution system, while the economic aspect is to analyze how much result energy of MHPP can be absorbed by PLN. Uultimately, it is described the financial feasibility of MHPP. The result is, legal aspect of development of MHPP of Palangai Hulu feasible to be continued to next process to sign contract of PPA. Technically, there is no negative impact of operating the PLN distribution system after the MHPP is connected to the PLN Rayon Balai Selasa system that analyzes using ETAP. Point of Connection (POC) as the selling point of electrical energy by DSE to the most economical PLN is on substation circuit of Balai Selasa. The energy production of MHPP Palangai Hulu can be absorbed 96.4% starting in 2018. The cost of POC is Rp. 11,577,000,000.

1 Introduction

Electricity demand in Indonesia always increases every year. During the five years ago in the period 2010 to 2014 there was an increase in electrical energy sales from 145.7 Tera Watt hours (TWh) to 197.3 TWh, the number of costumers increased from 42.2 million to 57 million and the electrification ratio increased from 66.2 % to 84%. PLN estimates that between 2014 and 2024 electricity consumption will increase to 464 TWh with an average growth of 8.7% per year [1]. The Government's effort to face the condition is to make regulation in the form of Law No. 30 of 2009 on electricity and the launching of acceleration program of 35,000 MW power capacity development [2].

Fulfilling thoses needs will be built power plant by PLN and Independent Power Producer (IPP). Electrical energy that generated by PLN and IPP power plants are integrated into a transmission and distribution network, before the energy gets to customers. The power generation of IPP before it constructed and connected to PLN network system must be done

* Corresponding author: hidayat@bunghatta.ac.id

grid study that undertaken by the electrical developer's. This study aims to determine the point of connection (POC), the amount of power generated IPP that can be absorbed by the PLN without causing negative impact to existing PLN network system, connection facility, equipment capacity upgrade after MHPP connected. PT. PLN has published a guidebook for the Connecting Center for New and Renewable Energy (PLT EBT) to the PLN Distribution System. These are used by PLN and PLT-EBT developers to ascertain and mitigate the risk of disruption to the system due to PLT-EBT connections and the basis for signing the Power Purchase Agreement (PPA) [3]. Development of Palangai Hulu MHPP with capacity of 2x4,9 MW owned by PT. Dempo Hydro Power (DHP) located in South Pesisir Regency of West Sumatera Province is planned to be connected to 20 kV distribution line of GH Balai Selasa [4].

Several studies that required to be done to achieve the objectives of the grid study include (a) load flow analysis in the distribution system; (b) short circuit analysis; (c) breaking capacity analysis (d) protection analysis and coordination set coordination (e) stability analysis, (f) harmonic analysis, using the ETAP 12.00 programming on peak load times (WBP) and out of peak load times (LWBP) in 2014.

2 Modelling of Load Flow Power System

The load flow system study is a very basic and important conducted in the grid study of the MHPP. The parameters are obtained such as voltage, current, power and power factor or reactive power each bus in an electrical network when existing and future condition [5]. The load flow power system model is based on a power system model consisting of 2 buses, although in reality the power system consists of several buses that will be interconnected with each other. The formula of load flow power system can be shown as;
The current formulation entering a bus is expressed by equation (1)

$$I_i = V_i \sum_{j=0}^{n} y_{ij} - \sum_{j=i}^{n} y_{ij} V_j \quad \text{, with } j \neq i \tag{1}$$

Power in bus,

$$P_i + jQ_i = V_i . I_i^* \tag{2}$$

$$I_i = \frac{P_i - jQ_i}{V_i^*} \tag{3}$$

Substituting equation (3) to (1) is obtained,

$$\frac{P_i - jQ_i}{V_i^*} = V_i \sum_{j=0}^{n} y_{ij} - \sum_{j=i}^{n} y_{ij} V_j \tag{4}$$

Based on equation (4) it can be seen that the power flow on the channel is not linear and must be solved by iterative numerical method. Some common methods used to solve the problem of power flow include Newton Rapson method, Fast Decouple, Gaus-Seidel, etc.

3 Grid Study of MHPP Palangai Hulu

3.1 Electrical System of South Pesisir

The Electrical system of South Pesisir district includies in the management of PT. PLN (Persero) West Sumatra Area Padang Rayon Painan and Rayon Balai Selasa. Currently, Electrical system of Rayon Painan and Rayon Balai Selasa got supplies from the 150 kV interconnection system of Sumbagsel through GI Bungus using two 20 kV feeders along 60 kms. One of the feeder is an xpreess feeder. In GI Bungus, there is a power transformer 30 MVA, 150/20 kV. Based on the review to the field, the quality of the electric voltage at the Painan Rayon is very low, ie at LWBP condition only 17.6 kV and WBP condition 14.8 kV. This condition is caused by voltage drop on the feeder due to the distance and the amount of power that was evacuated from GI Bungus to GH Painan. Therefore, PLN has done the addition of 150 kV substation (GI) Kambang which is still in preparation for operation. The total load of PLN system of Rayon Painan and Rayon Balai Selasa in 2014 amounts to about 25 MW which is partly supplied of PLTD Lunang and PLTD Lakuak. If MHPP Palangai Hulu operate, then PLTD Lakuak is expected stop operation because the cost of renting this diesel is relatively high (Rp. 3.500, - / kWh) [6]. PT. PLN (Persero) of West Sumatera region, that GI Kambang is already operated in 2018, so the PLN of Rayon Painan and Rayon Balai Selasa is supplied from GI Kambang. Therefore GH Painan, GH Lakuak and GH Balai Selasa will enter into GI subsystem. The location of GI Kambang is at GH Lakuak which has one power transformer 30 MVA, 150kV/20kV [1]

3.2 Load Flow Study

One of the studies is needed to connect the Renewable Power Plant to the PLN distribution network i.e. the load flow analysis. This review is based on PLN Directive No.0357.K/DIR/2014 to review the impact of changes in power flow on existing distribution systems due to the inclusion of power plants with power capacity up to 10 MW. Purpose of power flow study as follows;

1. Determine the operating conditions of the electric power system in a steady state
2. To know the operating voltage of the system is the standard range within 5% higher and 10% below the nominal value.
3. To identify power generation requirements, reactive power support requirements (VARs) and the placement of capacitors or reactors in order to maintain the system voltage and power factor (PF) as required 0.9 leading up to 0.8 lagging.
4. To know the technical shrinkage network voltage either branching.
5. For the purposes of controlling the condition of the system.

Load flow analisys of MHPP Palangai Hulu which is planned to be connected to GH Balai Selasa with distance 8.36 kms from MHPP Palangai Hulu. It is assumed that GH Balai Selasa has got supply from main station GI Kambang. Another thing to consider is the inclusion of other MHPP plants that have been previously PPA with PLN and incorporated in Rayon Balai Selasa subsystem [3].

The first step is to determine the bus bar voltage of MHPP Palangai Hulu when it will be parallel to PLN in 2018. It is obtained with simulation of power flow using ETAP application that obtain voltage at the point of connection of MHPP of Palangai Hulu at WBP condition 16,46 kV and LWBP condition 17,96 kV. The voltage values are outside the provisions of PLN to perform parallel, because the voltage drops below 10% of nominal 20 kV. Adjusting of the OLTC transformer in MHPP Palangai Hulu, obtained voltage point connections on the conditions WBP 19.72 kV and LWBP 20.70 kV as Table 1

Table 1 The Condition of MHPP Palangai Hulu Will Just Parallel to Distribution System

Condition	Status Load 2018	Voltage of Bus bar 20 kV MHPP (kV)	Noted	Distribution Line Looses (kW)
MHPP Palangai Hulu Connent to GH Balai Selasa	WBP	19,72	Normally	1020
	LWBP	20,7	Normally	1115

Base on these results, it can be concluded that MHPP Palangai Hulu can parallel with the PLN distribution system on condition of WBP or LWBP 2018 through setting of OLTC type generator transformer

The second step is to determine the power of MHPP generation in 2018 LWBP condition that can be absorbed by PLN without causing adverse impact to the existing system. By generating MHPP power through ETAP simulation we get voltage of 20 kV bus as shown in Fig. 1. The voltage of every bus 20 kV still in standard. Thus it can be seen that the MHPP is able to generate power according to its capacity without causing negative impact on the PLN system. The result of power flow analysis at LWBP condition in 2018, power generation of MHPP of Palangai Hulu can be absorbed by PLN equal to 96,9% without causing worse impact to PLN Rayon Balai Selasa system as shown in Table 2.

Due to LWBP condition of power generation of MHPP can be absorbed according to its capacity, it can be ascertained the same thing also happened in WBP condition. The power generating MHPP in the condition is a limitation of power that can be absorbed by PLN.

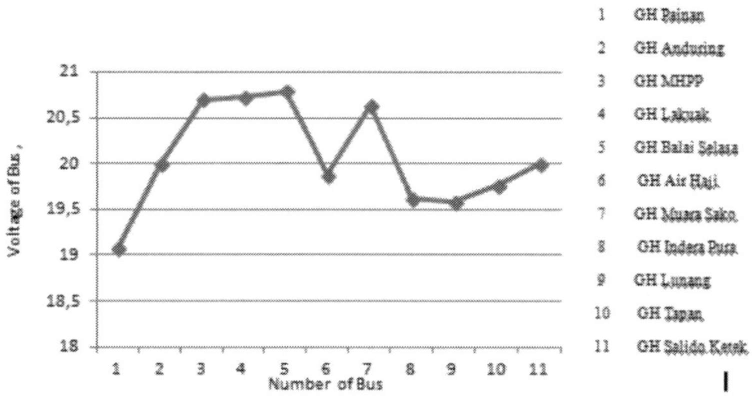

Fig. 1. Voltage of Bus in PLN Sektor Painan After MHPP of Palangai Hulu has been Connected

Table 2. Evacuate Energy of MHPP Palangai Hulu when WBP and LWBP Condition

Load Status in 2018	Energy Generated (GWh)	Energy in POC (GWh)	Losses (%)	Energy Evacuated (%)
LWBP	52.68	51.07	3.1%	96.9%
WBP	52.68	50.53	4.1%	95.9%
Average	52.68	50.80	3.6%	96.4%

3.3 Short Circuit Study

Short circuit study is required to prove that the rated capability of the circuit breaker equipment. The condition of the electrical system in the event of a short circuit taking into account the entry of the MHPP of Palangai Hulu to the distribution line 20 kV at the Rayon Balai Selasa, assuming that the WBP conditions in 2018 because the largest current is obtained at the maximum generation condition. The simulation of the operation pattern of the MHPP Palangai Hulu in WBP condition in 2018 and the short circuit study with 3-phase short circuit, phase and 1-phase short circuit to soil, obtained short circuit current to flow in evacuate distribution line from MHPP to POC as in Fig.2.

The results of the short circuit review wich the greatest short-circuit current of 3-phase noise at busbar location 20 kV in Bus Palangai Hulu is 1.33 kA. Based on distribution power regulation (Distribution Code) in 2009 determined the capacity of short circuit equipment on the voltage of 20 kV maximum 500 MVA or 14.4 kA. SPLN 118-4-1: 96 has also regulated the breaking capacity of cubicles for nominal voltage 20 kV ie 12.5 kA [7]

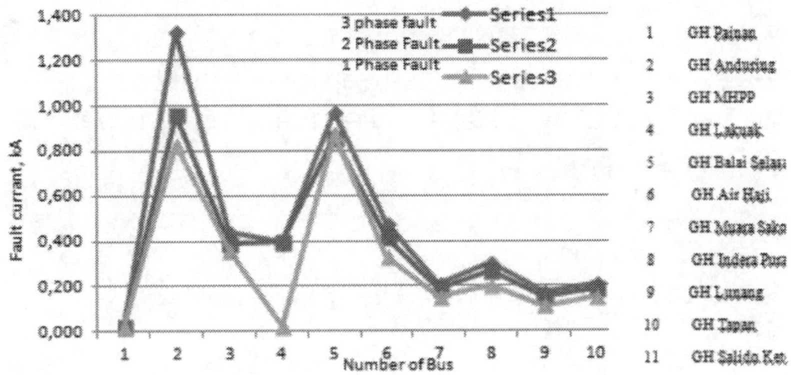

Fig.2. The Current Fault Flows in Evacuate Distribution Line from MHPP to POC

All power breaker equipment on the 20 kV side is required to have at least a breaking capacity of 12.5 kA and generally installed in the GI Kambang has a termination capacity of 25 kA whereas all breaker equipment installed and connectors network distribution has a breaking capacity of 16 kA or 12.5 kA so it can be confirmed with the largest short circuit in GI Kambang of 0.125 kA. The value is still below the circuit breaker capability installed both in the GI Kambang and the distribution network and substation, which means the fixed

4 Conclusion

Technical aspects such as analyzing the operational impact of the distribution system after the connections of MHPP to the PLN distribution system, while the economic aspect is to analyze how much result energy of MHPP can be absorbed by PLN. Ultimately, describe the financial feasibility condition. The result is legal aspect of development of MHPP of Palangai Hulu feasible to be continued to next process that is signing of contract of PPA. Technically, there is no impact on the operation of the PLN distribution system after the MHPP is connected to the PLN Rayon Balai Selasa system. The energy of MHPP Palangai Hulu can be absorbed 96.4% starting in 2018. The cost of POC is Rp. 11,577,000,000.

References

1. ESDM, General Design of National Electrical System (RUPTL) 2014-2024 (2014).
2. ESDM, Stipulate regulation in the form of Law on electricity and the launching of acceleration program of 35,000 MW, No. **30** (2009)
3. PLN, Guidebook for the Connecting Center for New and Renewable Energy (PLT EBT) power capacity development No. **0357**.K/DIR (2014)
4. Dempo Sumber Energy (DSE), FS of MHPP MW (2014)
5. J. S Hill, DOE Grid Study Is Finally Published - Supports Coal, Nuclear, Hydro, Not Renewables, Clean Tecnica, Agustus (2017)
6. PLN Area Padang, Operating System of PLN Rayon Painan (2015)
7. ESDM, Regulation of Sumatera Power Planing Network System, **No. 37** (2008)

MATEC Web of Conferences **248**, 02004 (2018) https://doi.org/10.1051/matecconf/201824802004
ESTIC 2018

Study of Capacitor Bank Switching Transient in Distribution Network

Indra Nisja*,, *Mirza* Zoni and Arnita

Department of Electrical Engineering, Bung Hatta University, Padang, Indonesia

Abstract. The power factor improvement has a great concern in the distribution system due to large motor installed in industrial system. The capacitor bank was installed to improve the power factor in the distribution system. Due to frequently capacitor bank switching has create transient phenomena in the distribution system and their effects the power quality. The aims of this research to study and simulate the effects of capacitor bank switching in the distribution system. The simulation has been conduct with PSIM software by modelled the system which contain of supply, load and three capacitor bank. The simulation was conducted with two conditions which are single bank and bank to bank switching. For single bank switching peak voltage of phase R reached about 60.9 kV and peak inrush current is reached about 6.7 kA. While for bank to bank switching the peak voltage of phase R reached as high as 59 kV and the peak inrush current is almost 8.2 kA.

1 Introduction

Along with the continuous increase in electricity demand in heavy industries caused by the installation of large motors in industrial systems involving increased reactive power consumption due to the typical power factor in steady conditions and as long as the motor is turned on. The result of this situation is the low power factor in the system. To overcome this problem, capacitor bank need install as a local source of reactive power. Three capacitors banks were installed in distribution substations system and these capacitors banks were designed to switch ON and OFF automatically based on power factor, voltage and volt ampere reactive in power system. The power factor itself will be fluctuation due to the load variations, so that the numbers of switching capacitor banks operations will be occur many times a day. Every switching event will cause the several impacts at distribution network like low frequency decaying, voltage and current distortion, resonance, inrush current and voltage transient [1]. The high magnitude voltage and current as well as high frequency transients can occurs during the switching of shunt capacitor banks, [2, 3, 4, 5]. Beside that, the capacitor banks paradoxically is an outlet source of harmonics. So that, the effect of capacitor bank switching transient on the distribution system must clearly understand. Voltage and current transients due to capacitor bank switching are often a concern for industrial and utility an engineer that are planning to install capacitors at distribution system, in this case is 20 kV.

* Corresponding author: drindra@bunghatta.ac.id

Depending on the size and rating, power capacitors are available installed at single phase or three phase circuit. Power capacitors are generally available in voltage rating up to 34,500 volts and 200 kVAR. Typically, the capacitor banks that applied in distribution substations consist of one to four banks. Individual units of power capacitor may be connected in series and multiples to provide banks of various capacities and voltage ratings [1].

This paper specifically was concentrated on the high frequency inrush currents and peak transient voltages that originated result of switching operation. The inrush current that occurs due to capacitor banks switching was becomes a big concern on these matters. Energizing of a capacitor bank that nearby fixed capacitor bank already in service produced a phenomenon called bank-to-bank switching [2]. It will be generated high frequency inrush currents in which the fixed capacitor bank charged the controlled bank. the According to IEEE Standard 18-2002, the shunt capacitors banks must withstand with 110% maximum continuous rms over voltage and 180% maximum continuous rms over current. This over voltage and over current includes both the fundamental frequency and any harmonic contributions. The standard also states that the VA rating of the capacitor cannot exceed 135%. [6]

2 Methodology

The system under study was included was included with the power source, three capacitor banks, three loads, feeder with overhead conductors and underground cables as shown in Fig.1. The capacitor banks are represented by cap. banks 1, 2 and 3 series with inductors. S_1, S_2 and S_3 are represents the circuit breaker to switching ON and OFF the capacitor banks. To simulate this model, PSIM software was used. For the initial condition all these three capacitor banks are opened. To determine the impacts of capacitor bank switching in distribution networks, a study of principle of power system operations, energy stored in a capacitor which included how a capacitor been charging and discharging are need. The primary concern of this paper is typically with how the capacitor switching transients will affect to voltage and current nearby industrial and commercial loads. The inrush current and voltage transient are occurs during the charging and discharging of the capacitor. This transient phenomena will cause current and voltage distortion at distribution network. Simulation by using PSIM software program and calculation of peak inrush current and the frequency for this transient given by: [6]

$$I_{peak} = \frac{V_{peak}}{Z_0} \quad (1)$$

$$Z_0 = \sqrt{\frac{L}{C}} \quad (2)$$

$$f = \frac{1}{2\pi\sqrt{LC}} \quad (3)$$

Where, I_{peak} = peak inrush current, V_{peak} = peak voltage, f = transient frequency

For a single bank switching, the amplitude of the inrush current will depend on the short circuit level at the point of common coupling (PCC) where the bank is connected. In this simulation the capacitor bank was switched into the feeder at the peak voltage of phase R to evaluate its effect on the feeder. To limit the frequency and magnitude of inrush current during the switching, the capacitor bank was equipped with 0.040 mH series transient inrush reactors. These reactors will increase the surge impedance magnitudes that will effectively reducing the peak inrush current. To study transient phenomena, the simulation has been done by single bank and bank to bank switching. For the single bank switching capacitor bank 1 has been turn ON at t = 10 ms, while for bank to bank switching the capacitor bank 2 has switching ON at t = 50 ms.

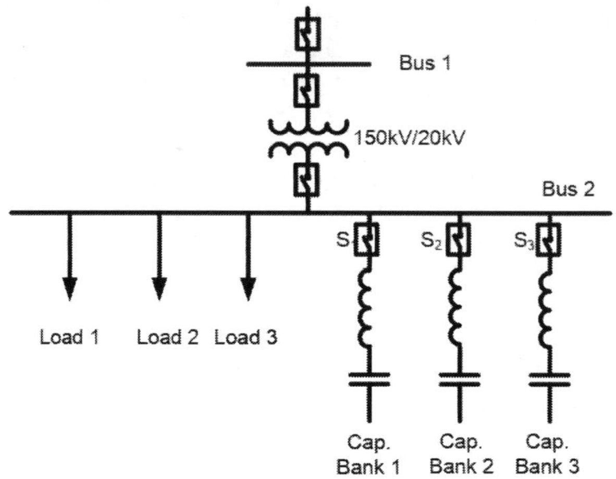

Fig 1: Typical utility substation showing a 4.5 MVAR capacitor bank and adjacent distribution loads.

2.1 Principle of energizing and de- energizing capacitor

If the capacitor is initially uncharged and there is no current flow to the circuit while switch is OFF. When the switch is closed at t = 10ms, charge is begins to flow and setting up a current flow. At the same time, the capacitor begins to energize. During energizing of capacitor, the charge is transferred between each plate until the capacitor is fully charged. Once the maximum charge is reached, the current is zero because the potential difference across the capacitor matched that supplied by the source. The common phenomenon in the process of de-energizing capacitor banks is reignition. An electric arc is produced in the circuit breaker while a capacitor bank is disconnected. To produce an effective de-ionization, the electric arc will be extinguished when the current wave is zero crossing. At that time the capacitor bank will be at its maximum voltage.

3. Results and discussion

3.1 Single capacitor-bank switching

Single bank switching is while the transient (voltage and current) occur for the closing of the first capacitor bank, while other capacitor banks are still not energized. To simulate this system, it does begin with energizing capacitor bank 1 by closing switches S_1. Fig. 2 (a) showed the three phase voltages distribution system for single bank switching. It shows the voltage before and after the switching operation. The frequency of the capacitor transient is equal to the system's natural frequency. Therefore, large capacitor banks will result in lower frequency decaying ring wave transients, while small banks will result in higher frequency ring wave transients. The capacitor bank switch is switching ON at peak value phase R voltage (t = 10 ms) with the peak voltage of phase R reached about 60.9 kV, more than it's twice steady state value before the energized. This transient voltage can happen several times a day depending on the system power factor. This is a transient overvoltage, since the controlled capacitor bank was originally discharged and the voltage oscillated over the system voltage as high as it started below. Fig. 2 (b) shows the voltage according to harmonic frequency. For the fundamental frequency of 50 Hz, the peak transient voltage is 60.9 kV. For 300Hz frequency the peak amplitude voltage in phase R and S are 6.2 kV. This means the magnitude of the 6th order harmonic voltage is 6.2 kV or 10.18% of fundamental frequency. The duration of the ring-wave transient is dependent upon the system X/R ratio at the capacitor bank. Transients associated with substation capacitor banks can last as long as at 10 ms to 30 ms.

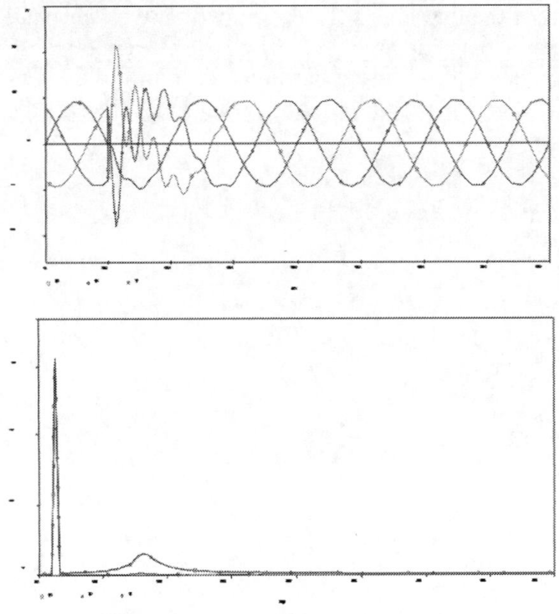

Fig.2: (a) voltage waveform (b) maximum peak voltage, for single bank switching

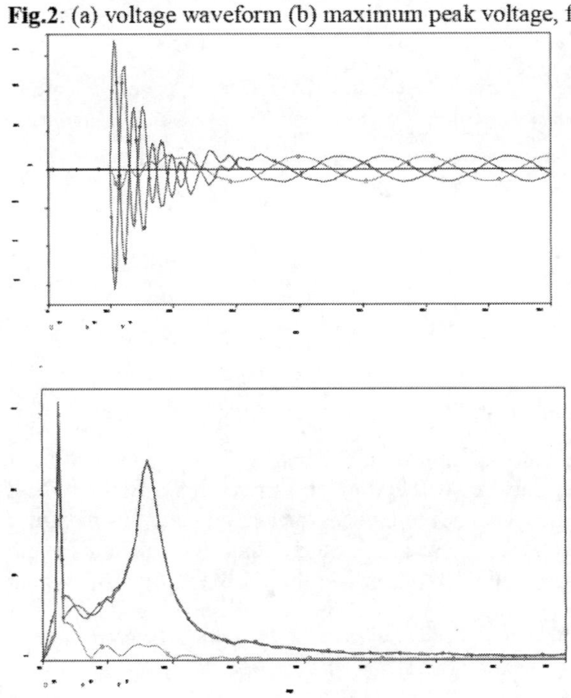

Fig. 3: (a) current waveform (b) maximum peak current, for single capacitor bank switching

The inrush currents that happen for single capacitor banks switching were also the big concern. Fig. 3 (a) shows the inrush current when by energizing the capacitor bank 1 by closing switches S_1. The peak inrush current value during of single bank switching is reached about 6.7 kA, more than eight times of their steady state value before the energized. The current was zero before the switch was closed at (t = 10 ms). After approximately 40 ms the transient was disappeared and the current back to it's steady state value. Fig. 3 (b) shows the current according to harmonic frequency. For the fundamental frequency of 50 Hz, the peak inrush current is 6.7 kA. For 300Hz frequency the peak amplitude of inrush current in phase S and T are 4.8 kA. This means the magnitude of the 6th order harmonic current is 4.8 kA or 71.64% of fundamental frequency. This is a high enough harmonics that will affect of the power system quality.

3.2 Bank-to-bank switching

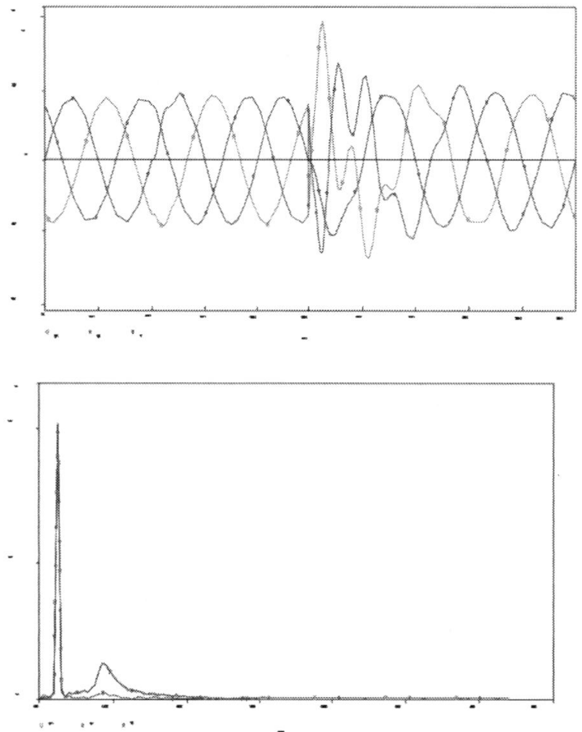

Fig. 4 (a) voltage waveform for bank, (b) maximum peak voltage for bank to bank switching

When a number of capacitor banks are used in parallel, the banks are not switched in at the same time. Some of the capacitor bank may be switched in later than others. Bank to bank switching is switching transients occur when a capacitor bank is energized in close proximity to capacitor bank that is already energized. During charging of the uncharged capacitor bank, the adjacent charged capacitor bank dumps a high frequency high and magnitude of current into the uncharged capacitor bank and this current is called inrush current. After capacitor bank 1 is closed and in steady state condition, then switch S_2 closed at time (t = 50 ms), so that the capacitor bank 2 will be charged. The result shows, the voltage transient happen at phase R, S and T. The phase R peak voltage increased as high as 59 kV is about twice of it steady state value. Voltage at phase S and T were increased as well, but not as high as the increase voltage in phase R as shown in Fig. 4 (a). Fig. 4 (b) shows the voltage amplitude phase for R and S for harmonic frequency of 150 Hz is 8 kV. This means the magnitude of the 3rd order harmonic voltage is 8 kV or 13.6% of fundamental frequency. Fig. 5 (a) shows very large inrush current flow due to bank to bank switching. The peak inrush current is almost 8.2 kA occurs for phase R, which more than 5 times its normal or steady state conditions. While the peak inrush current that occurs in phase S and T for frequency of 150 Hz as high as 7.8 kA. This means the magnitude of the 3rd order harmonic current is 7.8 kA or 95.1 % of fundamental frequency. So, the switching of capacitor bank besides causing transients will result in the emergence of current and voltage harmonics on the system as well, that will affect of the power system quality Furthermore, a high frequency transient component will arise, leading to further problems, especially with sensitive electronic loads.

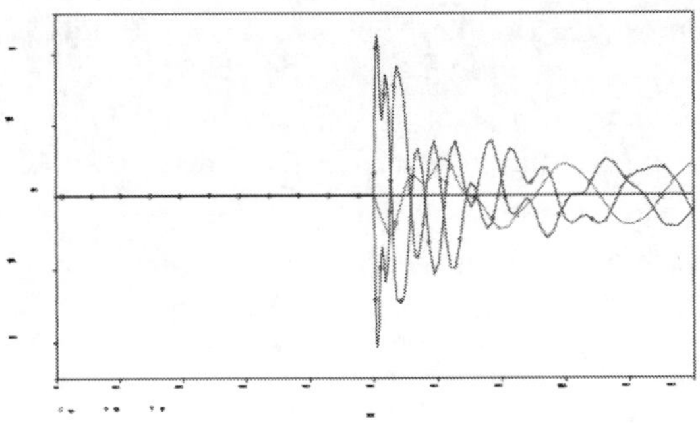

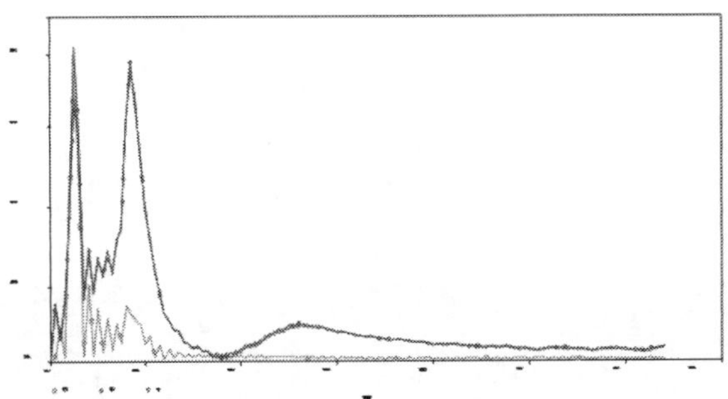

Fig. 5 (a) current waveform, (b) maximum peak current for bank to bank switching

4. Conclusion

After doing this research and analyzing it carefully, then can be made conclusions as follows: During the switching of the capacitor bank within the distribution substation, the transient voltage and inrush current occurs. The level of transient voltage that occurs in the distribution system due to capacitor bank switching, depend on the system itself. The switching transient can be occurs for single bank switching and bank to bank switching. Both switching has significant effects to the distribution system like voltage transient and inrush current. The resulting transient over voltage can reach a peak value about 60.9 kV and inrush current can reach about 8.2 kA. Due to this phenomenon happen in several times a day, it will have significant effect to the power system stability and quality. Besides the transient problems, capacitor banks switching will have other problems such as harmonics that will reduce the quality of quality.

References

1. Technote Series, *Practical Aspects of Shunt Capacitor Switching Transients,* Northeast Power Systems Inc. Queensbury, New York www.nepsi.com, (2012).

2. IEEE Standards 18-2012 (Revision of IEEE Std 18-2002), *Collection on Power Capacitors*, ISBN 1- 55937-416-0, (2012).
3. A. Greenwood, *Electrical Transients in Power Systems*, 2nd Edition, John Wiley&Sons Inc., (1991).
4. John H. Brunke, *Application of Power Circuit Breakers for Capacitive and Small Inductive Current switching*, Application of Power Circuit Breakers, IEEE Tutorial Course 93 EHO 388-9-PWR, (1993).
5. *Modeling and Analysis of System Transients Using Digital Programs*, IEEE PES Special Publication TP-133-0 (Ch.4 – Modeling Guidelines for Switching Transients).
6. M.F. Iizarry Silvestrini and T. E. Vélez Sepúlveda, *Mitigation of Back-to-Back Capacitor Switching Transients on Distribution Circuits*, Department of Electrical and Computer Engineering (2006): 1-8.

Online Position Control Performance Improving Applying Incremental Fuzzy Logic Controller

Dirman Hanafi[1,3,*], *Mohamed Najib* Ribuan[1], *Wan Hamidah*Wan Abas[2], *Hidayat*[3], *Elmy* Johana[1], *Herman* Wahid[4], *Rozaimi* Ghazali[5], *Hisyam* Abdul Rahman[1]

[1]Instrumentation and Sensing Technology (InSeT) Research Group, Faculty of Electrical and Electronic Engineering, Universiti Tun Hussein Onn Malaysia, Batu Pahat, 86400, Johor, Malaysia.
[2]Department of Electrical Engineering, Politeknik Kuala Terengganu, Terengganu, Malaysia
[3]Department of Electrical Engineering, Universitas Bung Hatta, Indonesia
[4]Process Tomography & Instrumentation Research Group (PROTOM-i), Faculty of Electrical Engineering, Universiti Teknologi Malaysia, Johor, Malaysia.
[5]Faculty of Electrical Engineering, Universiti Teknikal Malaysia Melaka, Melaka, Malaysia.

Abstract. This paper presents the online control system application for improving the DC motor performance. DC motor widely used in industries and many appliances. For this aim fuzzy logic controller is applied. The type of fuzzy controller use is an incremental fuzzy logic controller (IFLC). The IFLC is developed by using MATLAB Simulink Software and implemented in online position control system applying RAPCON board as a platform. The experimental results produced the best gains of the IFLC are 1.785, 0.0056955 and 0.01 for error gain (GE), gain of change error (GCE) and gain of output (GCU) respectively. Its produce smaller rise time, peak time, 0% overshoot and smaller settling time. Beside that the IFLC response also able to follow the set point. The controller response parameters values are also acceptable. It means that the IFLC suitable to be use for improving the position control system performance.

1 Introduction

DC motors widely used to improve performance, such as industry, home appliances, robot manipulators etc. because it has high reliability, flexibility and low cost. Most DC motor driver applications are in position control or in speed control systems [1].

Several controllers have been applied for position control system with DC motor as a driver such as Proportional Integral Derivative (PID) [1], Fuzzy Logic [2,3], Neural Network (NN) [4] and etc.

PID controller is commonly applied for controlling motor because of it has simple structure and comprehensive control algorithms [1,2]. Besides, it has been implemented in position control systems, but still suffer from poor performance due to non-linear parameters. The PID controller does not provide satisfactory results when the control parameters, loading conditions and motor itself change [5].

* Corresponding author: dirman@uthm.edu.my

Neural network (NN) controller is another kind of controllers that also has been applied for motor control. In the NN, there is abundant of architecture can be used to perform a variety types of functions. There are kind of neural network with high efficiency and strong function generalizing in terms of learning speed and simplicity of the structure. The NN need increasing the number input NN and add some delay to the input NN to improve the performance [4].

In this paper, the incremental fuzzy logic controller (IFLC) as an intelligent controller is applied. The IFLC does not need knowledge model of system, complex structure, learning process and etc. This controller work is based on the principle of human expert decision making in problem solving mechanism [2,3]. The Mamdani inference as a computational fuzzy inference type is used because its decision is more accurate.

2 DC Motor Modelling Approach

In armature control of separately excited DC motors, the voltage applied to the armature of the motor is adjusted without changing the voltage applied to the field. Figure 1 describes a separately excited DC motor equivalent model [1,6].

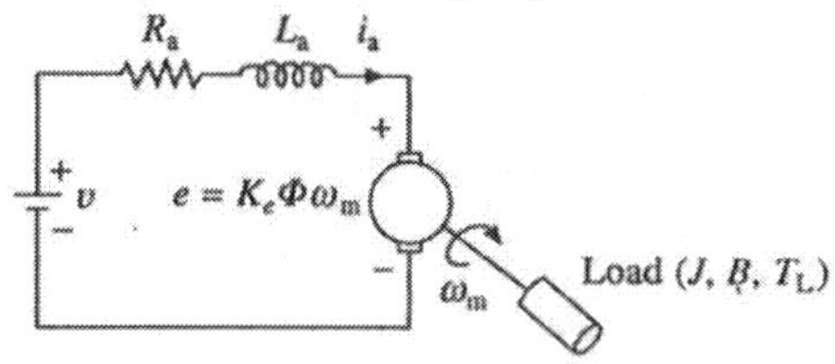

Fig.1. Separately excited DC motors

Where J is Moment of Inertia of the motor (kg.m^2), B is Friction coefficient of the motor(N.m.sec), $K_e$is Torque constant of the motor (N.m/A), $K_b$is Motor back emf constant (V/rad/sec), $I_a$is Armature current (A), $V_a$is Armature voltage applied (V), R_a is Armature resistance (ohms), and L_a is Armature inductance (H).

2.1 Dynamic Model

Voltage equation of the armature circuit under transient is given by [6]

$$v(t) = R_a i_a(t) + L_a \frac{di_a(t)}{dt} + K_e \Phi \omega_m(t) \quad (1)$$

From the dynamics of motor load system

$$J\frac{d\omega_m(t)}{dt} = T(t) - T_L(t) - B\omega_m(t) \quad (2)$$

Further

$$T = K_e \Phi i_a(t) \quad (3)$$

When field current is kept constant, flux remains constant. Replacing $K_e\phi$ by a constant K_e, yields

$$v(t) = R_a i_a(t) + L_a \frac{di_a(t)}{dt} + K_e \omega_m(t) \quad (4)$$

$$J\frac{d\omega_m(t)}{dt} = K_b i_a(t) - B\omega_m(t) - T_L(t) \quad (5)$$

$$K_b i_a(t) = J\frac{d\omega_m(t)}{dt} + B\omega_m(t) - T_L(t) \quad (6)$$

For no load previous equation become:

$$K_b i_a(t) = J\frac{d\omega_m(t)}{dt} + B\omega_m(t) \quad (7)$$

2.2 Transfer Function

Based on Laplace transforms of Eq. (1) and Eq. (6), the armatur current of motor is

$$I_a(s) = \frac{\omega(s)}{K_b}[J_m s + B] \quad (8)$$

Substitute Eq. (8) into Laplace form of Eq. (1) we get

$$V(s) = \frac{\omega(s)}{K_b}[L_a J_m s^2 + (L_a J_m + L_a B)s + (R_a B + K_b K_e)] \quad (9)$$

The relation between position and speed are $\theta(s) = \frac{1}{s}\omega(s)$, then the transfers function between shaft position and armature voltage without load are given as:

$$\frac{\theta(S)}{V(S)} = \frac{K_b}{L_a J_m s^3 + (L_a J_m + L_a B)s^2 + (R_a B + K_b K_e)s} \quad (10)$$

3 Incremental Fuzzy Control (IFLC)

An IFLC [3] adds a change in control signal Δu to the current control signal,

$$u(n) = u(n-1) + \Delta u(n)T_s \Rightarrow \Delta u(n) = K_p\left(\frac{e(n)-e(n-1)}{T_s} + \frac{1}{T_s}e(n)\right) \quad (11)$$

In Eq.(11)T_d is 0. For complete gain of PID, $\Delta u(n)$become as below

$$\Delta u(n) = K_p\left(e(n) + \frac{1}{T_i}\sum_{j=1}^{n} e(j)T_s + T_d\frac{e(n)-e(n-1)}{T_s}\right)(12)$$

Index n refers to the time instant. By tuning we shall mean the activity of adjusting the parameters K_p, T_i, and$T_d$in order to achieve a good closed-loop performance.

The IFLC as in Figure 2 is of almost the same configuration as the FPD controller except for the added integrator. The conclusion in the rule base is now called change in output (CU), and the gain on the output is accordingly GCU. The control signal $U(n)$ at time instant n is the sum of all previous increments,

$$U(n) = \sum_{j=1}^{n}(cu(j) * GCU * T_s) = \sum_{j=1}^{n}(f(GE * e(j) * GCE * e(j) * GCU * T_s)(13)$$

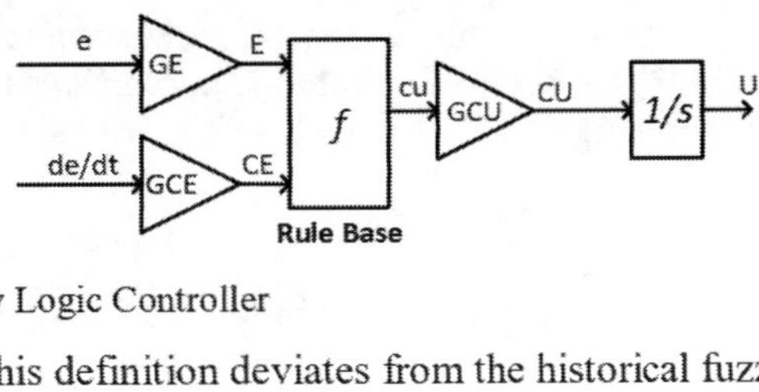

Fig.2. Incremental Fuzzy Logic Controller

Notice again that this definition deviates from the historical fuzzy controllers, where the sampling period $T_s$was left out. The function f is again the control surface of a PD rule base.

The mapping is usually nonlinear, but with the usual favorable choice of design is a linear approximation as Eq.(14).

$$f(GE * e(n), GCE * \dot{e}(n)) \approx GE * e(n) + GCE * \dot{e}(n) \quad (14)$$

Insertion Eq. (14) into Eq.(13) yields the control action,

$$U(n) \approx \sum_{j=1}^{n}\left(GE * e(j) + GCE * \dot{e}(j)\right) * GCU * T_s = GCU * \sum_{j=1}^{n}\left[GE * e(j) + GCE * \frac{e(j)-e(j-1)}{T_s}\right] * T_s = GCU\left[GE * \sum_{j=1}^{n} e(j) * T_s + GCE * \sum_{j=1}^{n}(e(j) - e(j-1))\right] = GCU * GCE\left[\frac{GE}{GCE}\sum_{j=1}^{n}(e(j) * T_s) + e(n)\right] \quad (15)$$

By assuming that $GCE * GCU = K_p$, $\frac{GE}{GCE} = \frac{1}{T_i}$ and $T_d = 0$, it is clear that the linear controller is a crisp PI controller.

Note that the proportional gain K_p now depends on *GCE*. The gain $1/T_i$ was determined by the ratio between the two fuzzy input gains, and the inverse of the derivative gain T_d in FPD control; the gains GE and GCE change roles in FPD and IFLC controllers.

4 Online Position Control Implementation

The online position control system with a DC motor as an actuator is implemented using computer, RAPCON board [7] and DC motor. Figure 3 (a) shows the hardware connection and (b) the IFLC controller block diagram.

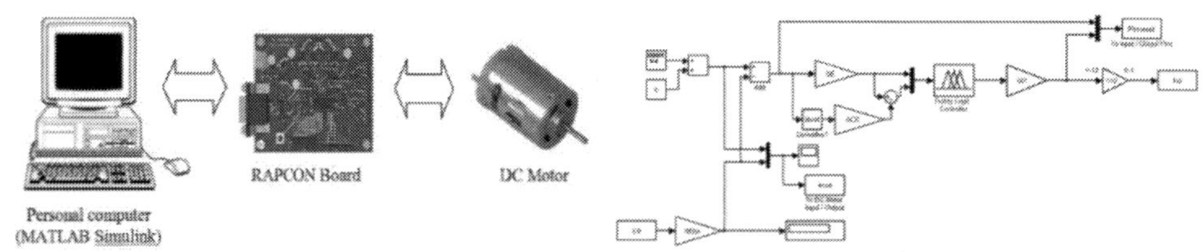

(a) (b)

Fig.3.(a) Hardware components connection, (b) the IFLC controller block diagram

5 Experimental Result and Analysis

The experiment for testing the controller performance is done in two ways. There are simulation and online scheme.

5.1 Simulation Scheme

The best settings of controller parameters values for simulation scheme are 2.721, 0.019 and0.259 for GE, GCE and GCU respectively. Figure 4 (a) represents the output response of IFLC controller and (b) its zooming.

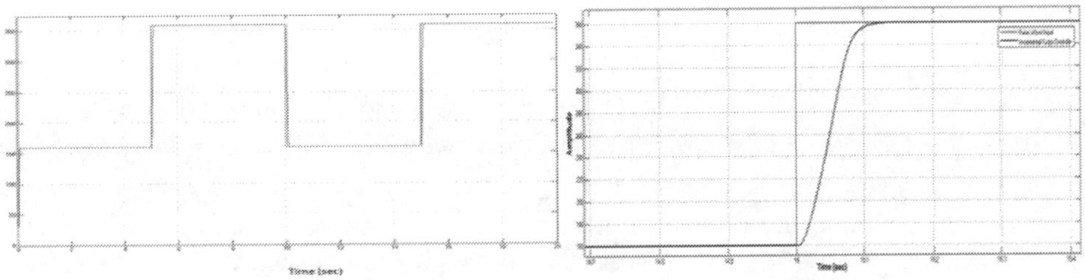

(a) (b)

Fig.4.(a) Output response of IFLC controller, (b) its zooming.

5.2 Online Scheme

The best controller parameter values for online are 1.785, 0.0056955 and 0.01 for GE, GCE and GCU setting respectively.Reference inputs in this case combination between Signal Generator and Constant value. Figure 5 (a) represents the output response of IFLC for online scheme experiment and (b) its zooming.

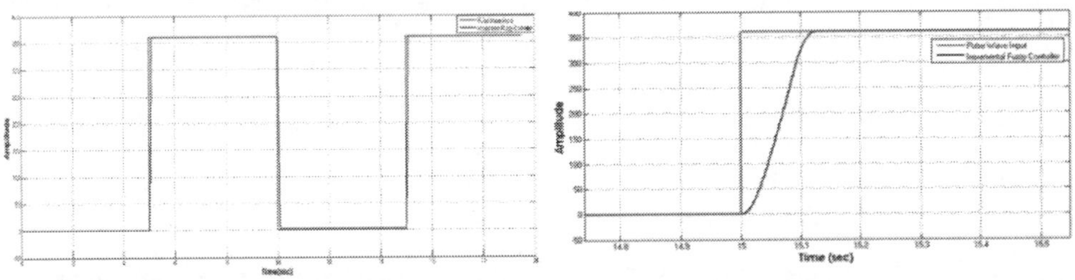

Fig.5.(a) Online IFLC response and(b) its zooming.

The IFLC responses specifications comparison between simulation and online are elaborated in Table 1.

Table 1. Specifications of IFLC response parameter

Parameter	Simulation	Experiment
Rise time, T_r	0.06s	0.085s
Peak time, T_p	0.12s	0.12s
Percent overshoot, %OS	0%	0%
Settling time, T_s	0.12s	0.12s

6 Conclusion

Based on the experimental results show that the IFLC is a suitable control for improving the position control system performance. Its produce smaller rise time, peak time, 0% overshoot and smaller settling time. Beside that the IFLC response also able to follow the set point.

This research was funded by Research Management Center (RMC) of Universiti Tun Hussein Onn Malaysia under GIPs grant with vot. no. U640.

References

1. V. Jambulingam, Position Control of DC Motor Using PI Controller, International Journal of Engineering Development and Research, **4(2)**, 1342-1345, (2016)

1. P. Maji, B. R. Jammu, S. K. Patra, K. Mahapatra,Design and Implementation of Online Fuzzy Logic Controller on FPGA, *Annual IEEE India Conference (INDICON)*, (2014)
2. G. Jayapalan, G. Agnihotri, .D. M. Deshpande, Real Time Implementation of Incremental Fuzzy Logic Controller for Gas Pipeline Corrosion Control, Advances in Fuzzy Systems, **2014**, 1-9, (2014)
3. M. Aamir, On replacing PID controller with ANN controller for DC motor position control, International Journal of Research Studies in Computing, **2(1)**, 21-29, (2013)
4. D. Hanafi, M. A. Sidik, M. Zoni, Hidayat, Time Base Firing Pulse Delay Control for Improving Single Phase Induction Motor Speed Performance Using Fuzzy Logic Control, ARPN Journal Engineering and Applied Sciences, **11(12)**, 7515-7521, (2016)
5. N. S. Nise, *Control System Engineering*,(7nd. ed. Addison-Wesley Publishing Company, Canada, 2014)
6. Zeltom LLC. RAPCON: Real-time Rapid Control Prototyping Platform for Matlab/Simulink, (2013)

Analysis of Characteristics Over Current Relay and Ground Fault Relay on Feeder Rayon Tabing

Arzul[1], *Ija Darmana*[2] , *Erliwati* [3], *Adiv Rama Salvayer* [4] and *Tris Safri Yetno* [5]

[1,2] Department of Electrical Engineering, University of Bung Hatta, 25243 Padang, Indonesia
[3] Department of Technique Elektromedik - Health Polytechnic Siteba Padang, Indonesia
[4] Electrical Engineering student of University of Andalas Padang, Indonesia
[5] Electrical Engineering student of University of Bung Hatta, 25243 Padang, Indonesia

Abstract. The reliability of power distribution of PT PLN (Persero) branch of Padang especially in Rayon Tabing most of the repeaters is still not optimal. This is due to an uncoordinated protection relay on each feeder. One effort made to improve the reliability of electric power is to rearrange the protection of overcurrent relays and ground disturbance relays on the feeder. So coordinated with each other well and is expected if there is interference in one of the repeater will not bring black out on other repeater.In this research we analyzed short circuit current at 20 kV side. From the analysis results obtained new relay protection settings based on the calculation of short-circuit current settings and compare the results with existing protection relays. By making efforts to improve the reliability of electric power is to rearrange the protection relay on each feeder. Based on short circuit current analysis on each repeater, the largest short circuit in 3 phase, 2 phase, and 1 phase disturbance is located at 1% disturbance location and the smallest disturbance is at 100% disturbance location. It can be concluded that the farther the location of the disturbance that occurs the smaller the disturbance, and vice versa, if the location of the disturbance that occurs closer then the greater the noise flow.

1 Introduction

In the access operation can not be used for error conditions (errors). One of the main tools in the protection system used is the current relay (overcurrent relay) [1] and soil relocation (Ground Faul Relay) [2, 3]. This relay instructs the breaker (CB) to disconnect if a short circuit occurs. Common problems in distribution systems include feeder discharge at 20 kV. caused by short circuit. If the protection release settings are not good. can cause total blackout.

To determine the protection system is already working in coordination with each other between the substation with another substation circuit required analysis of the current setting and working time overcurrent relays at distribution feeders Rayon Tabing.

* Corresponding author: zul22727@gmail.com

2 Literature Review

The voltage drop of the transmission network was first performed at the substation by a step-down transformer voltage [4, 5]. The resulting voltage is called voltage or voltage primary distribution medium. PLN primary distribution voltage is generally worth 20 kV. Network between power stations / plants and substations called the transmission network. while the network is out of the substation to the consumer / customer called with a distribution network.

2.1 Distribution system

Distribution is a system that serves to distribute electricity to the beneficiaries. The distribution system is divided into two (2) sections [6]. namely the medium-voltage distribution systems and low voltage distribution systems. Medium voltage distribution network begins Substation Electrical Center or on separate systems / isolated. At some point started from plants. Network form can be radial or closed (radial open loop) [7]. Single line diagram distribution system as in figure 1.

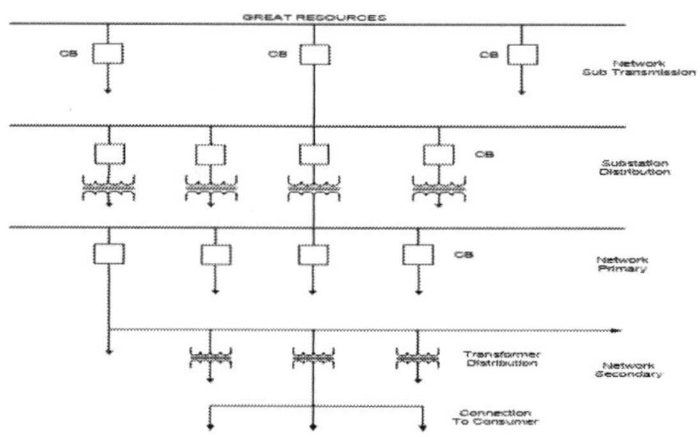

Fig. 1. Single Line Diagram Distribution System

2.2 Short-circuit

In a distribution system. rationing power through transmission network system that radial pattern that is by generator at the base. then large positive and negative sequence impedance of the system can be calculated by adding up all the positive and negative sequence impedance components ranging from the generator to subtation. While the zero sequence impedance is determined in other ways

2.2.1 Source sequence impedance

Sequence impedance source can be calculated by the equation

$$X_{1s} = X_{2s} = \frac{kV^2}{MVA_{hs-3\phi}} \quad (1)$$

$$X_{0s} = \frac{3kV^2}{MVA_{sc-1\phi}} - (X_{1s} + X_{2s}) \quad (2)$$

Impedance resources:

$$Z_{1s} = Z_{2s} = X_{1s}$$

$$Z_{0s} = X_{0s} \tag{3}$$

2.2.2 The sequence impedance transformer

For the calculation of the impedance transformer is taken reaktansinya. while tahananya ignored because they are small. In ohms. impedance is large is calculated by the equation:

$$Z_{1T} = Z_{2T} = +jz_1 \ x \frac{V_d^{\ 2}}{S_d} \tag{4}$$

2.2.3 Sequence Impedance Conductor

If an unknown type of conductor is used. then the conductor cross-sectional area sequence impedance obtained by viewing the table in the appendix. Line impedance value of positive and negative same sequence. ie X1 = X2 = XL. while the value of its zero sequence impedance is different. Based PLN Standart No. 64: 1985. the impedance value of the order of positive. negative and zero sequence in Annex IV. Sequence impedance so that the channel can be calculated by the equation:

$$Z_{n\,feeder} = Z_n x 1 \tag{5}$$

2.2.4 General Disorders Short-circuit equation

Short-circuit Disorders Three Phases: Large fault currents equal to the amount of current positive sequence is: $I_f = I_{a1}$

$$I_{hs\text{-}3\phi} = I_{a1} = \frac{V_f}{Z_1} \tag{6}$$

Short-circuited two phases: In this disorder is determined by the magnitude of the fault current impedance positive sequence and negative sequence alone. while komponn zero sequence does not exist. because at the moment there are no. Short circuited to the ground:

$$I_{hs-2\phi} = -j\sqrt{3}\frac{V_f}{Z_1 + Z_2} = \left|0{,}866 \, x \, I_{hs-3\phi}\right| \tag{7}$$

2.2.5 Adjustment Working Time Relay

Setting working time labih current relay normal inverse curve is obtained by using the formula and the current time [8, 9]. These formulas vary according to design manufacturer of relays. Setting working time overcurrent relay with inverse characteristics by "Standart BS.142.1996" and "Standard IEC 255-4" obtained:

$$\left(\frac{I}{I_s}\right) = \frac{I_{\text{shortcircuitat theendof theline}}}{I_{set}} \tag{8}$$

Multiple Time setting (TMS):

$$t = \frac{TMS x \beta}{\left(\frac{I}{I>}\right)^{\alpha} - 1} \quad \text{or} \quad TMS = \frac{t(s) x \left(\frac{I}{I>}\right)^{\alpha} - 1}{\beta} \tag{9}$$

In the overcurrent relay with inverse characteristics. the value α and β the inverse normal type can be seen in Table 1.

Tabel 1. Characteristics *Inverse*

Setting arus	α	β
Normal Inverse	0.02	0.14
Very Inverse	1.0	13.5
Extremely Inverse	2.0	80.0
Long-time Inverse	1.0	120.0

3 Results and Discussion

3.1 Data

Voltage = 20 kV
$MVA_{SC\ 3ph}$ = 1805,84 MVA
$MVA_{SC\ 1ph}$ = 504,91 MVA

Power Transformer Data
Brands = PAUWELS
Capacity = 30 MVA
Vector group = YNyn0
Tegangan = 150 / 20 kV
Impedance transformer = 12.4 %
Grounding (20 kV) = 40 Ω
Ratio CT = 1000/5 A

Data Conductor

Table 2. Data Feeders

Feeders	**Conductor type**	**Long Conductor (km)**	**Ratio CT**	$I_{load\ max}$ **(A)**
Air Pacah	A3C 3x150 mm^2	46.578	400/5A	150
J4.Exp – Lubuk Buaya	A3C 3x240 mm^2	6.559	400/5A	263

3.2 Results

From the figure 3 and 4 can be concluded that the data of the calculation already meets the characteristics of the more current relays, the characteristics that are achieved is the standard inverse.

1. Flow of the biggest upsets in 3-phase short circuit. the farther the distance from the location of the relay fault location. the smaller the perceived fault current relay.
2. The short circuit currents on the feeder Air Pacah. for 3-phase short circuit at fault location 1%. which amounted to 5685.09 Ampere. channel length = 46.578 km and the

load current = 150 Ampere (the second largest). Rated impedance channel and the channel length will affect the value of short circuit current.

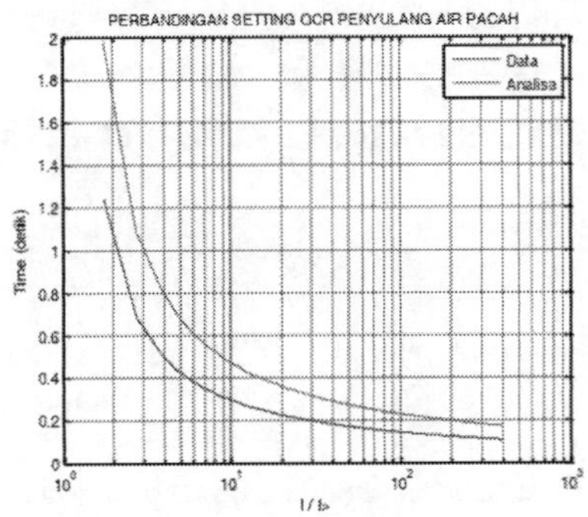

Fig. 3 Characteristics OCR for feeder Air Pacah

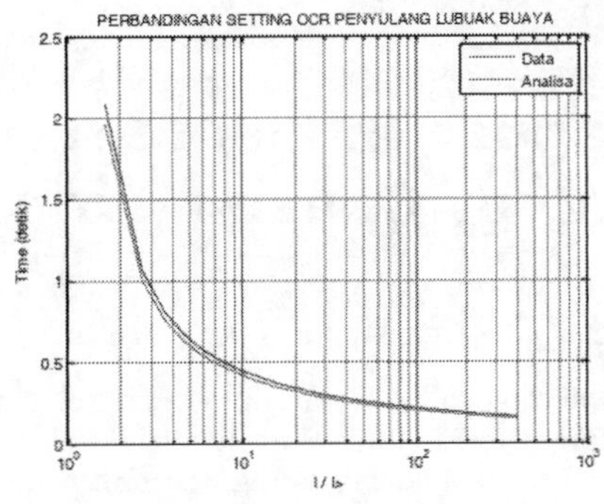

Fig. 4 Characteristics OCR for feeder Lubuak Buaya

References

1. A.R. van C. Warrington. *Protective Relays*. Their Theory and Practice. vol. 2. third edition. London, (1994)
2. Rones V.A.S., *Adaptive protection schemes for feeders with the penetration of Seigg based wind farm*, IEEE Xplore digital library, (2013)
3. S.M. Mousavi. H. Askarian Abyaneh. *Optimum setting and coordination of overcurrent relays considering cable damage curve*. IEEE Bucharest Power Tech Conference. June 28th - July 2nd. Bucharest. Romania, (2009).
4. Ija Darmana, *Implementasi Sistem Pentanahan Grid Pada Tower Transmisi 150 KV* (Aplikasi pada Tower SUTT 150 KV), Research of Applied Science and Education V9.i2 (185-194), Jurnal Ipteks Terapan, (2015)
5. Ija Darmana, Perbaikan Jatuh Tegangan Dengan Pemasangan Automatic Voltage Regulator, Research of Applied Science and Education Vol 8, No 4, Jurnal Ipteks Terapan, (2014)
6. PT. PLN (persero). *Design Criteria Enjinering Electricity Distribution Network Construction*, book I, (2010)
7. Gonen T. *Electric Power Distribution System Engineering*, Mc Graw Hill. Inc. (1986).
8. Khairizvan Edwar. *Analysis of Over Current Relay On 20 KV Distribution Channels Against Sympathetic Tripping Possibility*, Electrical Engineering Unand. (2014).
9. Rinan Al-Tawil. Seila Gruhonjic-Ferhatbegovic. *Application and settings inverse timerelay characteristics for feeders overcurrent protection*. 18th International Conference on Electricity Distribution, Juni (2005).

Gender Influence on the Household Electrical Energy Consumption Behavior

Maria Isfus Senjawati[1], *Lusi* Susanti[2,*],, *Hilma* Raimona Zadry[2], *and Prima* Fithri[2]

[1]Polytechnic of ATI Padang, Industrial Ministry, Indonesia,Student at Department of Industrial Engineering, Faculty of Engineering, Universitas Andalas, Padang, Indonesia
[2]Department of Industrial Engineering, Faculty of Engineering, Universitas Andalas, Padang, Indonesia

Abstract. Increased energy consumption can lead to depletion of fossil energy reserves that can pose a threat to energy supply in Indonesia. One way to increase the intensity of energy savings is to make changes in consumer behavior. The development of gender roles, abilities, and attention led most researchers to argue that women are more consistent in attention to environmental change than men. Women are more concerned with environmental quality and participate in environmental activities. Thus, this research tries to determine the influence of psychological factors on the behavior of household electrical energy consumption in term of gender differences. This study used questionnaires distributed using Google Docs to household electrical respondents several in Indonesia. Referring to ABC (Antecedents-Behaviour-Consequences) model of consumer behavior, the total of 808, questionnaires from 526 male respondents and 282 female respondents were successfully collected. The data were processed by the Partial Least Square Equation Modeling Model (SEM-PLS) using Smart-PLS software. The study concluded that,if given the consequences, women have higher influence to the behavior of electrical energy consumption than men.Overall psychological factors have a positive and significant influence on the behavior of electric energy consumption in Indonesia.

1 Introduction

One source of energy that still dominates the final energy consumption in various sectors is electrical energy. In 2013 and 2016, the national electricity demand is 190 Tera Watt Hour (TWh) and 232 TWh, respectively, with the household sector in both years is still dominating with a share of more than 41%, the highest among other consumer sectors such as industry(34%), followed by the commercial sector (24%), and transportation sector (0.1%) [1].

One way to increase the intensity of energy savings is to make changes in consumer behavior. Rani [2] states that consumer behavior is influenced by four main factors namely cultural, social, personal and psychological. Wijaya and Tezuka [3] states that one of the most significant efforts in encouraging people to behave in energy-efficient ways is to build energy-saving awareness.

* Corresponding author: lusi@eng.unand.ac.id

Research on energyawareness have been conducted by some previous studies i.e. study byWai [4], [11] which developed a conceptual model about building awareness of energy saving and study by Maag [5] which developed a Three Term Contingency model or also known as ABC (Antecedents-Behavior-Consequences). The model can be used to describe how to manage behavior and awarenessof energy saving. Study by Frederiks et al. [6], states that consumer behavior in terms of energy consumption is influenced by socio-demographic factors (gender, age, educational level, occupation and so on) and psychological factors (individual personality values, awareness, etc.) Psychological factors, especially social values for household energy needs together with personal values will result in a different awareness of energy consumption with different motivations [7]. Fithri et al. [8] also stated that gender influenced the behavior of electrical energy consumption. This study is conducted as an attempt to determine the effect of psychological factors on the behavior of household electrical energy consumption seen from gender differences.

2 Methods

2.1 Respondents

Respondents involved in this study were 808 household electrical energy consumers in Indonesia. The respondents consist of 526 men and 282 women from some of Java and Sumatra provinces.

2.2 Questionnaire Design

This research instrument used questionnaires which developed using data of information and behavior of household electrical energy consumption compiled from several literatures [7,9]. The questionnaires were filled online by respondents using google docs application which contains a list of questions or statements as follows:

a. Preface, contains the purpose of data collection, its benefits, and the main topic of this research.
b. The first part, contains a general information and socio-demographic data of respondents as documentation of the research.
c. The second part, contains the statements to obtain information about the psychological factors (45 statements), electrical energy saving behavior (17 statements), and the regulation factorsof electrical energy saving behavior (5 statements).
d. The third part, contains the amount and length of time the household electrical appliances are used.

2.3 Indicators and variables laten/construct

Table 1. Energy Saving Behavior Questionnaire Indicators Viewed from Psychological Factors.

Variabel	Indicator	Total
Values, Norms and Individual Personality (n)	n1, n2, n3, n4, n5, n6, n7,….., n36	36
Level of Awareness of Energy Problems (k)	k1, k2, k3, k4, k5, k6, k7,k8,k9	9
Behavior of Electrical Energy consumption(p)	p1, p2, p3, p4, p5, p6, p7, . . , p17	17
Regulation Factors (r)	r1, r2, r3, r4, r5	5

2.4 Data collection and analysis

The number of respondents obtained is 808 respondents with the description in Table 1.

Table 2 Recapitulation of Respondents

NO	Region	Male	Female	Amount
1	**Java**	276	132	408
2	**Sumatra**	250	150	400
	TOTAL	526	282	808

2.4.1 Outer Model (Measurement Model), the test is as follows:

A. ValidityTest

1) *Convergent validity*

Convergent validity test used to measure the internal positive correlation of the construct variable. Correlation value is valid if greater than 0.70. However, for early research stage , the values between 0.50 to 0.60 is considered sufficient [10]. After the re-estimation of 3 times, the value of outer loading on the data of women and men is already worth bigger than 0.5. Valid indicator data on psychological factors male is 23, behavior is 15, and regulation factorsis 5. While the female gender indicator valid on psychological factors is 25, behavior is 14, and regulation factorsis 5. Thus,convergent validity of factors psychological, regulation factorsand behavior are valid.

2) *Discriminant validity*

Discriminant validity is the unique test of each construct of a measurement model. This is done by comparing the loading constructs measured by loading other constructs. The overall constructs meet the criteria of discriminant validity, in which all constructs predict their indicators are greater than other indicators.

B. Reliability Test

Composite reliability (CR) is worth between 0 to 1 , the higher the value of Cronbach's alpha indicates the higher level of reliability of the measurement model. Variables have good reliability when CR value is greater than 0.7 and Cronbach's Alpha value is greater than 0.6 [10]. After testing for reliability measurement, the construct used in this model is acceptable and reliable in hypothesis testing.

2.4.2 Inner Model Evaluation (Evaluation of Structural Model)

The next step is to evaluate the structural model by looking at the path coefficient on the PLS output. Path coefficient obtained after bootstrapping on Smart-PLS software. The structural model evaluation can be seen in the original sample (O) contained in the output path coefficient. If the original sample is positive (+) then there is a positive correlation relationship, and vice versa if the original sample is negative (-) then indicates a negative linkage relationship. Inner model analysis can also be seen from the acquisition of R-Square on PLS output. R-Square is useful for viewing variable abilities in percentage form. The higher the R-Square value indicates a good model. The original values of the sample and R-Square can be seen in Table 2 and Table 3.

Table 3 Original Sample Value

	Male	**Female**
	Original Sample (O)	
Regulation factors-> Behavior	0.545	0.583
Psychological-> Behavior	0.293	0.213

Table. 3 shows that psychological variables and order/positive relationships to behavior (behavior). Women can calculate factors that lead to higher behavior than men. For psychological factors men who have psychological factor value to be higher than women.

Table 4 R-Square Value

	Male	Female
	R-Square	
Behavior	0.507	0.463

Table. 4shows that the factors and psychology can explain the effect for women by 46.3% and men by 50.7% against behavioral variables. So the factors that influence the electrical behavior of household electricity, men provide a greater role than women.

2.5 Analysis

Table 5 T-statistics and p-value

	Male		Female	
	T Statistics	P Values	T Statistics	P Values
Regulation factors-> Behavior	15.282	0	12.136	0
Psychological -> Behavior	7.966	0	4.5170	0

Table 5 shows a variety of behavioral, psychological factors having a greater t-statistic than women, a small p-value of 0.05; some elements have greater statistical t-statistics than women with a low p-value of 0, 05. Then it can be concluded that psychological variables give a positive influence on these variables, and male psychological factors are more dominant than women.

3 Conclusion

Psychological factor and the regulation factorson male gender can explain the effect of 50.7% on behavioral variables whereas in the female gender can explain the effect of 46.3% to behavioral variables behavior. This study concludes that psychological factors have a significant and positive influence on household electrical energy consumption behavior in terms of gender differences, where men have more influential to psychological factors than women do. The study suggest that (1) The government needs to make of energy conservation policy especially electrical energy with psychological factor approach such as communicative message and strong intervention and right target on the household electrical energy consumption behavior (2) Future study can investigate more psychological factors in other sectorson energy saving awareness behavior of household sector electricity.

This study is partly funded by Ristek-Dikti through PDUPT Grant No. 02/UN.16.17/PP.PDUPT.EET/LPPM/2018.

References

1. BPPT. *Indonesia Energy Outlook 2015. Journal of Chemical Information and Modeling* (Vol. **53**). https://doi.org/10.1017/CBO9781107415324.004. (2015)
2. Rani, P. Factors Influencing Consumer Behavior. *International Journal of Review Research and Academic Review.* Vol.**2(9)**: 52-61. (2014).
3. Wijaya, M. E., and Tezuka, T. Policy-Making for Households Appliances-Related

Electricity Consumption in Indonesia A-Multicultural Country. *Open Journal of Energy Efficiency*, *2013*(June), 53–64. https://doi.org/10.4236/ojee.2013.22009 (2013).

4. Wai, C. W. Energy Conservation: A Conceptual Framework of Energy Awareness Development Process. *3rd International Conference on Energy and Environment*, *74*(April 2006), 306–313. (2009).
5. Maag, J. W. *Behavior Management: From Theoretical Implication to Practical applications*. Belmont, CA: Thomson/Wadsworth. (2004).
6. Frederiks, E. R., Stenner, K., and Hobman, E. V. The socio-demographic and psychological predictors of residential energy consumption: A comprehensive review. *Energies*, *8*(1), 573–609. https://doi.org/10.3390/en8010573 (2015).
7. Vringer, K., Aalbers, T., and Blok, K. Household energy requirement and value patterns 1, *1* 553–566. (2007).
8. Fithri, P., Susanti, L., and Bestarina, K. Assessing Household Energy Savings and Consumer Behavior in Padang City. *Proceedings of the International Multi Conference of Engineers and Computer Scientists*, *2*(March), 18–22. (2014).
9. Rohana. *Analisis Perilaku Pelanggan PLN dalam Pemakaian Daya Listrik Studi pada PT PLN (PERSERO) Wilayah Sumatera Utara Cabang Medan*. Tesis Master. Universitas Sumatera Utara, Medan. (2008).
10. Ghozali, I Model Persamaan Struktural Konsep dan Aplikasi dengan Program Amos 16.0, Badan Penerbit UNDIP, Semarang. (2008).
11. Iwan, S., Hiroshi, M., & Lusi, S. Household lifestyle effect on residential electrical energy consumption in Indonesia: On-site measurement methods. *Urban Climate*, *20*, 20–30. (2017).

Intervention Selection to the Awareness of Energy-Saving Behavior in the Public Sector

Rozza Linda[1], * *Lusi* Susanti[2], and *Hilma* Raimona Zadry[2]

[1]Industrial Engineering Department, Faculty of Engineering, Ekasakti University, Padang, Indonesia
[2]Departement of Industrial Engineering, Faculty of Engineering, Andalas University, Limau Manis, Padang, West Sumatera, Indonesia

Abstract. Nowadays Indonesia is facing energy scarcity and still struggling with finding new energy resources. No other ways the employee have to use the currently available energy wisely. However wasteful behavior in energy use is still often seen in public sectors like government/private buildings, schools/campuses, hospital, and others. Various effort (such as energy saving campaign through mass media, pamphlet, etc) have been made by the building managers or owners to encourage users energy-saving behavior through various channels but found ineffective. Therefore, this study was conducted to select the alternative intervention to increase the awareness of energy-saving behavior in Indonesia by referring to the conceptual model of ABC (Antecedents-Behavior-Consequences) behavioral pattern. Several samples were taken from several public buildings, including government/private institutions, schools/campuses, hospital and mosques in Padang city. Data were collected through questionnaires filled by 30 respondents in each sector. The result shows that the phase of antecedent, the most appropriate type of intervention to build awareness of energy saving behavior for people is by making strong commitments in each building sector and the local government. In the consequence phase there are two criteria selected: reward and finance/legislation. The reward can be delivered for example, through providing a discount voucher or tax reduction. As for the consequence on finance/legislation can be done by creating and controlling the implementation of government regulations that are related to the electricity saving problem in public sector.

1 Introduction

Electricity still dominates the final energy consumption in Indonesia. The level of electricity demand is estimated to grow about 8.5% per year with the assumption of population grow of 1.1% per year [6]. If the trend of electricity consumption still follows business as usual (BAU) like today, Indonesia will experience electricity shortages shortly. This is exacerbated by the fact that Indonesia falls into the category of a country is wasted in electricity consumption in ASEAN. The problem of wasting electricity energy is generally caused by human factor (80%) and by technical factor 20% [5]. This wasteful behavior is increasingly being seen in public facility buildings, such as government/ private buildings,

* Corresponding author: lusi@eng.unand.ac.id

schools/campuses, hospital and others. One study conducted by Yang et al [8] The energy consumtion was calculated for each activity category (lighting, cooling, cooking, entertinment and other device) [7] states that energy consumption per square meter in office buildings spend about 70 – 300 KWH or equal 10 – 20 times greater than that consumption by households.

The public sector is still one source of energy waste in Indonesia. Efforts have been made by the goverment to encourage people's energy saving behavior through various channels such as public service announcements, posters, oral and written appeals, but the impact of the appeal is very small. Thus, it is needed an internal stimulus in arising the awareness/action which more concerned with the environment. Yik and Lee [9] stated that a lack of awareness of the importance of energy saving behavior is a major cause of energy wastage.

The Three Term Contingency theory is also known as the A-B-C model (Antecedences – Behavior – Consequences) [4]. This model explains how energy wastage occurs: Antecedences are the circumstances or situation that exist in the environment before a behavior is taken. Behavior shows how a person thinks, act and makes the decision that produce impact or consequences on the next process. While consequences is a situation that is instantaneous or in the long run is expected in accordance with the expected target behavior. If this behavior is reinforced, repeated that it becomes a habit, then the target to generate awareness of energy saving behavior can be expected.

So in encouraging energy-saving behavior in the public sector, it is necessary to conduct empirical studies of the types of interventions that are most appropriate to be applied based on the concept of Three-Term Contingency model, to produce the expected target behavior. This study aims to determine the most appropriate intervention for Indonesian society in building awareness of energy efficient behavior in the public sector.

2 Method

Respondent of the study are people related to the policy in the use of electrical energy as well as who understand about the conservation of electric energy in government/private institution, schools/colleges, hospitals and mosques.

2.1 Data collection and Analysis

Data collection was done by distributing questionnaires to predetermined public sector electricity users. The questionnaires were distributed as many as 120 pieces with 30 items for each sector (government/private, schools/colleges, hospitals and mosques) groups. Data processing is done on the questionnaires that have been collected. Data processing is done by AHP method using Expert Choice software version 2000.

The analysis was conducted on the result of data processing with AHP in the form of the most appropriate types of interventions to be conducted in the public sector in Indonesia to carry out energy-saving effort.

3 Result and Discussion

3.1 Questionnaire Distributions

Questionnaires were distributed to 120 respondents from government/private institutions, schools/colleges, hospitals, and mosques. The questionnaires were distributed for five months, from March 2017 to July 2017.

3.2 Pair Comparison

Pair comparisons are based on expert judgment which compares an element to other elements. Pair comparison of data using expert choice software. Priority is determined from the highest weight criteria. The elements selected in the priories are used as the most appropriate solution to the problem. The result found that the top three criteria selected in each agency is the commitment goal, reward and finance/legislation. Table 1 shows the alternative selected on the three criteria in each agency.

Table 1. Selected alternative for intervention of awareness to energy saving behavior

Sector	Selected Criteria	Selected Alternative	Assessment (%)
Government / private agency	Commitment goal	With municipality	43.9
		With employee	41.4
		With same institute	14.7
	Reward	Awards/prizes	85.9
		Tax deduction from land value tax	14.1
	Finance /Legislation	Legislation	55.4
		Price policies	28.9
		Subsidies	15.7
Schools/colleges	Commitment goal	With municipality	43.0
		With employee	40.2
		With same institute	16.8
	Reward	Awards/prizes	85.8
		Tax deduction from land value tax	14.2
	Finance /Legislation	Legislation	56.6
		Price policies	27.1
		Subsidies	16.3
Hospitals	Commitment goal	With municipality	42.8
		With employee	43.3
		With same institute	13.9
	Reward	Awards/prizes	86.1
		Tax deduction from land value tax	13.9
	Finance /Legislation	Legislation	56.8
		Price policies	27.4
		Subsidies	15.8
Mosques	Commitment goal	With municipality	41.3
		With employee	45.2
		With same institute	13.5
	Reward	Awards/prizes	85.3
		Tax deduction from land value tax	14.7
	Finance /Legislation	Legislation	57.2
		Price policies	26.8
		Subsidies	16.0

3.3 Analysis

3.3.1 Intervention on the Antecedent Phase

Goal commitment intervention is the selected intervention for antecedent phase. This intervention does not require high technology and great cost in the implementation and easy to follow up. Continuous commitment by an agency will change employee behavior if this is accompanied by continuous control by the authorities in an agency. To change the behavior of an individual/ employee in the public sector, it requires leadership with character and strict control system so that, the energy saving commitment can be implemented properly.

Selected alternative to goal commitment in government/private institutions, schools/colleges and hospitals are commitment by municipalities/government. This indicates that in the three institutions, the role of funders or owners of foundation in private institutions is very effective commit to realize energy saving behavior. As for the selected alternative mosque is a commitment to employees or in this case the offices/administrators of the mosque. It caused by the mosque funding are came from self-help community, so that the role of the employee is very dominant.

The form of commitment to government/private institution can be implemented by:

1. Forming a task force tasked with controlling energy use in the building.
2. Making the warning about the call to behave energy efficient
3. Creating written guidance on how to perform energy saving measure that is circulated to all sub-sections and also posted on strategic places in the office.
4. Conducting appeal and direction to all employees in every morning apple execution.

Commitment to schools/colleges can be realized in following ways:

1. Making the written guidance on how to conduct energy saving measure that is circulated the entire academic community and also tacked on strategic places in the schools/colleges.
2. Giving appeal to all students in the class to always behave energy efficient.
3. Improving the function of security officers to always control the use of energy in schools/colleges.
4. Providing the warning about the call to behave energy efficient.

Commitment to the hospitals can be done in the following ways:

1. Making written guidance on how to perform energy saving measures tailored to the sub-section of the hospital's work, because in the hospital, it has specialized tools whose use in understood only by certain people.
2. Creating written guidance on how to perform energy saving measures posted on the hospital's outpatient and strategic premises.
3. Providing the warning about the call to behave energy efficient.

Implementation of commitment to the mosque can be done in more simple way than three agencies above, because the energy user in the mosque is a congregation who come worship. Implementation of commitment to the mosque can be done by mosque officers who at the same time control the use of energy in the mosque.

3.3.2 Intervention in Consequence Phase

The selected consequence found are two intervention, namely reward interventions and finance/legislation intervention.

a) Intervention in the Form of Reward

Giving the reward to agencies that succeeded in making energy efficient effort in one of the most effective ways to be done in the public sector in Indonesia. This can provide motivation and stimulation of a person is behaving.

Asriani et al. [1] study indicated a significant influence between the provisions of reward to the employee to work discipline. However, this reward is only a short term effect [2]. Choosing the right form of reward is very important, but rewarding is a huge cost. Giving reward to the public sector in Indonesia can be realized for example in the form of discount voucher charging fuel to all employee who successfully makes energy-saving effort.

b) Intervention in the form of Legislation

The legislation is selected because government regulation is a force in a country to implement a commitment, so it is expected to effect on behavior change in the public sector. However, this method should be accompanied by clear sanction for agencies that do not implement these government regulation to make this regulation effective. The energy saving of the public sector in Indonesia is regulated by Minister of Energy and Mineral Resources No. 14 of 2012. In article 12 it is explained that the electrical energy saving in public sector can be done through: air system, light system, supporting-equipment , production process, and main energy utilization equipment. However, in the implement of the Minister's regulation to be effective, there should be sanction for agencies that do not implementation the regulation.

The sanctions that can be applied to government/private sector in the form of warning letters, enforce the high tax, reduce aid and revocation of operational permits. The sanction can also be applied to schools/colleges and hospitals. As for the mosques can by appealing to mosque officials to conduct energy saving measure or give a letter of reprimand to the mosque management who do not perform energy-saving measure.

4 Conclusion

This study found the most appropriate type of intervention to build the awareness of energy saving behavior in the public sector in Indonesia in the antecedent phase and the consequence phase. In the antecedent phase, it is suggested to make the strong commitment in every institution and local government to implement electrical energy saving measures. A form of commitment that can be done by each agency is to provide direction and appeal by the government and their respective leaders if all employees to perform energy-saving measures.

In the consequence phase, there are two criteria selected: reward and finance/legislation. One form of reward is given in the form of discount voucher fuel replenishment. As for the consequences on finance/legislation can be done by creating and controlling government regulations relating to electricity saving public sector in Indonesia.

Future research is advised to examine the effect of selected interventions on energy saving behavior in the public sector in Indonesia. This will provide more accurate information on alternative interventions in build awareness of electrical energy-saving behavior in the public sector in Indonesia.

The acknowledgement this study is partly funded by Ristek-Dikti through PDUPT Grant No. 018/UN.16.17/PP.PDUPT.EET/LPPM/2018.

Reference

1. Asriani, D. Gimin, Nas, S. Efect of Reward on PT.Citra Riau Sarana's employee work in Kuantan Singingi Regency. University of Riau (2014).
2. Geller, E. The Challenge of Increasing Proenvironmental Behavior, In:Bechtel, R., Churchman, A (Eds.), *Handbook of Environmental Psychology* (Wiley, New York, 2002)
3. Lestari, E. Analysis of Decision Support Systems for the Process of Increasing Position on PT X. **1/3/J-SI/2009**, (2009)
4. Maag, J.W, *Behavior Management : From Theoretical Implication to Practical Application* Belmont, CA : Thomson/Wadsworth, (2004)
5. ESDM Regulation No. 13 of 2012, Guidelines for Energy Savings in Government Buildings. Jakarta (2012).
6. PT. PLN (Pesero), PLN Statistics. PT PLN Jakarta Secretariat (2015)
7. Sukarno I, Susanti L, *Household Lifestyle Effect on Residential Electrical Energy Consumtion in Indonesia*. (2017).
8. Yang, L., Joseph, C., Lam, Tsang, C.L., Energy Performance of Building Envelopes in Different Climate Zones in China. Applied Energy **9**, 800-817 (2008)
9. Yik, F.W.H. and Lee, W.L., A Preliminary Inquiry into Why Buildings Remain Energy Inefficiency and Potential Remedy. Transaction, The Hongkong Institute of Engineers **1,** 2-3 (2002)

Designing Improvement of Procurement Business Process Reengineering Approach : A Study Case of Insurance Company

Almira Rhea Masayu and *M.* Dachyar*

Industrial Engieneering Department, Universitas Indonesia

Abstract. Streamlining procurement process is one of the strategies to reduce cost of a company, especially in the insurance industry. This research aims to propose design improvement of procurement process in insurance company by implementing Business Process Reengineering (BPR) method. Data collection regarding procurement process was collected in one of life insurance company in Indonesia. This research uses BPR method which consist of three steps namely mapping as-is process, analyze as-is process, and modeling to-be process. Mapping as-is process using iGrafx standard Business Process Modeling and Notation (BPMN). Analyze as-is process by its process time to indicate problems. Modeling process to-be by using BPR best practices that will be prioritized using AHP method and selected using pareto analysis. Scenarios for to-be process then created based on the selected best practices, that will be inputed into the simulation model iGrafx standard BPMN. This research produces three scenarios of to-be process for procurement in Indonesian insurance company that reduce 35-76% of the process time compared to as-is process.

1 Introduction

Insurance company is one of the most threatened industry in service industry [1]. Reducing cost has been the most popular strategy that CEO chose to improve their company's growth [1]. Reducing company cost can be supported by process efficiency [2]. Cost classification of insurance company shows that operational cost is the highest cost of insurance industry [1]. Procurement, which is a term of a process of purchasing goods or services, contribute to 75% of the total operational cost of insurance company [3].

Surprisingly, procurement in insurance industry has the lowest improvement process compared to other industries (such as FMCG, professional service, government, health, manufacture, and telecommunication). The main cause of the lack of improvement in procurement is from the quality of the data [4]. Procurement in insurance industry takes a long time of end-to-end process, because of the manual process of organizing administration in procurement process [5].

A company can improve its process efficiency by re-organizing and re-designing process [6]. Changing process flow needed a transformation of its business process [7]. This research

* Corresponding author: mdachyar@yahoo.com

aims to design improvement of the procurement process in one of the insurance company in Indonesia to increase its process efficiency with Business Process Reengineering (BPR) approach.

2 Literature Review

Business Process Reengineering (BPR) was first introduced by Davenport and Short and Hammer. BPR is a set of activities undertaken to achieve the results of the desired business process [8]. BPR were based on operational management theory, which is the design, activity, and improvement of systems that produce products or services. Operations Management focuses on the management of the entire system that produces finished goods as well as the process of delivering products to consumers [8]. BPR comes from the word "re-engineering". Reengineering is a fundamental way of thinking and a radical redesign of a business process to achieve dramatic improvement measured by its own measurement of cost, quality, service, and time [9]. Process innovation is one of the current innovation that organization concertrated on [8].

Data Flow Diagram (DFD) are visual tools for describing input and output models [10]. DFD consist of four notation, which are process, entity, storage, and flow [11]. The highest level of the DFD is a context diagram that provides an overview of a process. The next level is DFD level-0 which is the main process description. The level after level-0 is DFD level-1 which is the details of the process of each main process based on DFD level 0 [12].

Business Process Modeling Notation (BPMN) is a standard in capturing business processes in the early stages of system development [13]. BPMN represents graphic notations of business processes that facilitate understanding for business analysts and communication in business needs [14]. BPMN has four categories for graphical elements in building a model, namely Flow Objects, Connecting Objects, Swimlanes and Artifacts [15]. The application of BPMN is dominated by the main objectives such as documentation, redesign, continuous improvement, and knowledge management [16]

Analytical Hierarchy Process (AHP) is the most popular MCDM method used for research [17]. AHP is the first method introduced by Saaty, which is a method for problems with complex multi-criteria and takes into account both quantitative and qualitative aspects (Saaty, 1980).

3 Methodology

This research used BPR methodology approach. There are three main steps when conducting BPR method, which are modeling the as-is process, analyzing the as-is process, and modeling the to-be process. Preperation for step one –modeling as-is process- was done by creating procurement business process map using APQC framework to categorized process, SIPOC to see related stakeholders, and DFD to see flow of information on each process. Modeling the as-is process was created regarding the mapping of business process by APQC framework, SIPOC model, and DFD, by using simulation software iGrafx standard BPMN. As-is process analysis was observed by looking at the processing time in each of the activities in the as-is process model and benchmarking with the enablers to identify the problem. Modeling the to-be process are based on the best practices of BPR. Best practices were collected by literature review and then interviewing seven procurement expert in different insurance company in Indonesia to select best practice that are fit to the condition of insurance industry. Best practices then were prioritized by AHP method and selected by pareto analysis. The selected best practices will be incorporated into scenarios for the to-be process and simulated using iGrafx standard BPMN to see changes in processing time compared to the as-is process. Three scenarios were compared to see the lowest process time.

4 Results and Discussion

4.1 Procurement process map

Before making the as-is process, the process map of the procurement process is designed using the APQC framework, by dividing the process into several levels. The APQC framework is used to categorize processes into five process levels, which are categories, group process, process, activities, and assignment (APQC, 2015). Four category of processes are: Planning and Analysis, Procurement Strategy, Contract and negotiation, Approvals and document handover, SIPOC (Supplier, Input, Process, Output, Customer) method is used to see who is acting as a stakeholder in the procurement process. Four Stakeholders have a role in the procurement process in the insurer ie, third-party companies (suppliers), finance departments, legal departments, and departments requesting goods or services.

4.2 Modeling as-is process

DFD was first created to see the flow of the documents for each process group. There are three main entities for the procurement process with each entitiy has its own documents that flows in and out of the entities. Level-0 DFD of procurement process sees a few problem regarding the storage based of the entire tender documents.. BPMN have the activity point of view of the entire business process, which means that from BPMN we can see the flow of each process, but cannot see the flow of the documents. Many manual process from the procurement end-to-end process. BPMN also found many approval process that can takes time to finish and effecting the continuous of the next process, which caused a bottleneck problem.

4.3 Analysis of as-is process

The as-is process takes about 202 days to process. The parameters the company has for the whole process is 55 days. There are nine type of process that have a high process time compared to the others, which is RFP revision, completing CST, vendor profile checklist, and Due Diligence Requirement (DDR), Completing Risk Assessment Checklist (RAC), RAC approval circulation, tender proposal submission, proposal clarification, price negotiation, PA/AF approval circulation, and contract negotiation.

The activity that have high process time will then be improved by using the BPR process improvement targets as an alternative to the to-be process. Table 1 shows the problem description regarding the observer and enabler on each activity.

Table 1. Problem identification regarding process as-is by BPMN iGrafx

Process	Problem identification	Problem description
RFP revision	Bottleneck	The revision takes a long time, the bidding process cannot start if the RFP has not been finalized. 70% RFP through revision before final RFP
Completing CST, Vendor profile checklist, and Due Diligence Requirement (DDR)	lack of an automated system	Filling and retrieving document templates are done manually and in different places so it takes time to retrieve documents
Completing Risk Assessment Checklist (RAC)	lack of an automated system	Filling and retrieving document templates are done manually and in different places so it takes time to retrieve documents

Process	Problem identification	Problem description
RAC approval circulation	lack of an automated system	RAC hardcopy should be sent to several departments to get signatures from each department concerned
Tender proposal submission	lack of an automated system	The collection of proposals takes a long time, so it can not go to the next stage
Proposal clarification	lack of an automated system	Customized schedule adjustments for each related party
Price negotiation	lack of an automated system	Negotiations are done via email, missing information due to email piling up, lack of supplier response speed through email
PA/AF approval circulation	lack of an automated system	AF / PA hardcopy must be sent to several departments to get the signature from each department concerned
Contract negotiation	lack of an automated system	Negotiations are done via email, changes should be discussed with some managers from several departments concerned

4.4 Modeling To-Be Process

A literature study of BPR process improvement targets and adjustment of improvement targets on insurance conditions in Indonesia by appraising experts from several insurance companies in Indonesia. Literature study found 29 best practices of BPR. Seven (7) procurement insurrance expert in different insurance companies in Indonesia selected to 16 best practices by using Delphi Method.

AHP method is selected to find weighted importance from 16 BPR best practice and perform pareto diagram to prioritize business improvement. The weighted of importance of BPR best practices are Task Automation (13.1%), Resequencing (10.3%), Parallelism (10%), Technology Integration (9.6%), Triage (9.4%), Empowerement (6.8%), Task Elimination (6.4%), Specialist Generalist (5.2%), Task Composition (5%), Order Type (4.4%), Knock-down (4.3%), Contact Reduction (3.7%), Interfacing (3.7%), Exception (3%), Case Manager (2.8%), Extra Resource (2.3%).

Three scenarios which are scenario 1 (without automation) takes 147 days on Figure.1. Scneario 2 (half automation) takes 58,17 days, and scenario 3 (full automation) takes 50 days. Scenario three has the greatest impact on reducing processing time, which can decrease to 50 days.

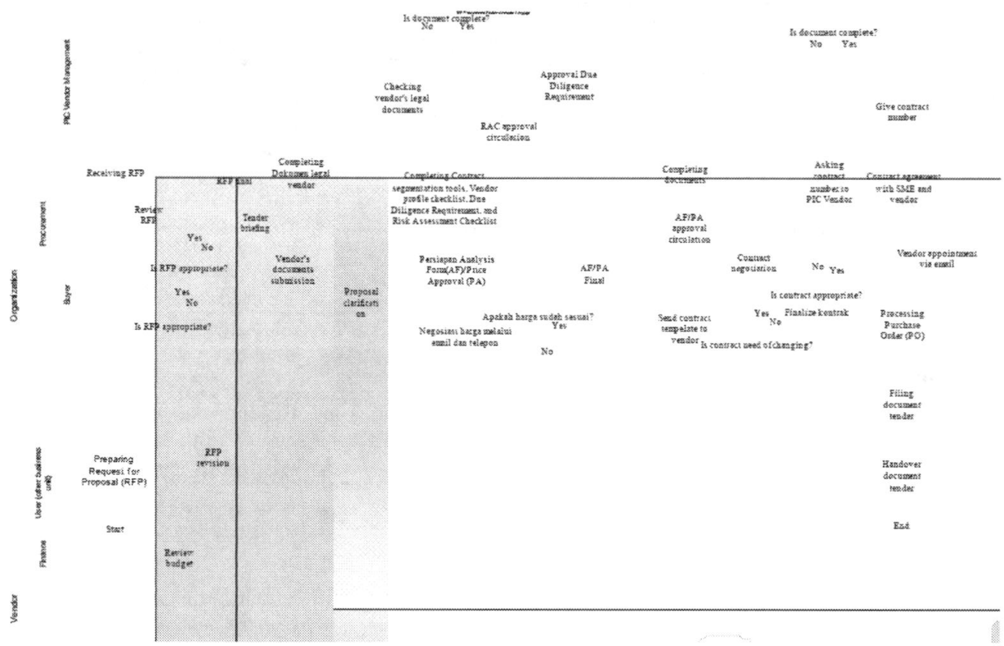

Fig. 1. Scenario 1 BPMN proses to-be

5. Conclusion

This research found the effective use of 16 best practices in procurement process of one of Indonesian insurance company, which are task automation, technology integration, parallelism, resequencing, order type, generalist-specialist, task composition, task elimination, empowerment, and triage.

Based on three scenarios, the best conditions are in the third scenario that is to automate the entire process. The third scenario resulted in a reduction in processing time of 76.48%. The third scenario shows that technology integration strategies can reduce processing time so as to achieve company standards.

References

1. PWC, (2017).
2. I. Cotter, (2015).
3. G. Sibanda, (2016).
4. Deloitte, *The Creative Studio* (2017).
5. V. Adhityahadi, (2017).
6. T. Munyard, J. Schade, (2017).
7. S. Kaniyar, P. Peters. (2015).
8. A. D. Cruz, E. E. V. Martinez, J. M. M. Hincapie i F. R. Torres, Intgbl Cap, (2016).
9. T. E. Erkan, Engr. and Tech. Int. J. of Econ. and Mgmt. Engr **3**, (2009).
10. P. Larsen, N. Plat i H. Toetenel, *Form. Asp. Compu*, (1994).

11. F. Schmalenberg i R. Vandenhouten, *International Symposium on Applied Machine Intelligence and Informatics,* (2016).
12. P. M. Jacob, I. M. H, J. Jose i J. Jose, International Conference on Information Science, (2016).
13. R. M. Dijkman, M. Dumas i C. Ouyang, Info. and S. W, Tech., (2008).
14. R. K. Ko, S. S. Lee i E. W. Lee, Bus. Prcss. Mgmt. J., (2009).
15. M. Chinosi i A. Trombetta, *Computer Standards & Interfaces,* (2012).
16. J. Recker, Bus. Prcss. Mgmt. J., (2010).
17. S. D. Pohekar i M. Ramachandran, *Renewable and Sustainable Energy Reviews*, (2004).

The Valuation of Geothermal Power Projects in Indonesia Using Real Options Valuation

Reza A. Bilqist, *M.* Dachyar*, and Farizal

Industrial Engineering Department, Universitas Indonesia

Abstract. Geothermal power project should be carefully evaluated, because not only the project is large-scale and high-risk, but also huge investments are needed to make the project profitable. The purpose of this study is to evaluate and analyze the present value of the geothermal power plant projects in Indonesia using net present value and real options valuation approach to obtain better project value that indicates profitable investments. Project value determined based on Net Present Value (NPV). The Monte Carlo Simulation calculated NPV based on discount rate, production volume, and operation and maintenance (O&M) cost. The risk of the project defined with 3 scenarios to obtain NPV and calculate Expected Net Present Value (ENPV). The result of this study showed Real Options Valuation obtained the highest project value indicating a higher return of the project that was previously undervalued using NPV and ENPV. Real Options Valuation approach improved the project value by incorporating scenario analysis and strategic option in the valuation process.

1 Introduction

Renewable energy in Indonesia only 6.2% of the total energy mix in 2015 that indicating Renewable energy in Indonesia reached 6.2% of the total energy mix in 2015 that indicating Indonesia has a large dependence on fossil energy [1]. Meanwhile, Indonesia has a great potential for renewable energy such as geothermal. Geothermal power generation is a promising substitute for the traditional-generating approaches like fossil energy [2]. Geothermal Power Plant that has been installed in Indonesia only 1640 MW about 4.9% of 29,544 MW total potential geothermal in Indonesia [1]. However, geothermal power generation should be carefully evaluated, because not only the project is large-scale and high-risk, but also huge investments are needed to make the project profitable.

2 Literature Review

Renewable energy power generation is a large-scale and high-risk project [3]. Uncertainty in a project of renewable energy power generation, have to take into consideration regarding their economic viability, which is linked to several uncertain variables that affect the evolution of the cash flows [4]. Valuation enables investors to determine asset value and take

* Corresponding author: mdachyar@yahoo.com

into consideration unlimited possible scenarios, as well as limiting losses in unfavorable scenarios. Valuation and redesigning business process are several methods that used in a project in order to get the best result or profit [5, 6], especially in a high-risk project such as oil and gas energy [7], and photovoltaic energy [8].

Geothermal is one of renewable energy project with huge potential, especially in Indonesia [9]. Maintenance management also affects operational activities in power generation, without proper management of operation and maintenance can lead to high costs in operating and maintenance activities and affect company performance [10]. Therefore, the value of the project must be evaluated financially to determine whether the project profitable or not, the main idea of valuation is to determine the financial value that closely resembles the true value of an asset or project [11].

Net Present Value or Discounted Cash Flow (DCF) is a traditional and simple valuation method, where incremental cash inflows and outflows for each period discounted and projected to present time when investment is made [12]. NPV methods assume that the investment is an all or nothing strategy and do not account for managerial flexibility that exists such that management can alter the course of an investment over time when certain aspects of the project's uncertainty become known (Mun, 2005). Project value has an important function as the basis of capital budgeting and is related to resource allocation to long-term project investments [13].

In the NPV approach, cash inflows and outflows are estimated over the life time of the project and are subsequently discounted to the present time. Meanwhile in Expected Net Present Value (ENPV), a number of scenarios are generated. The NPV for each of the envisaged project outcomes are determined and the ENPV is calculated by taking the sum of the product of the different NPVs and the probability of their occurrence [6].

The basic option theory in finance is a contract giving the buyer the right, but not the obligation, to buy or sell an underlying asset as stock or index, at a specific price on or before a certain date [14]. Real Options Valuation (ROV) refers to application of the option theory to the assessment of non-financial or "real" investments with learning and flexibility, such as multi-stage R & D, manufacturing plant expansion [15]. A project that can be terminated after a technical failure is an example of a real option. Initiation of a project generates the option, not the obligation, to continue the project [6].

Real Options Valuation not only enables investors to increase an asset value, due to the consideration of an unlimited number of possible future scenarios, as well as limiting losses in unfavorable scenarios, but also the most adequate when researches consider uncertainty scenarios aligned with managerial flexibility and the use of this approach can significantly change the value of an asset [16]. The implementation of Real Options Valuation were widely used in high risk project such as product development portfolio [15], oil and gas [7], mining [12], photovoltaic energy [3, 16], wind energy [17, 18, 19], water energy [20, 21], and nuclear [22].

Monte Carlo simulation utilized to obtain better actual value estimation of a particular project [11]. The Monte Carlo Simulation stands out because it enables the generation of a great variety of scenarios as well as multiple uncertainties [16]. As all those scenarios produce associated results, each scenario can have a forecast.

3 Methodology

The investment project is a geothermal power generation project in Sumatera with total capacity is 236 MW. The project divided into 2 units power plant, the Unit-1 power plant with 92 MW capacity and duration of project 10 years, the Unit-2 power plant with 144 MW project started after Unit-1 is successfully installed and power generated with duration of project is 8 years. After Unit-1 power plant started to operate for 8 years, the investors have right to defer or abandon the project Unit-2 power plant project if Unit-1 power plant had indicated not profitable. To determine the volatility of the projects, we must analyze the uncertainties that affect it. The uncertainties considered in this renewable energy project are discount rate, electric sales volume, and Operation and Maintenance (O&M) cost.

The project estimated with uniform sales volume derived from estimated electric produced and uniform Operation and Maintenance (O&M) cost. The electricity price on the basis of a feed-in tariff under the government's constant benchmarking is 13.8 cents $/KWh. Simple NPV analysis conducted to calculate project value based on Unit-1 and Unit-2 installed. The Monte Carlo Simulation conducted with 1000 trials to generate multiple project value using uncertain variables that influenced the project. The uncertain variables are discount rate, sales volume, and O&M cost.

In ENPV approach, three scenarios were generated in Unit-2 power plant project, which created based on the worst scenario, normal scenario, and best scenario of the Unit-2 project. The probability of occurrence for worst, normal, and best scenario is 20%, 50%, and 30% respectively.

The right to invest, defer, and abandon the Unit-2 power plant project become Real Options that must be considered in project valuation. The investors have the option to wait or defer investing until a later date for Unit-2 power plant project. That is, wait until uncertainty becomes resolved and then decide on the next course of action afterwards.

Investments will be made in the Unit-2 project if conditions indicate a good scenario and decide to abandon the project if the market condition is akin to normal or worst case scenario. In this study, Passive Wait and See Strategy were applied where we decide to wait one year assuming that we will gather more valuable information within time frame, and then decide to execute the project.

4 Result and Discussion

4.1 The Results of Monte Carlo Simulation for NPV, ENPV and ROV

The simulation results of NPV model for Unit-1 and Unit-2 power plant presented in Fig 1. Fig 1 showed the mean NPV for Unit-1 and Unit-2 is $92.25 million. The probability that Unit-1 project will break even or have NPV will be > 0 is 63.3% rather significantly increased compared to Unit-1 project executed.

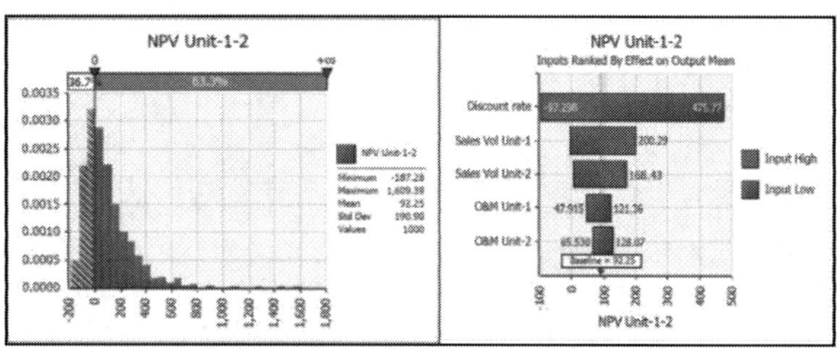

Fig 1. NPV simulation result and sensitivity chart for Unit-1 and Unit-2 projects

The NPV sensitivity chart after Monte Carlo Simulation shows that the Unit-1 and Unit-2 project is sensitive to the discount rate variation. The simulation results of ENPV model for Unit-1 and Unit-2 power plant presented in Fig 2. Fig 2 showed the mean NPV for Unit-1 and Unit-2 is $93.62 million. The probability that Unit-1 and Unit-2 project will break even or have NPV will be >0 is 64.6%. In ENPV model, distribution is less wide compared to

NPV model. This comparison can be seen in Fig 1 where range of value NPV model is -\$187 million to \$1609 million, but in Fig 2 of ENPV model the range of value is -\$173 million to \$1458 million.

The NPV sensitivity chart after Monte Carlo Simulation shows that the Unit-1 and Unit-2 project is sensitive to the discount rate variation. The simulation results of Unit-1 and Unit-2 power plant using ROV best scenario presented in Fig 3. Fig 3 showed the mean project value for Unit-1 and Unit-2 is \$130.86 million. The NPV sensitivity chart after Monte Carlo Simulation of ROV best scenario shows that the Unit-1 and Unit-2 project is sensitive to the discount rate variation.

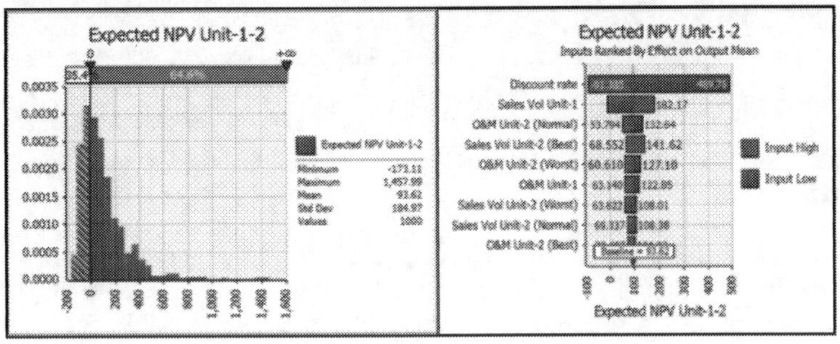

Fig 2. ENPV simulation result and sensitivity for Unit-1 and Unit-2 projects

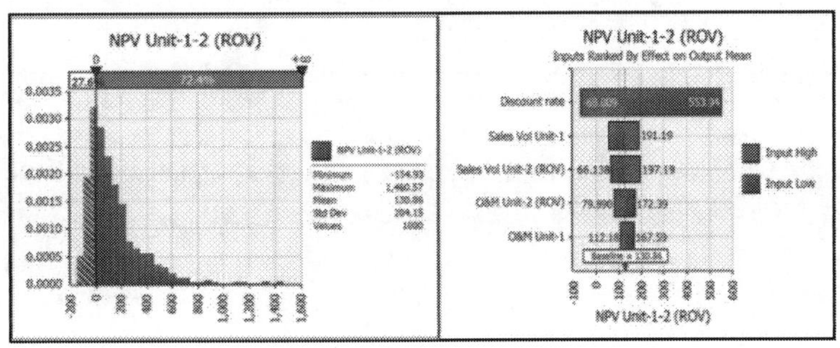

Fig 3. ROV simulation result of best scenario and sensitivity chart for Unit-1 and Unit-2 projects

4.2 Analysis for Monte Carlo Simulation

The result of Monte Carlo Simulation for NPV, ENPV and ROV presented as overlay graphical chart in Fig 4. Fig 4 shows ENPV approach obtained rather similar result with NPV, but ENPV provides a much tighter distribution although not really significant. Once we added ROV approach where we only execute the Unit-2 project if the best condition obtained, the risk has been reduced and the return of the project increased.

Based on the result obtained from NPV method through ROV method, we can define risk as uncertain variables that fluctuate in project value. All risks are mitigated in ROV approach since we do not execute the project if the worst or normal case scenario occurs. In this study also obtained that geothermal project really sensitive to discount rate and sales volume. Real Options approach enhanced project value since the project executed when the best case scenario occurs. This decision creates better valuation simply by having the right strategic optional available, acting appropriately, and valuing the project in terms of its "real" value, which includes this opportunity to make midcourse corrections when new information becomes available.

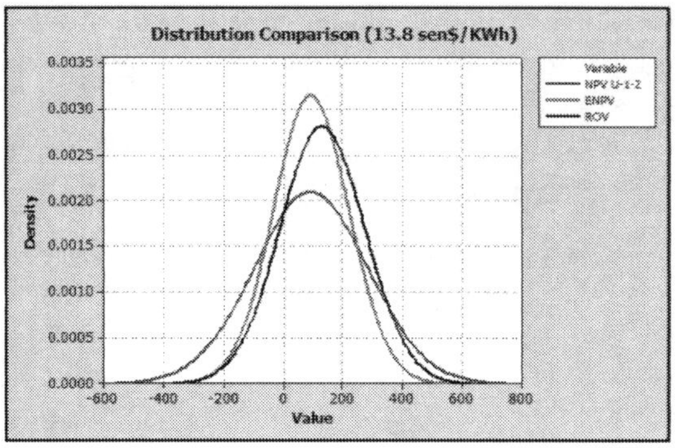

Fig 4. Overlay distributional comparison of NPV, ENPV, and ROV

5 Conclusions

This study showed that the valuation of geothermal power generation project using NPV method underestimates the value of the project rather than ROV. The ROV method assesses projects based on options in the implementation of a project. In ROV approach where we only executed the project if best scenario occurred, we obtain better value of the project and risk has been reduced. The result of the study showed discount rate is the most sensitive variable in the project followed by sales volume.

Since ROV applied option mechanism, which allows companies or investors achieve better overview and flexible management for project valuation. Therefore, ROV can help investor or companies to consider an option framework in the decision-making process. That is, to find strategic options that exist in projects or to create strategic options in order to increase the project's value.

References

1. Ditjen EBTKE, (2016)
2. N. A. Pambudi, Renewable and Sustainable Energy R. **81**, 2893-2901, (2018).
3. L. Tian, J. Pan, R. Du, W. Li, and G. Qibing, J. of App Energy **201**, 354-362, (2017)
4. M. M. Barroso and J. B. Iniesta, Energy Policy **55**, 335-352, (2013).
5. M. Dachyar and G. Novita, ARPN J. of Eng & App Sci **11**, 4539-4546,(2016).
6. B. J. A. Willigers and T. L. Hansen, J. of R&D Mgmt **38**, 520-537, (2008).
7. J. Guedes and P. Santos, Energy Economics **60**, 377–386, (2016).
8. K. Kim, S. Kim and H. Kim, IOP Conf. Series **40**, 1-8, (2016).
9. M. P. Hochstein and S. Sudarman, Geothermics **37**, 220-266, (2008).
10. M. Dachyar and Y. Tohir, ARPN J. of Eng & App Sci **13**, 2571-2580, (2018).
11. J. Mun, *Real Option Analysis Tools and Techniques*, (John Wiley & Sons, New Jersey, (2005).
12. N. A. Gardner, *McGill University Annal*, (2015).
13. M. Tsui, *Valuing Innovative Technology R&D as a Real Option*, (MIT, Boston, 2005).
14. M. Amram and N. Kulatilaka, *Real options: managing strategic investment in an uncertain world*, (Harvard Business School Press, Boston, 1999)
15. M. Jin and X. Tian, JIEM **8**, 72-84, (2015).

16. G. Esteves and I. Leite, IEEE Xplore, 1-6, (2017).
17. D. Loncar, I. Milovanovic, B. Rakic and T. Radjenovic, Renew & Sustain Energy Rev **75**, 354–367, (2017).
18. M. M. Barroso and J. B. Iniesta, Journal of Energy **77**, 422-433, (2014).
19. D. Angelopoulos, R. Bruckmann, F. Jirous, I. Konstantinaviciute, P. Noothout, J. Psarras, L. Tesniere and B. Breitschopf, Energy & Environment **27**, 82-104, (2016).
20. K. Kim, H. Park and H. Kim, Renew & Sustain Energy Reviews **75**, 918–926, (2017).
21. J. Marques, M. Cunha, D. Savic and O. Giustolisi, Procedia Eng **89**, 856-863, (2014).
22. L. Tian, H. Shan and N. Zhu, Energy Procedia **104**, 299-304, (2016).

Determining Criteria of Human Resource Information System that Affect Human Resource Performance in Companies Using DEMATEL-Based ANP Method

Eveline Siregar and *M.* Dachyar*

Industrial Engineering Department, Universitas Indonesia

Abstract. Decision making in HR activities is delayed because of insufficient information. Human Resources (HR) and information system are the important roles in helping company HR works. Objective of this study is toidentifycriteria of Human Resource Information System (HRIS) that affect human resources in companies by using DEMATEL-based ANP (DANP) method.HRIS criteria are collected from several studies to be validated by experts in HR of manufacturing companies. The survey was conducted with 5 experts and their judgments are calculated in limit supermatrix. The limit supermatrix of DANP showed that 18 criteria in HRIS. This study obtained five most important criteria of HRIS in HR performance are high quality data presentation, quick and precise, accessible, information need in time, and fulfill needs of HR.

1 Introduction

Human Resource (HR) affect the development of an organization/company.HR problems that are often encountered such as the retention of human resources, the lack of human resources development, poor employee relations, traditional administrative system, manual payroll system, and lack of integrated system. Information system (IS) is a combination of information technology that use technology to support operational and management activity [1]. There are two important sources in the organization, human resources and information systems (IS).HR and IS require significant management to affect the overall performance of the business. Insome developed countries, HRIS weremanaged independently rather than using outsourcing services. The purpose of this study is to identify the criteria of HRIS (Human Resource Information System) that affect the performance of human resources in companies using DANP method in manufacture companies. A research conducted in UK showed modules of HRIS in the UK have been used to help facilitating the performance of human resources in the company[2].

* Corresponding author: mdachyar@yahoo.com

2 Literature Review

Human Resource Information System (HRIS) was defined as information system, within the traditional functional areas of the organization, and designed to support planning, administration, decision-making and Human Resource Management (HRM) control activities [3]. In the other hand, Kavanagh [4] defined HRIS as a system that used to acquire, manipulate, store, retrieve and distribute related information about the organization's human resources.There are many advantages of using HRIS such as faster process information, greater information accuracy, better program planning and development, improved employee communication, reduced cost of data storage in HR, more transparency in the system, more meaningful career planning and counseling at all levels, and better ability to respond to environmental changes [5].

Company performance is the company capacity indicator to achieve its goal and performance which include financial and non-financial action [6]. Financial size includes economic factor and non-financial include an indicator of achievement, as market share, quality, satisfaction and market effectiveness. In the other hand, the most important thing to develop and achieve organization's goal is employee's contribution. A potential and skilled employee could help to achieve competitive advantage over their competitors [7].

The DEMATEL-Based ANP (DANP) method combines other hybrid techniques. Analytic Network Process (ANP) has interactions and feedback within (inner dependence) and between clusters (outer dependence) [8].The DANP method are used to obtain form of a comprehensive unweighted supermatrix by constructing a direct effect matrix. This pairwise comparisons are not only done in groups, but also to the whole system.The total intergroup relationship matrix is used to measure the appropriate portion of unweighted supermatrixto obtain weighted supermatrix.This following steps are the process of DANP method[9]. There are seven steps in DANP process: 1. Calculate the direct effect, 2. Normalize the direct-effect matrix,3. Build up Total-Influence Matrix T, 4. Analyze the results, 5.Build up an unweighted matrix, 6. Create a weighted matrix, 7. Limit supermatrix.

3 Methodology

The questioner in this studywas reviewed by 5 expert judgments. All experts are HR manager with more than 10 years of experience in their area. Geometric mean value of criteria obtained from expert judgments that has more than 3.75 are accepted. There are 4 dimension and 18 criteria selected by experts, as follow: System quality dimension (D1) consists of four criteria: accessible (C1), quick and precise (C2), easy to use (C3), and multipurpose (C4). Information quality dimension (D2) consists of five criteria: output in useful format report (C5), comprehensive information (C6), up-to-date information (C7), information need in time (C8) and accurate information (C9). User satisfaction dimension (D3) consists of three criteria: high quality data presentation (C10), fulfill needs of HR (C11), and satisfaction (C12). Information system success dimension (D4) consist of six criteria: enhance human resource planning (C13), online system (C14), career improvement (C15), salary advice (C16), improve the assessment and training needs (C17) and enhance performance management (C18). The second queationaire of this research were distributed to 10 HR experts in different manufacturing companies and used Likert scale from 1 to 5.

The result of the respondents were generated to be an averaged matrix (A) as shown in Table 1.

Table 1. Averaged Matrix (A)

	C1	C2	C3	C4	C5	C6	C7	C8	C9	C10	C11	C12	C13	C14	C15	C16	C17	C18	Total
C1	0.00	3.50	2.20	1.70	1.50	2.40	2.90	3.30	3.00	3.20	1.50	1.10	1.00	3.40	0.40	0.80	0.50	0.50	32.70
C2	2.90	0.00	3.40	3.30	3.40	3.40	3.20	3.30	3.10	3.30	3.10	0.70	1.00	3.60	1.10	0.60	0.80	0.80	40.80
C3	3.40	1.90	0.00	1.90	0.50	0.50	0.50	0.60	0.50	0.40	1.10	0.40	0.40	3.10	0.40	0.30	0.30	0.30	16.50
C4	3.40	2.40	1.30	0.00	2.10	3.40	2.80	3.10	2.90	2.90	3.20	0.30	1.70	2.90	0.30	0.30	0.20	0.20	34.20
C5	2.20	3.30	2.70	3.00	0.00	3.30	3.20	3.40	3.40	3.50	3.30	1.00	0.30	3.20	0.70	0.70	0.90	0.70	38.80
C6	3.40	2.90	0.50	0.30	1.80	0.00	3.30	3.40	3.50	1.10	0.70	0.60	0.80	3.30	3.50	0.50	0.50	0.50	28.60
C7	2.10	3.00	3.00	2.60	2.80	3.50	0.00	3.40	3.40	3.50	3.40	3.40	0.90	3.30	0.50	0.40	0.60	0.60	40.60
C8	3.20	2.30	3.00	2.30	3.00	2.60	3.40	0.00	2.70	3.20	0.50	0.60	0.30	3.00	0.60	0.60	1.00	1.10	34.60
C9	2.80	3.30	1.10	1.10	3.40	3.40	3.30	3.40	0.00	3.40	0.40	0.30	0.80	2.30	0.50	0.50	0.50	0.50	31.20
C10	2.20	3.40	2.80	3.20	3.40	3.40	3.30	3.50	3.40	0.00	0.80	0.40	0.30	2.70	0.20	0.30	0.40	1.10	35.10
C11	2.90	3.20	3.40	2.90	2.90	3.00	3.30	3.50	3.20	3.30	0.00	1.40	1.10	3.40	0.30	0.40	0.30	0.50	39.00
C12	3.50	2.60	3.20	3.30	3.50	3.50	3.40	3.50	3.40	3.40	3.40	0.00	3.10	3.70	3.10	3.30	3.60	3.70	58.90
C13	3.20	1.80	0.60	0.70	0.50	3.50	3.50	3.80	3.60	3.00	1.90	1.00	0.00	3.50	3.20	3.30	3.30	3.60	44.00
C14	3.60	2.60	3.30	3.70	2.70	2.90	3.00	3.30	3.60	3.10	1.60	0.80	0.70	0.00	0.70	0.60	0.50	0.70	36.90
C15	3.20	2.00	0.30	2.90	0.60	3.50	3.20	3.30	3.40	3.20	2.90	0.80	3.10	3.40	0.00	3.10	3.20	3.00	46.50
C16	3.40	3.70	0.80	0.80	3.40	3.80	3.40	3.60	3.40	3.40	0.70	1.30	1.00	3.60	2.00	0.00	2.00	3.10	42.00
C17	3.20	3.40	0.70	0.30	0.30	2.80	2.70	3.30	2.70	2.90	0.70	3.00	3.00	2.60	1.40	3.20	0.00	3.20	38.10
C18	3.30	3.50	3.10	3.40	3.20	3.50	3.40	3.50	3.40	3.20	3.30	3.70	3.10	3.60	3.10	3.60	3.30	0.00	57.10
Total	51.90	53.60	35.70	37.40	38.20	52.00	52.80	54.90	52.00	49.90	31.80	20.80	22.60	55.80	17.80	23.00	21.90	34.10	

The averaged matrix (A) are processed to limit supermatrix to obtain global weight. This following steps show the process:

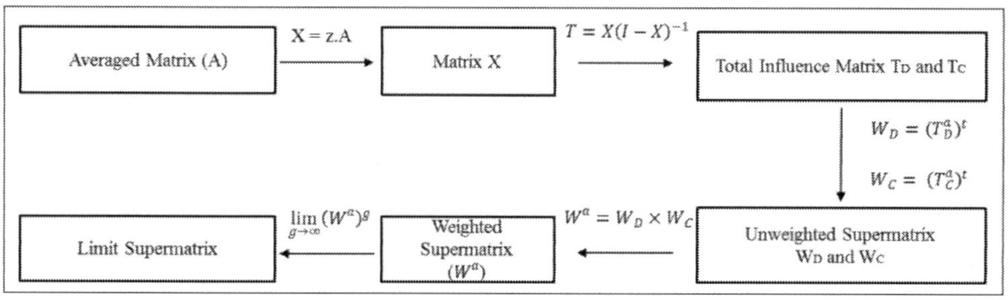

Fig 1. DANP Process

After generating average matrix (A), we form Matrix (X) which is the result of an overall matrix of direct-indirect influence. Then, we would like to find the Total Influence Matrix (T)from the multiplication of Matrix X. Furthermore, we form unweighted supermatrix (W_D and W_C) to generate local weight and weighted supermatrix (W^{α}). The last step is to form a limit supermatrix to obtain a stable supermatrix and generate a global weight for the criteria.

4 Result and Discussion

The result of DANP method is the limit supermatrix and the global weight for the criteria. Limit supermatrix process needs normalization to make sure that the unweighted supermatrix (W_D and W_C) have same unit.Table 2 summarizes all the weight by DANP method. Information quality (D2) has the greatest local weight dimension (0.3268) while information system success (D4) has the smallest local weight (0.1570). Moreover, high quality data presentation (C10) also has the greatest global weight while career improvement (C15) has the smallest global weight (0.0152). The greater weight of local dimension and global weight meana higher level of importance than the other dimension dalam rangka mengambil keputusan dalam bidang HR berdasarkan sistem informasi yang ada.

According to the result of DANP method, information quality dimension helps HR to improve their performance, such as an up-to-date information will make all activities in HR like selection of new employee, promotion of employee, general affair activities, transportation and lodging expenses, and payroll system becomes easier. Information quality generates accurate data presentation that helps HR to have a valid data, so the human errors in HR activities will be minimized.Quick decision making in HR activities would improve corporate's performance.

Table 2. Weight by DANP

Dimension	Local Weight Dimension	Criteria	Local Weight Criteria	Global Weight
D1	0.2990	C1	0.2855	0.0834
		C2	0.2982	0.0884
		C3	0.2063	0.0646
		C4	0.2100	0.0645
D2	0.3268	C5	0.1612	0.0547
		C6	0.2049	0.0664
		C7	0.2087	0.0680
		C8	0.2183	0.0708
		C9	0.2069	0.0672
D3	0.2173	C10	0.4957	0.1089
		C11	0.3098	0.0691
		C12	0.1945	0.0410
D4	0.1570	C13	0.1339	0.0202
		C14	0.3936	0.0638
		C15	0.1017	0.0152
		C16	0.1184	0.0168
		C17	0.1185	0.0172
		C18	0.1338	0.0196

5 Conclusions

As a conclusion, there were four dimensions at this research, that were system quality dimension with 4 criteria, information quality dimension with 5 criteria, user satisfaction with 3 criteria, and information system success with 6 criteria. The most important dimension was quality information (0.3268). The five biggest global weight from 18 criteria were high quality data presentation (0.1089), quick and precise (0.0884), accessible (00834), information need in time (0.0708), and meet your needs (0.0691).

References

1. Silver, M. Lynne, M. M. Beath and Mathis, *The Information Technology Interaction Model: A Foundation for the MBA Core Course* (1995).
2. K. S. Ball, Pers. Rev., 677-693 (2001)
3. G. DeSanctis, MIS Quart., **10,** 15-26 (1986).

4. M. J. T. M. &. J. R. D. Kavanagh, *Human Resource Information System : basics, application and future directions* **2**, (2012).
5. B. Gupta, IOSR J. of Bus. and Mgmt., 41-46 (2013).
6. N. Venkatraman and V. Ramanujam, Acad. of. Mgmt. Rev., 801-808 (1986).
7. C. Zehir, Y. Gurol, T. Karaboga and M. Kole, *12th International Strategic Management Conference, ISMC 2016*, 372-381 (2016).
8. M. Dachyar and Yupita, Adv. Sci. Let., 2244–2247 (2014).
9. C.-C. Hsu and J. J. Liou, J. of Air Trans. Mgmt., 40-46, (2013)

Analysis of Outpatient Service Queue of Public Hospital in Jakarta

M. Dachyar*, Farizal, and *M. Mansur* Yafi

Industrial Engineering Department, Universitas Indonesia

Abstract. In recent years Public hospital services are becoming a burden in many large cities. Many patients are needed to be served, but it was constrained by the number of doctors. Even though the resources are limited, the efficiency of public hospital services is important. The patient's queues can be shortened by service quality enhancement of outpatient. This study has applied *Business Process Reengineering* (BPR) method to improve the efficiency of public hospital services. Igrafx software is used to visualize the improvement design. The result has suggested a solution to accelerate the outpatient health services of a public hospital in Jakarta.

1 Introduction

The quality of public hospital has to be improved as patients' demand grows [1]. The services quality improvement of the public hospital is different from the private one, especially in term of goals where private hospital are more profit-oriented [2]. The acceleration of outpatient services procedure, the public hospital can also increase patient satisfaction [3].

The public hospitals in a big city in Indonesia like Jakarta should get serious attention from the government in term of improvement. The patients could have been queuing since 5 am. This research is aimed to reduce the queues of outpatient service in public hospital in Jakarta. Based on the observation conducted between March and June 2017, the patient queues from 9 am to 10 am the queue on internal disease clinic around 110 patient, the queue on cardiac clinic around 104 patient and pulmonary clinic 60 patient.

2 Literature Review

Outpatient service process includes a registration process, medical examination, health treatment in the examination room, further examination, and received prescriptions [4]. Patient waiting time describes access to health services that become an indicator of health service quality. The waiting time potentially affects patient dissatisfaction with services, which closely related to the image of the hospital. The reason for giving the least attention to waiting time in the hospital is acceptable where service improvement is not a priority.

Queue occurs when objects to be served is confronted with delays of busyness service system. It happened when there is no balance between the serviced and services [5]. These

* Corresponding author: mdachyar@yahoo.com

are the visualization of long queue where easily can find such as in teller transaction in the Bank, at the airport when passengers came in the check-in desk, in the supermarket queuing on paying goods, in a car wash where cars are queuing to be washed. In the service sector queuing is bored. Consequently, the costumers will leave. This is considered as a loss for the organization [6]. The best service among others is to provide quick service to the customers, so they don't wait too long. However, the impact of this quick service delivery will incur costs for organizations such as hospitals, as they have to add service facilities and technologies that are expensive.

Business process re-engineering (BPR) is a fundamental evaluation with a radical business process change for obtaining the improvement in main and contemporary performance indicator such as cost, quality, service, and speed [7]. In addition, according to Schonsleben business process re-engineering is the improvement of a business process by redesigning the process fundamentally [8]. Du Plessis also explained that BPR is a fundamental analysis and design of each process and radical business process including the business operation, the management system, the division of tasks, the structure, the principle, and the organizational behavior. The goal is to improve the performance in the interest of fulfilling business development demand also information technology development as the main stimulant of this process [9].

There were several companies that have implemented BPR, such as milk company [10], pharmaceutical company [11], shipping company [12].

3 Research Methodology

This research uses Business Process Re-engineering (BPR) method to map out outpatient service for reducing the outpatient service queue of a public hospital in Jakarta. BPR contains three (3) steps, which are the preparation for outpatient service improvement, the mapping and analysis of outpatient service process' initial condition, and the designing of outpatient service process improvement.

There are three main sub-processes in the outpatient service procedure, which are the registration, the polyclinic service, and the drug service. Based on the analysis using a *Pareto diagram*, three clinics were chosen as the research object. They are internal disease polyclinic, cardiac polyclinic, and pulmonary polyclinic.

4 Results and Discussion

4.1 Prepare for Outpatient Service Process Reengineering

The preparation phase is a stage for comprehending the initial condition of outpatient service process in public hospital and knowing the purpose of its business process.

4.2 Mapping and Analysis of Outpatient Service Process Initial Condition

Mapping the Initial Condition. In this phase, *time study* technique is used for achieving the standard time of each hospital service. After identifying the initial condition with *process chart,* the author creates the as-is process.

The patient registration process is to get the queue number for short time if they come with a complete document. Medical records were sought manually which takes about 30 minutes, then delivered to polyclinic. The next process is blood pressure measurement and body weight measurement is common to all the polyclinic. The patients will be examined by a doctor and prescribed for medicine at available pharmacies in the hospital. The process of taking the drug requires a little queue time because some drugs are not yet available and must be mixed manually first. The patient can go home and return as scheduled by the doctor. The patient queue occurs from 9-10 am at every clinic and pharmacy, after that the queue will be reduced. Based on queueing data simulation was developed, the average wait time for internal disease clinic is 6 hours 51 minutes, the average wait time for the cardiac clinic is 6 hours 46 minutes, the average wait time for the pulmonary clinic is 5 hours 45 minutes. The last step of analyzing phase is conducting an interview with experts to identify the root problems that need to be improved, such as:

1. The patient takes the queue number twice, for registration and the polyclinic
2. The prescription is still written manually.
3. The prescription copies are still written manually and inefficiently.
4. There is a long queue on polyclinic service.
5. Without verification on registration phase, the patient should automatically be registered in each polyclinic.
6. The pharmacists cannot access the patient's medication database directly.
7. The doctors and nurses are ready at the clinic but the medical records have not yet delivered to the clinic, because of the manual search of medical records.

4.3 Design New Outpatient Service Process

To reduce patient waiting time needs some improvement steps. Registration changed into electronic registration. The prescription process can be accelerated if the doctors input the data from every clinic. Increasing the number of doctors is one of the solutions to reduce queue time. Table 1 shows the suggested improvement that can be applied in the new design.

Table 1. Outpatient Service Improvement Suggestions

No	Improvement	Problem
1	Online registration	1
2	Printing the prescription	2
3	Printing the prescription copies	3
4	Increasing the number of doctors	4
5	Integrating the hospital information system with national health insurance information systems	5
6	Installation of prescription information system	6
7	Installation of medical records information system	7

Table 2 shows stepwise improvement in three polyclinics.

Table 2. Four Steps of Service Improvement in Three Polyclinics

Polyclinic	Stage of improvement			
	1	2	3	4
	Improvements Combination			
Internal disease clinic	1,3	2,5	6	7
Cardiac clinic	1,3	2, 4, 5	4,6	7
Pulmonary clinic	1, 3	2, 5	4, 6	7

The new service design is produced by combining some improvement suggestions. The service improvements consisted of several phases; the capability on implemented the technical capacity and financial capability of the organization.

The simulation shows the greatest efficiency of each improvement stage as shown in Figure 1.

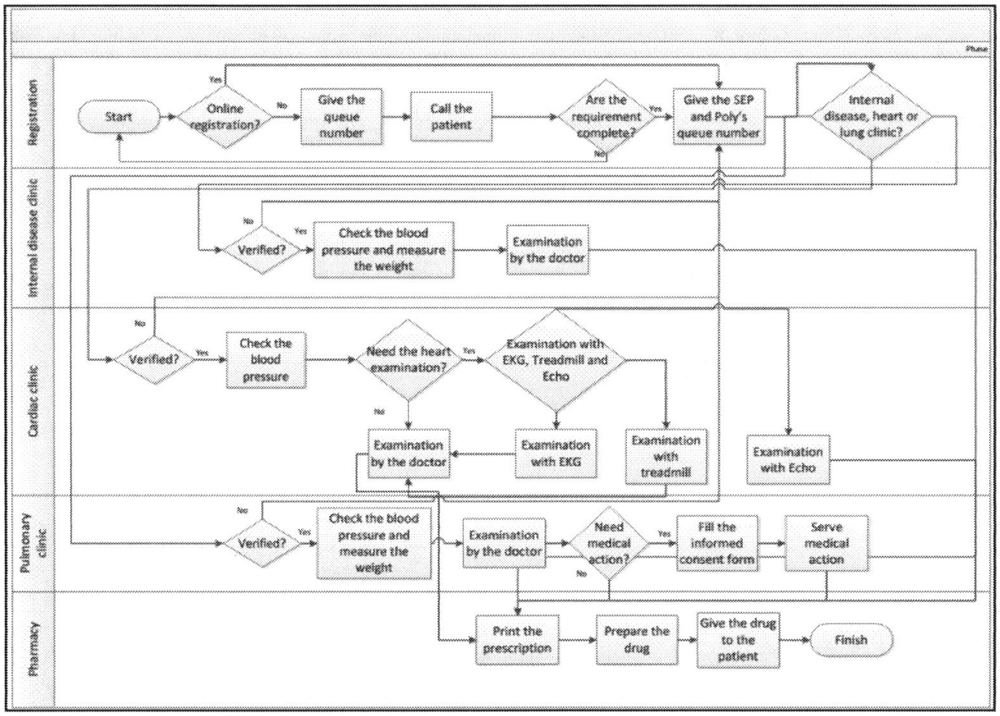

Fig. 1. Improved Outpatient Procedure

By implementing this improvement, internal disease clinic gained 21.7% efficiency time service process, cardiac clinic gained 22.5% and pulmonary clinic gained 21.8%.

Table 3 shows the effectiveness of the service improvement, the initial condition and service improvement are simulated using the Igrafx software. The simulation result shows the outpatient service time reduction through the efficiency percentages Improvements are made in stages to determine the efficiency that occurs when the hospital can not yet make the comprehensive improvements over a short period of time due to hospital inadequacy in terms of technical, cost, and personnel. The service time in the hospital since the patient arrived to finish getting the drug takes more than 4 hours on average.

Patients who first come to the hospital when the total time of their service is not up to 4 hours, because waiting time in the polyclinic and in the pharmacy is still short, but this can happen if come to the hospital before 6 am, but this can be longer than 4 hours since later in the day.

Table 3. Process Improvement Simulation Result

Polyclinic	Current condition result Avg Cycle (Hour)	Stage	Improvements result Avg Cycle (Hour)	Efficiency
Internal disease clinic	7.34	1	6.83	6.90%
		2	6.54	10.90%
		3	6.22	15.30%
		4	5.75	21.70%
Cardiac clinic	7.33	1	6.63	9.50%
		2	6.36	13.20%

Polyclinic	Current condition result Avg Cycle (Hour)	Stage	Improvements result Avg Cycle (Hour)	Efficiency
		3	6.14	16.20%
		4	5.68	22.50%
Pulmonary clinic	6.09	1	5.57	8.50%
		2	5.28	13.30%
		3	5	17.90%
		4	4.76	21.80%

5. Conclusion

This research resulted in the recommendation of a gradual improvement for three clinics on outpatient service of a public hospital in Jakarta. Internal disease polyclinic service time has been reduced from 6.9%-21.7%. Cardiac polyclinic service time has been reduced from 9.5%-22.5%. Pulmonary polyclinic service time has been reduced from 8.5%-21.8%. In order to reach less than 4 hours of service time, further research is needed.

This research funded by Universitas Indonesia – PITTA 2017

References

1. A. Shabbir, S. A. Malik, Int. J. of Q. & Rel. Mgmt, 538-557 (2016).
2. J. S. Harrison & S. M. *Strategic Management of Healthcare Organizations.* New York: Business Expert Press (2015).
3. Z. Lulu, M. Li, F. Ye, T. Ding, & P. Kang, *An Investigation Report on Large Public Hospital Reforms in China.* Singapore: Springer (2016).
4. S. Chand, H. Moskowitz, J. B. Norris, S. Shade, & D. Willis, H. C. Mgmt. Sci., 325-340 (2009).
5. L. J. Krajewski, L. P. Ritzman, & M. K. Maholtra, *Operations Management: Processes and Supply Chains.* Essex: Pearson Education (2013).
6. J. A. Fitzsimmons & M. J. Fitzsimmons, *Service Management: Operations, Strategy, Information Technology.* New York: McGraw-Hill (2010).
7. M. Hammer & J. Champy, *Reengineering The Corporation: A Manifesto for Business Revolution.* Perfectbound (2001).
8. P. Schönsleben, *Integral Logistics Management: Operations and Supply Chain Management Within and Across Companies.* Boca Raton: CRC Press (2012).
9. S. Mohapatra, *Business Process Reengineering.* London: Springer (2013).
10. M. Dachyar & E. Christy, J. of Phys. (2014).
11. M. Dachyar & G. Novita, ARPN J. of Engr. and App. Sci. (2016).
12. M. Dachyar & B. Praharani, *Improvement of Procurement Business Process (Procure to Pay) in Indonesia Shipping Company.* Knowledge, Service, Tourism & Hospitality. CRC Press (2016).

Improvement Priorities: Public Hospital Service Quality

M. Dachyar*, Farizal, and *Alicia* Ti

Industrial Engieneering Department, Universitas Indonesia

Abstract. Patients' satisfaction in public hospitals in Jakarta is decreased due to the increase participant of outpatients using national health insurance. One of the principle influenced satisfaction is services. The purpose of this research is to get priority improvement of service quality dimension. ServQual method is used to obtain the patient's perception and expectation gap on hospital services. Analytical Hierarchy Process (AHP) is used to obtain the dimension importance weight. Five experts were involved on choosing the appropriate service dimension. Questionnaires were given to 400 patients. Results showed that the largest gap on overall services dimension and sub-dimension are cleanliness, comfort of check room and bathroom. The weight of importance is included, results show the priority improvement of trust dimension and sub-dimension doctor is checking the patient's condition correctly.

1 Introduction

Indonesia seeks to realize a global commitment to develop Universal Health Coverage (UHC) for its entire population through the National Health Insurance (BPJS) program which has started since January 1, 2014 which aims to provide health protection for the citizens.

Based on data from social security system, there was an increase of 17% national health insurance participants from 133 million people in early 2015 to 156 million by the end of 2015. This indicates that 62% of Indonesians have become participants. Data from social security service, satisfaction decreased in 2014 to 2015 from 81% to 78.9%.

2 Service Quality

Service quality is a concept that has attract many interest and debates in literature research because it is difficult to define and measure it in the absence of the overall consensus published for both [1]. Service quality can help an organization to achieve or exceed customer expectations [2]. Service quality is what Customer receives (technical dimension) and how customers receive it (functional dimension) [3].

Measuring the quality of hospital services and understanding patients' perceived quality is important [4]. Fulfilling the needs of patients and making health care standards is an important thing to achieve high quality [5], so identifying and understanding what is meant

* Corresponding author: mdachyar@yahoo.com

by quality dimension is important to the customer becomes important for the quality of service.

3 Patients' Satisfaction

In the Donabedian quality measurement model, patient satisfaction is defined as the assessment of patient report results, while the process and structure of the service can be measured by patient experience [6]. Jenkinson C et al. (2002) and Ahmed et al. (2011) indicates that patient satisfaction largely represents attitudes toward service or aspects of service [7, 8]. Satisfaction itself is an important factor affecting loyalty [9]. However, according to Dachyar (2011), if there are complaints from customers, it does not affect the loyalty of these customers [10].

4 Methodology

To measure the quality of service and determine the priority of improvement, used SERVQUAL method and Analytical Hierarchy Process (AHP).

Service quality can be measured using SERVQUAL which is a tool to measure service quality [11]. This tool is used to measure the quality of service by using the concept of gaps and service quality dimensions. The original SERVQUAL consists of 22 statements using the Likert scale with seven points to the service quality dimensions of tangibles, assurance, empathy, assurance, reliability [2], but to measure the quality of public hospital services, there are tools developed by Aagja (2010), which is called PubHosQual.

The PubHosQual Difference with ServQual consists of the validity of the content. 4 of the 5 ServQual dimensions are included in PubHosQual and there is also another dimension independent of the ServQual dimension (excluding empathy related social responsibility). The advantage of PubHosQual is with the inclusion of considerations regarding services in the hospital sector.

Analytical Hierarchy Process (AHP) is an effective method used in complex multi-criteria decision making, in which this method can help decision makers in setting priorities and making best decisions [12]. AHP is a measurement theory used to solve the ratio scale of discrete and continuous pairwise comparisons [13].

5 Data Collection

The original theory of ServQual created by Parasuraman consists of 5 attributes, namely tangibles, reliability, responsiveness, assurance, and empathy. ServQual has been applied in various sectors but influential factors may vary from one sector to another, so it is necessary to re-examine the factors for the perception of service quality in each sector [14]. In hospitals, patients often find it difficult to assess the technical quality of medical services accurately so that many researchers create a quality service framework that can measure the quality of care in hospitals. From various literatures, collected the dimensions of service quality which then conducted an assessment by experts to find out whether the dimensions obtained from the literature in accordance with the conditions or dimensions of service quality in Indonesia, especially regional public hospitals in Jakarta.

Expert assessment is done using questionnaire using Likert scale 5 (five) points to get service quality dimensions that are suitable for public hospitals in Indonesia and is screened using geometric mean. From the results of expert assessment, 20 sub-dimensions are produced by experts which means in accordance with the conditions of hospitals in Indonesia. 20 Sub-dimensions received from experts come from 7 dimensions of hospital service quality, namely; Registration, medical services, overall service, administrative completion, social responsibility, trust and treatment process. The dimensions that are used can be seen in table 1.

5.1 Pilot Testing

In the early stages of data retrieval, pilot testing was conducted with 40 respondents representing or having similarities with the target respondents to check the validity of the questionnaire that has been designed. After the initial questionnaire was distributed, the results of the questionnaire tested its validity and reliability. Validity test aims to determine whether the questionnaire is valid or not and reliability test aims to determine whether the scale used consistently or not.

The questionnaire can be declared valid if the KMO value (Kaiser-Meyer Olkin) reaches at least 0.5 [15]. For this study, each dimension in the questionnaire is more than 0.5 so it can be declared valid. Questionnaire can be stated reliabel; Reliable, consistent, and relevant if it has a Cronbach's Alpha value of more than 0.6 or greater because it can be said to be sufficient to measure scale and reliability [16]. All of the questions came out reliable and valid.

5.2 Data Sample

In this study, the determination of the number of samples is done using convenience sampling method. The determination is based on the total population. Patients using 95% confidence level. Based on the number of outpatient patients in hospital age, the minimum sample size was 385. The questionnaire was distributed to 432 outpatients at a local general hospital located in Jakarta. Respondents consisted of 74% women and 26% men. with an average age of 39 years. The majority of respondents were patients who had visited more than 5 times (66%).

6 Result and Discussion

Data processing calculated the gap value between patient's perception and expectation in outpatient department (Servqual Dimension). There are seven service quality dimensions with 20 subcriterias, it was found that all subcriteria is of negative value with subcriteria hospitals have clean washrooms, clean rooms/wards without foul smell being the largest negative gap score which is -1.17. The large negative value of the gap between perception and expectation in the sub-dimension hospitals have clean washrooms and speed of service and make patients spend hours for queueing. Improvement priorities can be seen in table 1.

Table 1. Improvement Priority Rank of Dimension

Dimension	Gap Score (A)	Weight (B)	Weighted Gap Score (A*B)	Rank
Trustworthiness	-0.73	0.25	-0.18	1
Clinical Care Process	-0.68	0.23	-0.15	2
Overall Service	-0.91	0.12	-0.11	3
Social Responsibility	-0.76	0.12	-0.09	4
Medical Service	-0.61	0.13	-0.08	5
Discharge	-0.75	0.08	-0.06	5
Registration	-0.79	0.07	-0.05	6

In data processing, we need AHP to calculate the weight of service quality dimension (PubHosQual) and ranking of each subcriteria. The weighting results show that the most important dimension is the trustworthiness but this dimension has the main priority to be improved with the gap value of -0.18.

The weighting results show that the most important sub-dimension is confidence in the doctor who treated me in hospital but this sub-dimension has the main priority to be improved with the gap value of -0.10. Improvement priorities for sub-dimension can be seen in table 2.

Table 2. Improvement Priority Rank of Dimension

Dimension	Sub-dimension	Gap Score (A)	Weight (B)	Weighted Gap Score (A*B)	Rank
Registration	Hospitals will give prompt admissions to their patients	-0.96	0.04	-0.04	7
	The employees handling admission in hospitals should be polite	-0.61	0.02	-0.01	18
Medical Service	Hospitals will have knowledgeable and experienced physicians	-0.47	0.10	-0.05	6
	Hospitals will have knowledgeable and experienced nurses	-0.75	0.04	-0.03	11
Overall Service	Materials associated with the service (such as pamphlets or statements) will be visually appealing in a hospital	-0.74	0.01	-0.01	20
	Hospitals have clean washrooms, clean rooms/wards without foul smell	-1.18	0.01	-0.01	19
	When patients have problems, excellent hospitals will show a sincere interest in solving it	-0.83	0.02	-0.02	17
	Hospitals will give prompt service to customers	-1.02	0.03	-0.03	10
	Hospitals will always be willing to help patients	-0.79	0.03	-0.02	15
	Hospitals have the patient's best interests at heart	-0.92	0.02	-0.02	14
Discharge	Patients of hospitals will be given prompt discharge	-0.73	0.03	-0.02	13
	Employees of Hospitals will explain the discharge process to the patients and family members	-0.77	0.05	-0.04	9
Social Responsibil i-ty	Hospitals should provide equal treatment stemming from the belief that every one, big or small, should be treated alike	-0.74	0.03	-0.03	12
	Hospitals should provide good service at a reasonable cost, but not at the expense of quality	-0.66	0.03	-0.02	16
	Employees of hospital should have a sense of responsibility, i.e. they are regular, punctual, sincere, and without going on strikes	-0.87	0.06	-0.05	5
Trustworthi -ness	Hospital provided services as promised and on time	-0.80	0.10	-0.08	2
	Confidence in the doctor who treated me in hospital	-0.65	0.15	-0.10	1
	Explanation provided by doctor about health status, medical tests	-0.71	0.07	-0.05	4

Dimension	Sub-dimension	Gap Score (A)	Weight (B)	Weighted Gap Score (A*B)	Rank
Clinical Care Process	Description offered by the doctor about treatment procedures and outcomes	-0.66	0.10	-0.06	3
	Medical advice and instructions provided by doctor	-0.67	0.06	-0.04	8

7 Conclusion

In this study, we got 7 dimensions of hospital service quality and 20 sub-dimensions suitable with the condition in Indonesia, especially in regional public hospital.

Hospital service experts consider that the dimension of trustworthiness is the most important dimension in the quality of hospital services and Confidence in the doctor who treated me in hospital to the most important sub-dimension. From the results obtained, the overall service dimension is the dimension with the greatest negative gap value for hospital service quality. Hospitals have clean washrooms, clean rooms/wards without foul smell is a sub-dimension with the greatest negative gap value and for the result of the combined gap and weight value, it is found that trustworthiness is the priority dimension of improvement and confidence in the doctor who treated me in hospital becomes a sub-dimension of priority improvement.

This research funded by Universitas Indonesia – PITTA 2017

References

1. M. Wisniewski, Man. S. Q. **11**, 380-388 (2001).
2. Parasuraman, V. A. Zeithaml and L. L. Berry, J. of Ret, 12-40 (1988).
3. Gronroos, Eur. J. of Mkt., 36-44 (1993).
4. K. Braunsberger and R. H. Gates, J. of Consr. Mkt. **19**, 575-590 (2002).
5. Ramachandran and N. Cram, J. of C. Engr. **30,** 219-228 (2005).
6. Donabedian, J. of the Amer. Med. Assn, 1743-1748 (1988).
7. Jenkinson, A. Coulter, S. Bruster, N. Richards and T. Chandola, Q. S. H. C. **11**, 335-339 (2002).
8. Ahmed, N. Allah, K. Shadiullah, K. Habibullah, A. Muhammad and H. K. Muhammad, Gomal J. of Med. Sci. **9**, 183-188 (2011).
9. M. Dachyar and E. S. Athory, J. of Phys.: Conf. Series, 1-8 (2015).
10. M. Dachyar and Fatkhurrohman, Int. J. of Soc., Behav., Edu., Econ., Bus. and Ind. Engr., 370-374 (2011).
11. R. Ladhari, Man. Serv. Q., 65-86 (2008).
12. T. L. Saaty, *The Analytical Hierarchy Process*, (1980).
13. R. W. Saaty, Math. Model. **9**, 161-176 (1987).
14. P. Aagja and R. Garg, Int. J. of Phar. and H. C. Mkt. **4,** 60-83 (2010).
15. Field, *Discovering Statistics Using SPSS*, (2009).
16. M. Loewenthal, *An introduction to psychological tests and scales*, (2004).

Determining Criteria for Food SME's Supplier Selection Using DEMATEL-Based ANP Method

Yasmin Ramadhani, *M.* Dachyar* and *Farizal*

Industrial Engineering Department, Universitas Indonesia

Abstract. SMEs' contribution to Indonesia's national GDP is quite high. However ineffective supply chain management causes the competitiveness of food SMEs company is low. The purpose of this research is to determine the appropriate criteria for supplier selection to improve the competitiveness of food SMEs company. This research method using a combination of DEMATEL and ANP called DEMATEL-Based ANP. This method is used to generate supplier selection criteria and their weight, which are obtained from the judgment of seven experts. This research resulted in the 20 criteria considered important from the results of the expert judgment. Criteria with the greatest weight is the communication openness with the weight value of 14%.

1 Introduction

SMEs' contribution to Indonesia Gross Domestic Product (GDP) in 2017 reached 61,41%. The non-oil and gas industry that gives the greatest role to the national GDP is the food and beverage industry which holds the largest percentage in the role of national GDP. Association of Indonesian Food and Beverage Entrepreneurs stated that Indonesia is the fifth most competitive food industry in ASEAN.

One of the most effective ways for a company to be competitive is implementing Supply Chain Management or SCM [1], [2]. Supplier selection is crucial and considered as a strategic way in the management of an adequate supply chain [3].

2 Literature Review

SCM is used to manage the supply chain to enhance the company's performance by making improvements in planning and management activities such as material planning, inventory management, capacity planning, and logistics with suppliers and customers [4]. Food supply chain management is created to explain the activities or operation of production, distribution and consumption to maintain the safety and quality of various food effectively and efficiently [5], [6].

Multi-Criteria Decision Making (MCDM) is one of the most commonly used methods to resolve the complex problem that exhibits high uncertainty, conflicting goals, different interests, and from multiple perspectives [7].

* Corresponding author: mdachyar@yahoo.com

The integration of Decision Making Trial and Evaluation (DEMATEL) method with Analytical Network Process (ANP) called DANP is an appropriate method to include interaction and interdependence which exists between dimension and criteria that appear in real life problems [8]. ANP is one of multiple criteria decision making, which most widely used because it allows more complexes relationship between elements by changing hierarchy into networks [9]. The DANP method is used to overcome difficulties in giving evaluation for pairwise comparison performed by the expert by modifying the question in the pairwise comparison. DANP solves the problem of 9-scale questions for the usual pairwise comparison which confuses the expert in their judgment by using total influence matrix [10].

3 Research Methodology

3.1 Criteria Assessment

In this research, researcher collects data three times. Data processing begins with identifying criteria from the study of literature. Judgment of 7 experts will validate the criteria from the literature which are 35 criteria and seven dimensions through the first questionnaire. The results will be processed using the geometric mean to eliminate the valuation criteria. The results which are 20 criteria and seven dimensions will be used for the next data collection to find the linkage between relationships judgment criteria to the expert. The final data collection will be processed by DEMATEL-Based ANP method to find the global weight of the criteria.

The results of dimensions and criteria that have been eliminated are seven dimensions and 20 criteria. The first dimension (D1) is economic, consist of price (C1), the speed of delivery (C2), and quality (C3). The second dimension (D2) is the response speed, consist of quick response time (C4), communication skills (C5), and service performance (C6). The third dimension is capability (D3), consist of problem-solving (C7), capacity (C8), and correct quantity (C9). The fourth dimension is customer relations (D4), consist of mutual trust (C10), honesty and truth (C11). The fifth dimension is supplier strategic (D5), consist of certificates (C12), guarantee (C13), and sales and marketing ability (C14), after sales service (C15), and support service (C16). The sixth dimension is flexibility (D6), consist of flexibility to respond to unexpected demand changes (C17), availability of resources (C18), and willingness to improve the product and process continuously (C19). The seventh dimension is social (D7) with the criteria of communication openness (C20).

3.2 Global Weight Calculation

The first step in data processing with DANP method is to create the direct-influence matrix (G) in the form of a 20x20 matrix because the number of criteria is 20, from the expert judgment. The results of the calculation of the direct-influence matrix (G) will be proceed with the calculation of Equation (1) to obtain the normalized direct-influence matrix from the direct or indirect effect, i.e., matrix $\boldsymbol{X}$.

$$\boldsymbol{X} = v\boldsymbol{G} \tag{1}$$

The value of v is obtained based on Equation (2) where the maximum value of the row sums and the maximum value of the column sums of the matrix components.

$$v = min_{ij}\left\{\frac{1}{\max i \sum_{j=1}^{n} g_c^{ij}}, \frac{1}{\max j \sum_{i=1}^{n} g_c^{ij}}\right\}, \qquad i,j \in \{1,2 \dots, n\} \tag{2}$$

The next step after obtained the Matrix X is to get the total-influential matrix $\boldsymbol{T}_c$ from Equation (3) below

$$\boldsymbol{T}_c = \boldsymbol{X}(\boldsymbol{I} - \boldsymbol{X})^{-1} \tag{3}$$

where $\boldsymbol{I}$ is the identity matrix. The next step after getting Matrix $\boldsymbol{I}$ is to calculate matrix $(\boldsymbol{I} - \boldsymbol{X})$. Matrix $(\boldsymbol{I} - \boldsymbol{X})$ will be inversed to obtain matrix $(\boldsymbol{I} - \boldsymbol{X})^{-1}$. The next step after getting the inversed matrix $(\boldsymbol{I} - \boldsymbol{X})$ is to multiply the Matrix $\boldsymbol{X}$ with the result of the Matrix $(\boldsymbol{I} - \boldsymbol{X})^{-1}$ to obtain the total-influential matrix $\boldsymbol{T}_c$. After calculating the total-influential $\boldsymbol{T}_c$ is calculating the total-influential matrix $\boldsymbol{T}_D$. Matrix $\boldsymbol{T}_D$ is obtained from the sum of the total-influential matrix $\boldsymbol{T}_c$ based on each dimension.

The next step after getting the total-influential matrix for the dimension is creating Influential Network Relation Map (INRM). INRM is obtained from the sum and reduction of r and s value. The value of r and s will be compared with the threshold value to get the relationship between dimensions. Fig 1 shows the relationship between each dimension.

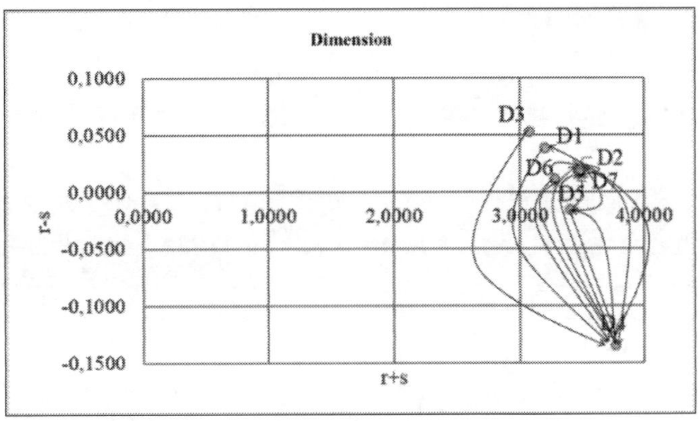

Fig. 1. Influential Network Relation Map of Dimensions

To make unweighted supermatrix for the dimensions, Equation (4) is used

$$\boldsymbol{W}_c = (\boldsymbol{T}_c^{nor})' \tag{4}$$

where to get matrix $\boldsymbol{T}_c^{nor}$, the first step is doing normalization for matrix $\boldsymbol{T}_c$. Normalization is calculated by dividing each element of the row with an average of each element in the criteria row of the same dimension in the matrix of criteria. To make the weighted supermatrix $\boldsymbol{W}_c^*$ through Equation (5) by doing the multiplication of each element $\boldsymbol{T}_D^{nor}$ with each division matrix $\boldsymbol{W}_c$. The normalized matrix $\boldsymbol{T}_D$ in the form of $\boldsymbol{T}_D^{nor}$ is calculated by dividing each element of the row into the matrix dimensions.

$$\boldsymbol{W}_c^* = \boldsymbol{T}_D^{nor}\boldsymbol{W}_c \tag{5}$$

The final step is to form the limit weighted supermatrix to obtained stable supermatrix through Equation (6). Limit supermatrix shows global weight for criteria that may indicate the importance of each criterion in the selection of suppliers.

$$\lim_{\varphi \to \infty} (\boldsymbol{W}_c^*)^{\varphi} \tag{6}$$

Table 1. Limit Super-matrix

	C1	C2	C3	C4	C5	C6	C7	C8	C9	C10	C11	C12	C13	C14	C15	C16	C17	C18	C19	C20
C1	0,044	0,044	0,044	0,049	0,049	0,049	0,042	0,042	0,042	0,055	0,055	0,048	0,048	0,048	0,048	0,048	0,046	0,046	0,046	0,049
C2	0,042	0,042	0,042	0,046	0,046	0,046	0,040	0,040	0,040	0,052	0,052	0,045	0,045	0,045	0,045	0,045	0,043	0,043	0,043	0,046
C3	0,047	0,047	0,047	0,052	0,052	0,052	0,045	0,045	0,045	0,058	0,058	0,051	0,051	0,051	0,051	0,051	0,049	0,049	0,049	0,051
C4	0,045	0,045	0,045	0,049	0,049	0,049	0,043	0,043	0,043	0,055	0,055	0,049	0,049	0,049	0,049	0,049	0,046	0,046	0,046	0,049
C5	0,045	0,045	0,045	0,049	0,049	0,049	0,043	0,043	0,043	0,055	0,055	0,048	0,048	0,048	0,048	0,048	0,046	0,046	0,046	0,049
C6	0,044	0,044	0,044	0,048	0,048	0,048	0,042	0,042	0,042	0,054	0,054	0,047	0,047	0,047	0,047	0,047	0,045	0,045	0,045	0,048
C7	0,049	0,049	0,049	0,054	0,054	0,054	0,047	0,047	0,047	0,061	0,061	0,053	0,053	0,053	0,053	0,053	0,051	0,051	0,051	0,054
C8	0,040	0,040	0,040	0,044	0,044	0,044	0,038	0,038	0,038	0,050	0,050	0,043	0,043	0,043	0,043	0,043	0,042	0,042	0,042	0,044
C9	0,044	0,044	0,044	0,048	0,048	0,048	0,042	0,042	0,042	0,054	0,054	0,047	0,047	0,047	0,047	0,047	0,045	0,045	0,045	0,048
C10	0,069	0,069	0,069	0,075	0,075	0,075	0,066	0,066	0,066	0,085	0,085	0,074	0,074	0,074	0,074	0,074	0,071	0,071	0,071	0,075
C11	0,065	0,065	0,065	0,071	0,071	0,071	0,062	0,062	0,062	0,080	0,080	0,070	0,070	0,070	0,070	0,070	0,067	0,067	0,067	0,071
C12	0,026	0,026	0,026	0,028	0,028	0,028	0,025	0,025	0,025	0,032	0,032	0,028	0,028	0,028	0,028	0,028	0,027	0,027	0,027	0,028
C13	0,023	0,023	0,023	0,026	0,026	0,026	0,022	0,022	0,022	0,029	0,029	0,025	0,025	0,025	0,025	0,025	0,024	0,024	0,024	0,026
C14	0,029	0,029	0,029	0,031	0,031	0,031	0,027	0,027	0,027	0,035	0,035	0,031	0,031	0,031	0,031	0,031	0,030	0,030	0,030	0,031
C15	0,027	0,027	0,027	0,029	0,029	0,029	0,026	0,026	0,026	0,033	0,033	0,029	0,029	0,029	0,029	0,029	0,028	0,028	0,028	0,029
C16	0,029	0,029	0,029	0,031	0,031	0,031	0,027	0,027	0,027	0,035	0,035	0,031	0,031	0,031	0,031	0,031	0,030	0,030	0,030	0,031
C17	0,046	0,046	0,046	0,051	0,051	0,051	0,044	0,044	0,044	0,057	0,057	0,050	0,050	0,050	0,050	0,050	0,048	0,048	0,048	0,051
C18	0,042	0,042	0,042	0,046	0,046	0,046	0,040	0,040	0,040	0,052	0,052	0,045	0,045	0,045	0,045	0,045	0,043	0,043	0,043	0,046
C19	0,045	0,045	0,045	0,050	0,050	0,050	0,043	0,043	0,043	0,056	0,056	0,049	0,049	0,049	0,049	0,049	0,047	0,047	0,047	0,049
C20	0,133	0,133	0,133	0,146	0,146	0,146	0,128	0,128	0,128	0,165	0,165	0,144	0,144	0,144	0,144	0,144	0,138	0,138	0,138	0,146

4 Results and Discussion

The result of DANP method is global weighting to criteria as seen in Table 2. The global weighting for each supplier criterion is obtained from the diagonal value of limit super-matrix from Table 1.

Table 2. Dimension and Criteria Weights

Dimension	Criteria	Global Weight
D1	C1	0,04437
	C2	0,04202
	C3	0,04703
D2	C4	0,04922
	C5	0,04902
	C6	0,04788
D3	C7	0,04718
	C8	0,03845
	C9	0,04204
D4	C10	0,08493

Dimension	Criteria	Global Weight
D4	C11	0,07979
D5	C12	0,02800
	C13	0,02537
	C14	0,03086
	C15	0,02905
	C16	0,03094
D6	C17	0,04796
	C18	0,04317
	C19	0,04675
D7	C20	0,14592

The global weight is defined the importance weight of criteria. The highest weight criteria is communication openness, and the lowest weight is the guarantee (C13). The importance weight of criteria can be considered to select the right supplier.

The data are validated using geometric mean. The results of judgement 7 experts value will be processed using the geometric mean to eliminate the valuation criteria less than 3,4375. The results are 20 criteria and seven dimensions.

5 Conclusion

This research resulted in the 20 criteria considered important for supplier selection in food SMEs from the results of the seven experts judgment. The results are calculated using geometric mean to eliminate the criteria according experts' judgment. Criterion with the greatest weight is the communication openness with the weight value of 14% from DANP method calculation. Criterion with the lowest weight is the guarantee with the weight value of 3%. The communication openness is considered important and the guarantee is considered less important for supplier selection in food SMEs company. The company can consider the criteria for selecting the right supplier.

This research funded by Universitas Indonesia – PITTA 2018

References

1. S. Li, S. S. Rao, T. S. Ragu-Nathan, and B. Ragu-Nathan, J. Oper. Manag., **23**, 618–641 (2005).
2. D. J. Ketchen, W. Rebarick, G. T. M. Hult, and D. Meyer, Bus. Horiz., **51**, 235–243 (2008).
3. C.-N. Liao and H.-P. Kao, Expert Syst. Appl., **38**, 10803–10811 (2011).
4. C. Chandra and S. Kumar, Ind. Manag. Data Syst., **100**, 100–114 (2000).
5. T. Marsden, J. Banks, and G. Bristow, Sociol. Ruralis, **40**, 424–438 (2000).
6. J. Blandon, S. Henson, and J. Cranfield, J. Int. Dev., **21**, 971–984 (2009).
7. J.-C. Pomerol and S. Barba-Romero, **25** (2000).
8. W. Y. Chiu, G. H. Tzeng, and H. L. Li, Knowledge-Based Syst., **37**, 48–61 (2013).
9. M. Dachyar and Yupita, Adv. Sci. Lett., **20**, 2244–2247 (2014).
10. I. Gölcük and A. Baykasoılu, Expert Syst. Appl., **46**, 346–366 (2016).

Strategy Design to Improve the Implementation of Supply Chain Management in Food SMEs

Agustina Windaryanti, *M.* Dachyar*, and Farizal

Industrial Engineering Department, Universitas Indonesia

Abstract. Food and beverage industry serves as one of the primary industries by the Ministry of Industry for industrial development plan until 2035. To improve the competitiveness of food SMEs, the strategy of implementing supply chain management is designed. The purpose of this research is to obtain strategy recommendations for the implementation of supply chain management in food SMEs. The literature study was conducted to find out the criteria, sub-criteria, and strategies that need to be implemented in improving the supply chain management of the food SMEs. Analytical Network Process (ANP) method is used to find the weight of criteria, sub-criteria, and strategy of implementing supply chain management. Criteria, sub-criteria, and strategies are judged by seven experts. This research obtained five criteria, twelve sub-criteria, and twenty-three strategies for the implementation of supply chain management. The result of data processing shows strategy that has the highest weight is to coordinate the logistics department with other functions in the company.

1 Introduction

Best supply chain management practice in food and beverage industry will increase the industry's competitiveness level. Problems encountered in the food and beverage industry is the supply chain. Losses due to inefficient supply chains in the food and beverage industry reached 86.6 trillion rupiah [1].

2 Literature Review

Operations management (OM) is a management area that handles supervision, design, production, and reproduction of products and services that improves the company's efficiency [2]. A supply chain consists of all entities involved in the process of meeting the needs of consumers, either directly or indirectly. Supply chain members are not just manufacturing and supplier companies, but also retailers and even consumers themselves [3]. The food supply chain is different from the supply chain of other products. The fundamental difference between the food supply chain and other supply chains is the persistent and significant change in the quality of food products throughout the entire supply chain to the point of final consumption [4].

* Corresponding author: mdachyar@yahoo.com

The previous study, research involving SMEs in Turkey, the implementation of SCM can lower inventory levels, decrease production time, increase flexibility, demand forecast accuracy, cost savings and accurate resource planning [5].

Multi-criteria decision making (MCDM) has classified into two categories: multi-attribute decision making (MADM) and multi-objective decision making (MODM) decisions [6]. The MADM technique involves very much human participation and judgment. MADM consists of methods of ELECTRE, Technique for Order of Preference by Similarity to Ideal Solution (TOPSIS), AHP, ANP, Preference Ranking Organization Method for Enrichment Evaluation (PROMETHEE), MAUT, and others. MODM consists of goal programming and genetic algorithm.

Analytical Network Process (ANP) is a framework used to solve problems related to decision making without having to make assumptions about the relation of dependence and the interrelation of elements with different levels and also the reliance of elements in the same level [7]. An important aspect of pairwise comparisons is mutual property. When one element has a ratio x times more dominant than the other concerning the given property, the inverted comparison do where the lower element has a reciprocal value of 1 / x [8]. The consistency validation of pairwise comparisons is required. Consistency ratio value should not be more than 0.1 for pairwise comparison is said to be consistent [8]. ANP has been used previously in the area of fuel subsidy policy elections [9], supplier selection, and location selection [10].

3 Research Methodology

Data is processed by using the ANP method steps, as follow: (1) identification of criteria, sub-criteria, and strategies; (2) criteria, sub-criteria, and strategies selection; (3) determination of the interdependence relationship between criteria, sub-criteria, and strategies; (4) model construction based on dependency relationship; (5) construct pairwise comparisons questionnaire; (6) construct supermatrix; (7) obtaining the weight of strategies.

Seven experts judged criteria, sub-criteria, and strategies. Results of judgment calculated using geometric mean and selected five criteria, twelve sub-criteria, and twenty-three strategies affecting the improvement of SCM implementation. The criteria are food traceability [11], collaboration [12], information [13], flexibility [12] [14], and internal organization [15]. Twelve sub-criteria are regulation, quality and food safety [11], logistics and distribution planning, supplier management, inventory management, demand planning, production planning [12], information sharing [13], product flexibility [14], human resources management, knowledge management [15], top management commitment [16].

The twenty three strategies are registering to The National Agency of Drug and Food Control of Republic of Indonesia (A1), registering to halal certificate (A2), providing clear traceability information (production date and expiration date) until it arrives to the consumer (A3), performing hygienic supply chain management (A4), considering the quality of the packaging (A5) [17], to coordinate the logistics department with other functions in the company (A6) [18], to conduct assessment and monitoring to suppliers based on cost, quality and delivery performance (A7) [19], building long-term relationships with suppliers based on trust (A8), collaborating with suppliers for better inventory management (A9) [20], implementing Just In Time (JIT) to reduce inventory (A10) [12], using information technology throughout the supply chain to inform demand and production information to reduce the number of finished goods (A11), identifying factors (promotions, discounts, etc.) that affect the number of requests and assess the influence of each factor on future consumer demand (A12), comparing and analyzing real demand and demand forecasts to improve the level of accuracy of future demand forecasts (A13), using a production planning system (A14), sharing information regularly (A15), relevant, accurate, and complete information

(A16), ability to standardize the product (A17) [21], investing in Research and Development process for product innovation (A18) (Tummala & Schoenherr, 2011, employee training programs (A19), cooperation among employees (A20), knowledge sharing in the various functions of the supply chain (A21), the acquisition of knowledge from people who have experienced in a part of the supply chain (A22), creating Standard Operating Procedures (SOP) throughout the supply chain (A23) [22]. Experts judged the linkages from one element to others. The construction of ANP model shown in Fig.1.

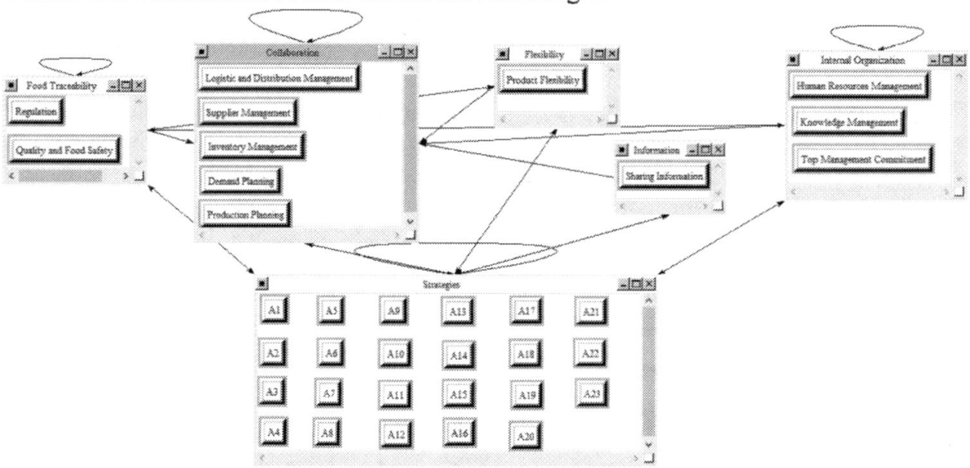

Fig. 1. The ANP Model of Strategy to Improve the Implementation of SCM

The inconsistency value of all priority vectors in this study is less than 0.1 so that the ANP model is consistent and valid. The transformation from unweighted supermatrix, to weighted supermatrix, limited matrix generated the criteria, sub-criteria and strategies weight.

4 Results and Discussion

4.1 Criteria and Sub-Criteria Weight

The highest to lowest criteria's weight in a row namely, collaboration (0.562), food traceability (0.282), flexibility (0.076), internal organization (0.049) and information (0.031). The highest weight indicates that collaboration is the most important criteria in the improvement of the implementation of SCM system in food SMEs because collaboration starting from raw material purchase to the distribution of finished goods is the essential thing for SMEs to survive. Food traceability is the second most important criteria because the food industry has characteristics such as a relatively short lifetime and direct effect to human health and safety, so it is essential to know the date of expiration for the consumer.

The highest weight of sub-criteria is quality and food safety (0.232), followed by logistics and distribution planning (0.207), production planning (0.136), inventory management (0.096), product flexibility (0.076), demand planning (0.076), regulation (0.05), supplier management (0.047), top management commitment (0.029), knowledge management (0.002), and human resources management (0.018).

4.2 Strategies Weight

Coordinating the logistics department with other functions in the company has the highest weight / A6 (0.183). The increasing of demand results in the highest strategy weight is A6 (0.242). The challenge to innovate the food products results in of the highest strategy weight is A6 (0.191).

5. Conclusion

This research obtained five SCM criteria, twelve sub-criteria, to find the best SCM strategy in Food SMEs. The most important criteria for implementing SCM is the collaboration, and the most critical sub-criteria is food traceability. Collaboration with other companies or suppliers will increase the competitiveness to lead the market. To improve SCM performance, the food SMEs management can implement these priorities: (1) coordinate the logistics department with other functions in the company has the largest weight, (2) use production planning system, (3) register to The National Agency of Drug and Food Control of Republic of Indonesia; (4) consider the quality of the packaging; and (5) provide clear traceability information (production date and expiration date) until the product arrives at the consumer.

This research funded by Universitas Indonesia – PITTA 2018

References

1. R. A. Mulyo, *Food Loss and Food Waste Rice Comodity in Indonesia* (2016).
2. R. K. Jain, H. Nemade, G. H. Yadav, and P. Sehgal, Int. J. Adv. Eng. Technol., (2013).
3. S. Chopra and P. Meindl, *Supply Chain Management : Strategy, Planning, and Operation* (2015).
4. R. Akkerman, D..V.Meer, and D.P. Van Donk, Int. J. Prod. Res., **48**, 3475–3492 (2010).
5. S. C. Lenny Koh, M. Demirbag, E. Bayraktar, E. Tatoglu, and S. Zaim, Ind. Manag. Data Syst., **107**, 103–124 (2007).
6. C.-L. Hwang and K. Yoon, *Multiple Attribute Decision Making: Methods and Applications A State-of-the-Art Survey,* 58–191 (1981).
7. T. L. Saaty, J. Syst. Sci. Syst. Eng., **13**, 129–157 (2004).
8. T. L. Saaty and M. Sodenkamp, Int. J. Appl. Decis. Sci., **1**, 24 (2008).
9. I. Hardiati, M. Dachyar, and Yadrifil, Int. J. Appl. Eng. Res., **10** (2015).
10. M. R. Khadivi and S. M. T. Fatemi Ghomi, Waste Manag., **32**, 1258–1265 (2012).
11. T. Bosona and G. Gebresenbet, Food Control, **33**, 32–48 (2013).
12. V. M. R. Tummala and T. Schoenherr, Int. J. Logist. Syst. Manag., **8** (2011).
13. E. Hassini, C. Surti, and C. Searcy, Int. J. Prod. Econ, **140**, 69–82, (2012).
14. J. H. M. Manders, M. C. J. Caniëls, and P. W. T. Ghijsen, J. Purch. Supply Manag., **22**, 181–195 (2016).
15. P. Beske, A. Land, and S. Seuring, Int. J. Prod. Econ., **152**, 131–143 (2014).
16. R. . Singh, R. Kumar, and R. Shankar, Int. J. Manuf. Res., **7**, 165–180 (2011).
17. M. Bourlakis, G. Maglaras, E. Aktas, D. Gallear, and C. Fotopoulos, Int. J. Prod. Econ., **152**, 112–130 (2014).
18. C. Basnet, Manag. Res. Rev., **36**, 153–172 (2012).

19. N. Banaeian, H. Mobli, B. Fahimnia, I. E. Nielsen, and M. Omid, Comput. Oper. Res., **89**, 337–347 (2018).
20. S. S. Padhi, R. K. Pati, and A. Rajeev, J. Clean. Prod., (2018).
21. R. K. Singh, Meas. Bus. Excell., **17**, 80–97 (2013).
22. C. Lin, C. Kuei, and C. Kang-Wei, Int. J. Oper. Prod. Manag., **33**, 347–370 (2013).

Scheduling for Indonesia's Aircraft Wing Structure Design Project with Critical Path Method and Resource-Constrained Project Scheduling

Melissa Putri Hidayat, *M.* Dachyar*, and *Farizal*

Industrial Engineering Department, Universitas Indonesia

Abstract. As an archipelagic country, Indonesia has high needs for its aviation development. About 270 million people are predicted to fly from and within Indonesia in 2034, which is three times larger than current market conditions. Due to the complexity of aircraft development projects, management of a well-designed Indonesian aircraft design project needs to be implemented. This research aims to analyze the scheduling of aircraft wing structure design project made in Indonesia. Critical Path Method (CPM) and Resource-Constrained Project Scheduling (RCPS) are used to get better scheduling with a shorter duration to complete the project. Results from resource allocation using RCPS can yield a shorter project duration (730 working days) compared to the company's normal resource allocation (950 working days).

1 Introduction

Indonesia can be regarded as a highly developed market for air travel. This can be seen from the 28.4% increase of air transportation passengers from 27,421,235 passengers in 2009 to 96,529,128 passengers in 2016 [1]. Indonesia's domestic flights are ranked fourth in the world for the growth of the aircraft passenger market with an increase of 14.3% in 2016 [2].

Each project has different constraints. However, there are some general limitations in project management, such as quality, scope, scheduling, resources, budget, and risks. Through these project management constraints, the viability of a project can be analyzed [3].

Aircraft development projects are complex. Based on the PMI report, 50% of aviation projects experience delays [4]. Therefore, management of a well-designed Indonesian aircraft design project is necessary.

2 Literature Review

Project management is the application of knowledge, skills, tools, and techniques to meet project needs. Project management can be done with the appropriate application and

* Corresponding author: mdachyar@yahoo.com

integration of the process in the project to execute it effectively and efficiently [3]. Implementation of project management can be done by various methods, such as using the conventional method like Critical Path Method [5], Program Evaluation Review Technique [6], or alternative methods like Critical Chain Method [7], Business Process Reengineering [8], and various other methods.

Critical path is the path of activities sequences that determine the duration of the project. The purpose of the critical path is to identify the sequence of activities to be finished in time to achieve project deliverables [9]. Critical Path Method (CPM) is used for all project types, such as construction, facility maintenance, or software development [10].

Scheduling for projects with limited resources can be done with Resource-Constrained Project Scheduling (RCPS) to allocate resources to activities and obtaining the shorter duration of critical activity so that the project duration can be reduced after knowing the critical path from CPM method [11].

3 Research Methodology

3.1 Data Collection

Duration and dependency relationship between each activity are collected so activity could be scheduled. In this project, there are 59 activities to achieve project deliverables.

This project has ten human resources with the same competence and qualifications to be able to work on all activities on the aircraft wing design project. Based on company's normal allocation, resource allocation for each activity is carried out based on at the duration length for each activity. Shorter duration of activity means fewer workers are allocated. Conversely, the longer the duration of the activity, the more workers that should be allocated to the activity.

Table 1. Company's Normal Resource Allocation

Activity Duration	Num. of Resources
>= 2240 hours	5 people
>= 1600 hours	4 people
>= 960 hours	3 people
>= 320 hours	2 people
< 320 hours	1 people

3.2 Project Scheduling Results with Critical Path Method

Critical Path Method (CPM) is used to find the critical path that exists in the project. Scheduling of resources is carried out with Microsoft Project software. At first, the resources are allocated according to the normal resource allocation set by the company in Table 1. By allocating resources based on company's normal allocation, the total duration of the project is 950 working days as shown in Fig. 1.

Project duration could be shortened by allocating more resources to activities that are on a critical path so that the overall duration of the project can be shortened by the Resource-Constrained Project Scheduling (RCPS) allocation. By prioritizing resources on activities that are on a critical path, the duration of the critical activities will be shorter, which results in a shorter duration of the project. The overall duration of the project after allocating resources with RCPS is 736 working days as shown in Fig 2.

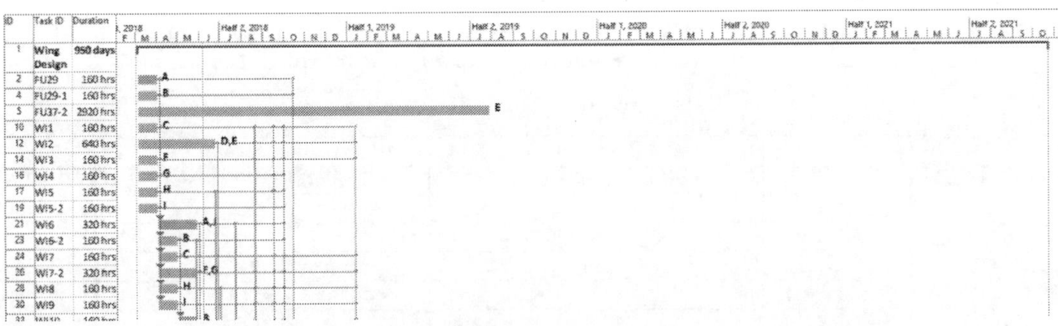

Fig. 1. Part of Gantt Chart for Project Scheduling with Company's Normal Allocation

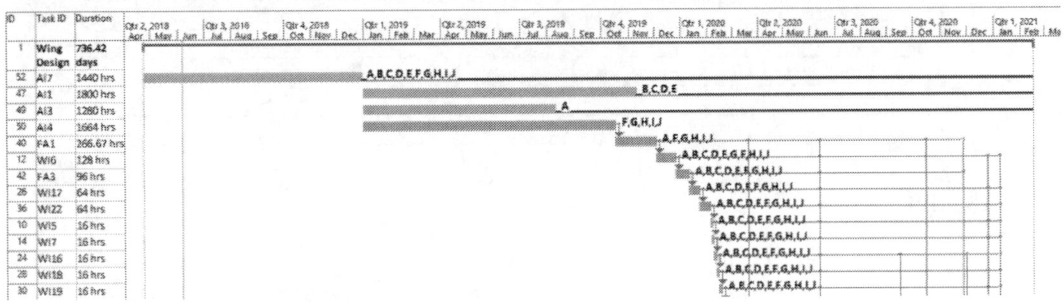

Fig. 2. Part of Gantt Chart for Project Scheduling with RCPS

4 Results and Discussion

Each project has different constraints. However, there are some general limitations in project management, such as scheduling, resources, budget, and risks. Through these constraints, the viability of a project can be analyzed [3].

4.1 Resources and Scheduling

By looking at the critical paths of the project, a resource could be allocated better with RCPS allocation. Activities that are on critical paths are the primary focus for resource allocation. Several activities also have higher priority for the RCPS allocation [11].

It was found that there is a reduction in the duration of the project by 214 working days from the normal resource allocation compared to the allocation of resources with RCPS. The project duration with RCPS allocation is 736 working days.

The CPM method with RCPS allocation takes account of the resource limitations as well as the duration of each activity in the allocation of its resources. It is suitable for the aircraft wing structure design project that has limited resources.

4.2 Budget

Budget to be spent by the company are viewed from the resources length of work. Longer working hours indicates that higher budget needs to be spent on the project.

The result shows that resource allocation on a critical path with RCPS did not change the total working hours of the resource compared to the company's normal resource allocation. Working hours of the resource with critical path allocation have a range of 4,700-5,900 hours. Working hours for resources with normal allocation has a range of 4,300-6,880 hours. From the results, it could be seen that the division of work for each resources using RCPS is more evenly distributed than the company's normal allocation as shown in Table 2.

Table 2. Working Hours for Each Resource

Resource	Working hours with normal allocation (hours)	Working hours with RCPS allocation (hours)
Resource 1	6.360	5.526
Resource 2	4.300	5.360
Resource 3	4.660	4.920
Resource 4	4.770	5.516
Resource 5	4.760	5.385
Resource 6	6.880	5.675
Resource 7	6.840	4.758
Resource 8	4.400	4.748
Resource 9	5.840	5.399
Resource 10	4.300	5.825
Totals	53.110	53.110

4.3 Risk

The sequence of project activities depends on the relationship between activities (predecessor and successor). Critical paths have an important role in the duration of the project with CPM method. The longer and the higher number of critical paths in a project, the greater the risk of the project. The project risk is shown from the stability of project duration. The longer and larger the number of critical paths, the greater the likelihood of adding to the overall duration of the project due to the activities on the critical path that might be delayed.

Table 3. Critical Path Comparison Results

Resource Allocation	Critical Path	Critical Activity
Normal	3 Critical Path	8 Activities
RCPS	67 Critical Path	23 Activities
Difference	63 Critical Path	15 Activities

Resource allocation affects the path and critical activities of the project. The risks of the design of aircraft wing structure project become higher when resources are allocated by RCPS method as shown in Table 3. Special attention from the company is required in carrying out the project if the resource allocation is done with RCPS allocation.

To avoid unpredictable additional duration of the project, the company can make regular scheduling improvements to ensure the current conditions of the project. By regularly scheduling the design activities, company found the best sequence of activities, duration of the project, allocation of resources, and the estimated costs to be spent.

5 Conclusion

Project management for the design of aircraft wing structure made in Indonesia with CPM using RCPS allocation resulted in a shorter duration (736 working days) compared to the company's normal allocation without using RCPS (950 working days). Risk of unpredictable delay from the project using RCPS allocation will increase compared to the normal allocation scheduling of the company. Addition of activity on the critical path can increase the overall project duration in case of delays happened on activity completions.

This research is funded by Universitas Indonesia – PITTA 2018

References

1. The World Bank, (2016). [Online]. Available: https://data.worldbank.org/indicator/IS.AIR.PSGR?end=2016&locations=ID&start=2009.
2. IATA, *Annual Review* (2017)
3. Project Management Institute, *A Guide to the Project Management Body of Knowledge (PMBOK® Guide)*, (2017)
4. K. Venema and A. Gueniche, Benef. Microbes, **5**, 97–98, (2014)
5. S. Zareei, *Renewable and Sustainable Energy Reviews*, **81**, 756–759 (2018)
6. G. Bergantiños, A. Valencia-Toledo, and J. Vidal-Puga, Discret. Appl. Math., **243,** 11–20 (2018)
7. A. Izmailov, D. Korneva, and A. Kozhemiakin, Procedia - Soc. Behav. Sci. **235**, 189–197 (2016)
8. M. Dachyar and Z. A. H. Sanjiwo, Indian J. Sci. Technol., **11**, 1–8 (2018)
9. C. Carson, P. Oakander, and C. Relyea, *CPM Scheduling for Construction: Best Practices and Guidelines* (2014)
10. J. Kim, C. Kang, and I. Hwang, Int. J. Proj. Manag., **30**, 264–272 (2012)
11. J. R. Meredith and S. J. Mantel, *Project management — A managerial approach*, (2009)

Scheduling of Aircraft Design Project: A Comparison of Critical Path Method, Design Structure Matrix and Genetic Algorithm Approaches

Ariani Amalia, *M.* Dachyar*,and Farizal

Industrial Engineering Department, Universitas Indonesia

Abstract. The purpose of this study to minimize the overall duration of aircraft design project in Indonesia. This study proposes a combination method of Design Structure Matrix (DSM) and Genetic Algorithm (GA) for scheduling aircraft design project activities. DSM is used to identify the information flow between project activities. GA use reworks data, which was the result output of DSM, and activity duration to optimize the task sequence by considering the human resource constraints. A comparative analysis was conducted between the schedule based on DSM and GA and schedule based on Critical Path Method. Combination of DSM and GA obtained 63 days shorter project durations compared to CPM. This research is convenient for design projects scheduling with limited resources that can be obtained by using alternative methods other than traditional methods.

1 Introduction

In the next 20 years, IATA [1] predicts that the number of passengers in Indonesia will reach 220 million passengers and ranks five in the world. The aircraft production in Indonesia establishes connectivity between regions, especially in remote areas. The aircraft project is a product development project that is influenced by uncertainty factors and requires high technology [2]. Currently, aircraft projects in Indonesia have never been completed and require additional time for the design and production process. Delays in design projects often occur due to legal, technical, design changes and improper planning [3]. The complexity of the information flow may lead to rework or iteration of activities that related to project delays [4]. The aircraft design project requires optimal project scheduling that can help overcome the delays caused by rework, resource and precedence constraints [5].

2 Literature Review

Project management is an application of knowledge, skill, tools, and techniques to undertake project activities to meet project requirements[6]. The project management processes consist of initiation, planning, implementation, monitoring and control, and closure. It is important

*Corresponding author: mdachyar@yahoo.com

for every organization to have good project management [7]. The scope of a project must be clearly defined and limited. The project scopes include the number of systems implemented, the number of business units involved, and the necessary business process re-engineering. Any proposed changes then evaluated against business benefits, and if possible, will be implemented in the next phase[6]. The timing of the project should be well managed so that the available deadlines and budget can be met. Projects that fail to reach the targeted time, budget costs or meet the needs of stakeholders would have a negative impact on business, government, and individuals themselves [8]. In project planning, the project team may face uncertain or unexpected things [9]. Uncertainty factors can lead to performance variability that affects the project planning and scheduling process [10]

Planning and scheduling process are important factors that affect project success. There are several methods used in project scheduling, including Critical Path Method (CPM), Program Evaluation and Review Technique (PERT), Critical Chain Project Management (CCPM), Design Structure Matrix (DSM), and metaheuristic methods such as Genetic Algorithm (GA). Some studies suggest that traditional methods, such as CPM, PERT, and CCPM, have limitations in project scheduling that have complex information flows [2, 11]. This suggests that alternative methods should be proposed in addressing complex project information flow issues.

In contrast to traditional methods, DSM more effective for managing feedback information on construction projects, engineering design, and product development [4]. DSM could manage project activity efficiently especially on complex projects. GA have been widely used in project scheduling compared to other metaheuristic methods, particularly in DSM analysis [5]. DSM and GA used to optimize the project scheduling model[5]. DSM utilized to identify complex inter-activity relationships. Meanwhile, GA modeling used DSM framework to obtain decent initial solution candidates.

Combination of DSM and GA not only can be applied to projects with complex information flows, resource constraints, overlapping and iterative activities,but also identify the relationships between project activities, task sequence optimization, and resources allocation. Efficient task sequencing are required to obtain optimal project scheduling and resources allocation. Therefore, the purpose of this research is to develop scheduling of aircraft design project to get the shortest duration using Design Structure Matrix and Genetic Algorithm and compare it with Critical Path Method. The combination of both methods is expected to overcome the challenges of the complexity of the information flow and simultaneously optimize project scheduling with several constraints considered.

3 Methodology

In this study, an aircraft design project data from an aircraft design company in Indonesia were used. The data consist of 128 activities including activity duration, activity dependency, and human resources. The data that were verified and validated by project manager. Validated data processed using CPM and the combination of DSM and GA.

Data were processed by submit into DSM matrix. The interaction between project activities could be clearly seen. Project activities weredivided into three types: parallel, sequential, and coupled activities. Partitioning and tearing process utilized to minimize coupled activity on DSM matrix. Partitioning and tearing process obtained new task sequence. The new task sequence becomes the input on GA modeling.

The initial parameters used in GA modeling are population number, chromosome length, generation number, crossover rate, and mutation rate. The initial population is a set of chromosomes where each chromosome represents a solution vector composed of genes. In this study, chromosomes are a sequence of activities and genes are the identity of a project activity. The chromosome generated in each population must comply with the predecessor and rework constraints.

Each generated chromosome is checked for its fitness and improved in each iteration. The function of the iteration is to obtain the best chromosome from each iteration process. The new chromosomes generated after the crossover and mutation process are examined based on their constraints and its fitness. Iteration stops after fullfilling the stopping criteria.

GA simulation used two scenarios obtained from DSM. Both scenarios were determined based on partitioning and tearing results on DSM.Comparative analysis conducted between schedule result of DSM and GA combination and CPM.Microsoft project were used to evaluate the effectiveness of the proposed method.

4 Result and Discussion

4.1 The Result of GA simulation: Scenario 1

Scenario 1 based on activity data obtained by DSMpartitioning. Duration of each rework activity identified in scenario 1 assumed to be 0.25 of initial duration. Iteration activity in scenario 1 assumed to occur only once. The result of the solution obtained for each combination of modeling parameters shown in Figure 1. Based on Figure 1, the total project duration for scenario 1 is 966 working days. The minimum total duration obtained in combination 9 with the crossover rate is 0.9 and the mutation rate is 0.1.

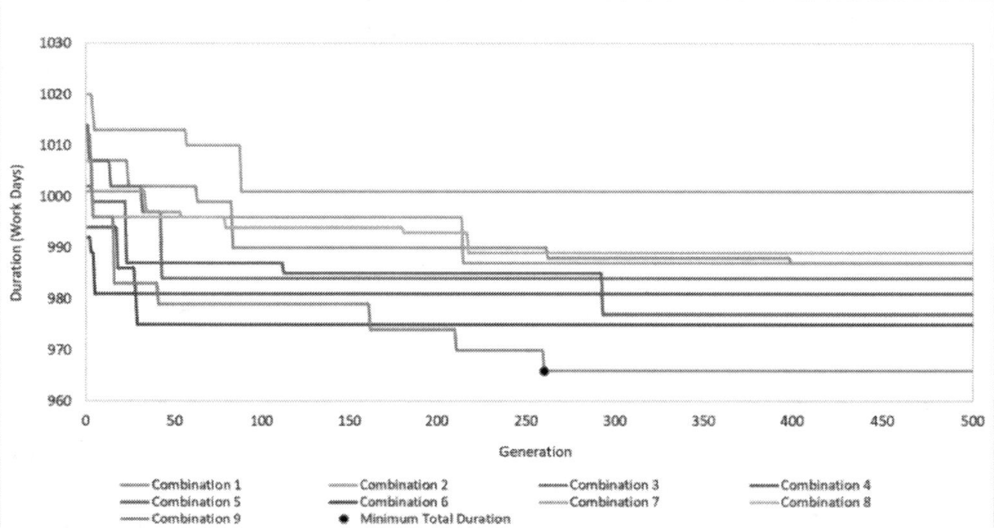

Fig1. GA Simulation Result: Scenario 1

4.2 The Result of GA simulation: Scenario 2

Scenario 2 based on activity data obtained by DSM tearing. The duration of each rework activity in scenario 2 divided into two, which are 0.25 and 0.15 of initial duration. The difference of rework factor due to feedback information received by some rework activities fewer than other rework activities. Iteration activity in scenario 2 assumed to occur only once. The result of the solution obtained for each combination of parameters is shown in Figure 2. Based on Figure 2, the total project minimum duration for scenario 2 is 941 working days.

The minimum total duration obtained in combination 5 with the crossover rate is 0.8 and the mutation rate is 0.05.

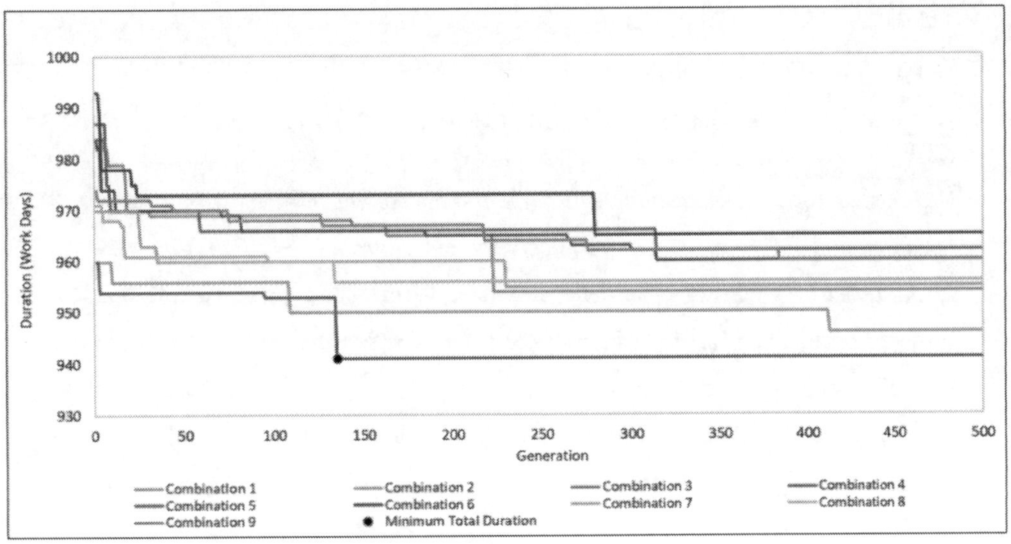

Fig2. GA Simulation Result: Scenario 2

4.3 Comparative Analysis

The use of DSM and GA in aircraft structure design projects resulted in a shorter total project duration.Based on CPM method, the total project duration obtained is 1.004 days. In scenario 1, project duration is 38 days faster than CPM. Meanwhile in scenario 2, the project duration is 63 days faster than CPM.CPM has a disadvantage in identifying rework activities. Rework of unidentified activities resultedin less optimal total duration.

The duration difference in both methods can be due to several things. In CPM method, the sequence of activities and the allocation of human resources applied to the project are processed manually. By using DSM, the complexity of the information flow can be identified. Complex projects such as an aircraft development, an information flow that unidentified lead to increase in the complexity of the project. Therefore, accurate activity identification is essential in project implementation.

Total duration of project is related to project cost and the sequence of activities related to the scope of the project. By using DSM and GA, the total project duration could be minimized by limiting the use of human resources. The sequences of activities obtained using the DSM and GA are more variated than CPM. This sequence of activity variations can result in optimal project schedule by fulfilling project constraints such as predecessor and rework activities.

5 Conclusions

This research combines DSM and GA to shorten the duration of the overall design project by considering the complexity of the information flow and human resources. Combination of DSM and GA obtained63 daysshorter project durations compared to CPM. Variations ofactivity sequences and the allocation of human resources areconvenient to illustrate the project scheduling scenarios in aircraft project design.

This research funded by Universitas Indonesia – PITTA 2018

References

1. IATA, *Annual Review*, (2016).
2. B. Shekar, R. Venkataram and M. Satish, Concr. Eng: Res. and Apps., **19**, pp. 283-294, (2011).
3. McKinsey, *A Risk Management Approach to A Successful Infrastructure Project*, 2013.
4. I. Gunawan and K. Ahsan, Int. J. Project Org. and Mgmt., **2**, 311-327, (2010).
5. A. A. Yassine, O. Mostafa and T. R. Browning, Com. & Ind. Eng., **107**, 39-56, (2017).
6. Project Management Institute, *Practice Standard for Scheduling*, (2017).
7. F. F.-H. Nah and J. L.-S. Lau, Bus. Process Mgmt., **7**, pp. 285-296, (2001).
8. M. Dachyar and Z. Sanjiwo, Ind. J. of Sci. and Tech., **11**, 1-8, (2018).
9. Project Management Association of Japan, *A Guidebook of Project & Program Management for Enterprise Innovation*, (2005).
10. M. Dachyar and N. R. Pratama, *Journal of Physics: Conference Series,* **495**,1-9, (2014).
11. I. M. Srour, M.-A. U. Abdul-Malak, A. A. Yassine and M. Ramadan, Auto. in Construct., **29**, 1-11, (2013).

Scheduling of Empennage Structure Design Project of Indonesia's Aircraft with Critical Path Method (CPM)

Arin Wulandari, *M. Dachyar** and *Farizal*

Industrial Engineering Department, Universitas Indonesia

Abstract. Production of the aircraft becomes very important to facilitate many passengers of aircraft in Indonesia in the future. The design of a very complex aircraft requires proper project management. This study aims to perform the best scheduling project of empennage structure design of Indonesia's aircraft with limited resources. Critical Path Method (CPM) is used to schedule the project based on activity and resource. CPM scheduling results in completion time for activity data for 400 working days. CPM scheduling after assigning human resources results in a longer completion time for 1023.47 working days. CPM scheduling is carried out by reallocating of human resources to produce a shorter completion time, and it results in completion time for 955 working days.

1 Introduction

Indonesia's need for connectivity with air transport is increasingly high in connecting resources, communities, and industries to protect the country's economy [1]. According to the Air Transport Statistics report of 2015 and 2016, in 2011 to 2016 the number of passenger on domestic destinations has an increasing trend [2], [3]. IATA report (2016) states the number of passenger aircraft in Indonesian is predicted to reach 220 million people in2034[4].

The design of an aircraft is a complex and difficult undertaking to obtain the optimal end product [5]. The complex aircraft design requires proper project scheduling to achieve the project's success. Critical path method (CPM) is a traditional method of project scheduling that results in completion time based on activity and resource.

2 Literature Review

A project is a temporary undertaking to create a unique product, service, or outcome [6]. Project management is the universally accepted standard to deal with businesses that are

* Corresponding author: mdachyar@yahoo.com

temporary, unique, and has a specific purpose [7]. The success of project management can be measured through the "iron triangle" model that includes time, scope, and cost. The model of the iron triangle has a point of view that the project can be successful even though project management was not successful because it has higher goals and long-term[8].Project success should be measured within project scope, scheduling, cost, quality, resource, and risk constraints in term of completing the project [9].

Critical Path Method (CPM) is a method that used to estimate minimum project duration and determine the amount of scheduling flexibility on logical network paths in the schedule model[6]. CPM uses networking forms in project planning to coordinate activities, schedule, and monitor project progress[10]. In the use of CPM, if resources complete an activity before the planned completion date, time can not be disseminated to the next activity[11].

Calculation of critical path includes two stages which areforwardpass calculation and backward pass calculation. The forward pass is a critical path calculation starting from the beginning of the node to the last node in the network. The backward pass is the inverse of the forward pass, the critical path calculation that starts from the end of the node to the initial node in an activity network diagram.The critical path is determined when the forward pass and backward pass has zero slack time.

In the process of network replanning, there is a resource allocation method which is also called as limited resource planning. Resource allocation is an attempt to find the critical path as short as possible based on available or permanent resources. The problem with this method is that employees may not be technically qualified to perform on more than one activity within a network. Unfortunately, not all CPM networks can perform easy resource scheduling. The project manager must make an effort to reallocate resources with the aim of reducing the critical path, provided that the slack time is not planned deliberately as a safety valve[9]. The method to reduce critical path based on the resource is called resource reallocation.

3 Research Methodology

Data is collected by deriving secondary data which is a document of the company. The data taken are activity data and human resource data. Activity data consist of project task or activity, duration of the activity, and predecessor of activity. The activity data of the empennage structure design project consists of 55 activities that are divided into seven main activities. The human resources data of the company have requirements for allocation in project scheduling, the distribution of human resources depends on the duration of the project activity.

Data is calculated by critical path method (CPM) to find scheduling of aircraft empennage design project. First, CPM is scheduledwith only activity data by calculating early start (ES), early finish (EF), late start (LS), and late finish (LF) to obtain slack time. The calculation of zero activity slack time is a critical pathon the project. The recent CPM scheduling is then assigned with human resource data to obtain the new critical path. CPM scheduling with activity and human resource data followed by network replanning of human resource reallocation to reduce the critical time of the project. Thecompletion timeof the CPM scheduling can be determined from Gantt chart or the bar form graph. The CPM scheduling results are then analyzed based on the project constraint to measure the project's success. Project constraints used in this study are scheduling based on human resource, cost, and risk.

4 Results and Discussion

4.1 CPM Scheduling based on Activity Data

Critical activities are determined by calculating the slack time of each activity. Critical activities are obtained when the slack time is equal to zero. The project completion time ofaircraft empennage design project on the activity data is displayed on the part of Gantt chart as in Figure 1.

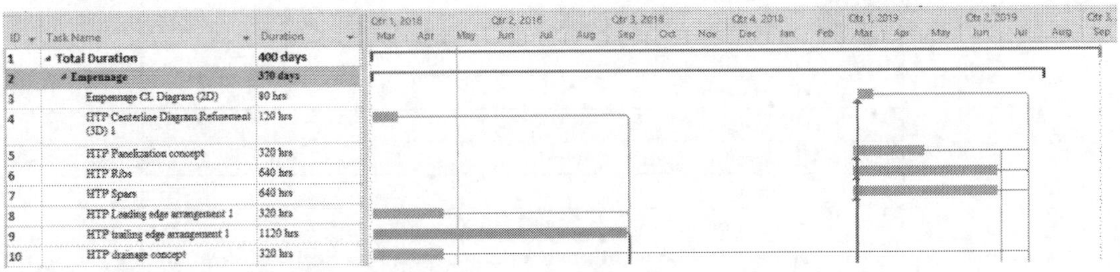

Fig. 1. Gantt Chart 8 Activities of 55 Activitiesin Activity Data

The completion time for CPM scheduling based on activity data is 400 working days. This completion time does not consider the number of human resources working in each activity.

4.2 CPM Scheduling based on Activity and Human Resource Data

The data of activity and human resource interconnection is obtained by assigning human resources to each activity. Human resource is assigned to each activity with the requirements that have been described in the data collection phase. Figure 2 illustrates the part of Gantt chart display for activity and human resource activity data.

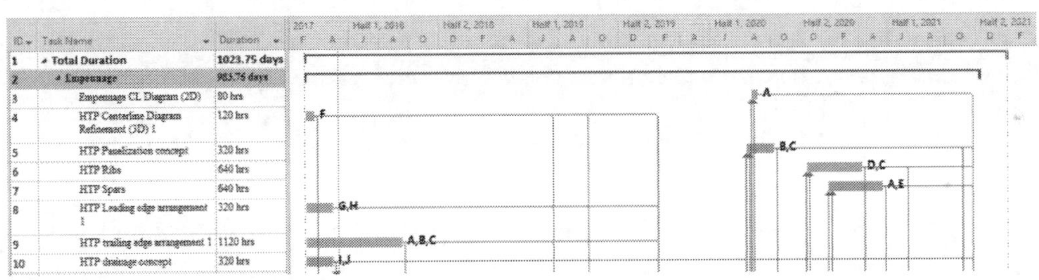

Fig. 2. Gantt Chart 8 Activities of 55 Activities in Activity and Human Resource Data

By assigning human resources on the activity data, there is a change in the order of work activities that cause the completion time of the overall activity is longer which is 1023.75 working days. The number of critical activity on CPM scheduling based on activity and human resource data is 20 activities. Human resources working in CPM scheduling based on activity and human resource data has 64,290 working hours.

4.3 CPM Scheduling with Human Resource Reallocation

The completion time of the CPM scheduling results can be shortened by setting up human resource based on resource reallocation. The resource is reallocated by adding human resources to the critical path to reduce time. The number of activity that carried out of human resource reallocation isfive activities. The results of human resource reallocation are three non-critical activities, and two activities that remain critical. Two activities that remain critical are not a problem because the goal of human resource reallocation is not to convert critical activity to non-critical, but to reduce critical time so that the total time of completion can be reduced. Table 1 summarizes the critical activities with human resource reallocation and then described the Gantt chart in Figure 3 to see the project completion time.

Table 1. Summary of Critical Activities with Human Resource Reallocation

No.	Task Name
1	HTP trailing edge arrangement 1
2	HTP drainage concept
3	HTP Hardpoint
4	HTP hoisting point definition
5	VTP trailing edge arrangement 1
6	VTP hoisting point definition
7	Section 538 Subassy & Panels Breakdowns
8	Section 538 Frames/Stringers/Longerons Cross Section & Datum
9	Empennage Major Assembly Sequence
10	Catia Filing Database: Empennage
11	Design Development Management
12	Design Data Base Control
13	Structure Design Standards
14	R80 Aircraft Assembly Sequence Document
15	3D System modelling
16	Material Selection & Design Allowables

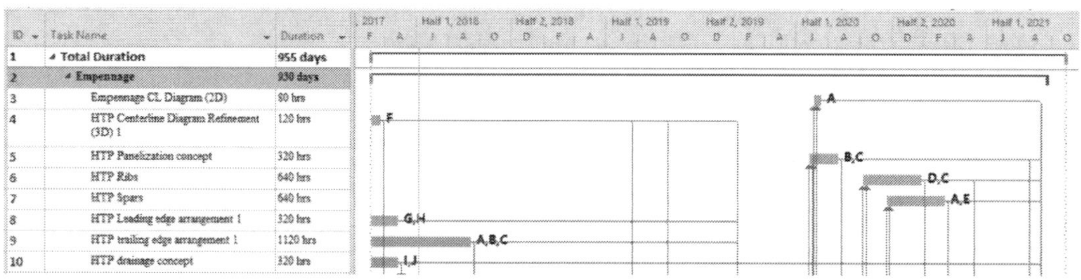

Fig. 3. Gantt Chart 8 Activities of 55 Activities of Activity and Human Resource Data withHuman Resource Reallocation

CPM scheduling with human resource reallocation has a shorter completiontime which is 955 working days. The number of critical activity on CPM scheduling after resource reallocation is 16 activities. Human resources working in CPM scheduling after resource reallocation has 64,290 working hours.

4.4 Project Constraint

Project constraints that used in this study are scheduling based on human resource, cost, and risk. Project constraint analysis is performed on CPM scheduling based on activity and human resources data before human resouce reallocation and after human resource reallocation. Before the human resource reallocation,human resources are assigned on CPM scheduling with activity data based on company requirements. CPM scheduling after human resource reallocation process is resetting human resources and it results in the faster completion project time than CPM scheduling before human resource reallocation. Constraint cost in this research is the salary of human resource which seen from the amount of human resource work hour. The incurred cost by CPM scheduling before human resource reallocation is the same as the CPM scheduling after human resource reallocation which is a salary for 64,290 working hours. Constraint risk can be seen from the number of critical activities. CPM scheduling before human resource reallocation has a greater number of critical activities than CPM scheduling after human resource reallocation, thus CPM scheduling before human resource reallocation has a higher risk in project implementation.

5. Conclusion

Critical Path Method (CPM) is a traditional management project in scheduling that is commonly used for various project. This studyshows three phases of calculating CPM schedulingin empennage design structure projectwhich are CPM scheduling based on activity data, activity and human resource data, and activity and human resource data after resource reallocation.The results of this studyare CPM scheduling based on activity data with completion time for 400 working days, activity and human resources data with completion time for 1023.75 working days, and activity and resource data after resource reallocation with completion time for955 working days. CPM scheduling has a longer completion time when activity data is assigned with human resources because the following activities can be done when the human resources in the previous activity are finished working. The results show that CPM scheduling after human resource reallocation results in faster completion project timeand has a lower risk due to reduced critical activities. The same constraint of CPM scheduling before human resource reallocation and after human resource reallocation is the cost of payroll for 64,290 working days.

Acknowedgement : This research funded by Universitas Indonesia – PITTA 2018

References

1. U.S. Embassy Jakarta Economic Section, (2012).
2. BPS, (2016).
3. BPS, (2017).
4. IATA, (2016).
5. M. Vignesh, V. K. Morris, and C. Aravind, *Springer*, (2014).
6. PMI, *A Guide to the Project Management Body of Knowledge - PMBOK Guide*. (2013).
7. M. Dachyar and Z.A.H. Sanjiwo, Indian J. Sci. Technol.,**11**,1-8, (2015).
8. M. Radujković and M. Sjekavica, Int. J. Proj. Manag., **196**, 607–615, (2017).
9. Project Management Institutes,*A guide to the project management body of knowledge (PMBOK guide)*. (2017).
10. S. Zareei, *Renewable and Sustainable Energy Reviews*, **81**. 756–759, (2018).
11. J. Feng, L. Li, N. Yang, Y. Hong, M. Pang, X. Yao, L. Wang,Water Sci. Eng., **4**, 225–236, (2011).

Implementation of Risk Management in Manufacturing of Wellhead and Christmas Tree Equipment (Risk management framework)

Abdul Hamid[1, 2, *], *Ishak* Bin Baba[1], *Sulaiman* Bin Haji Hasan[2], *Agung Setyo* Darmawan[3] and *Nushatisah* [2]

[1]Faculty Technology Engineering Universiti Tun Hussein Onn Malaysia, 86400 Johor, Malaysia
[2]Faculty Engineering, Mechanical Engineering Department, Universitas Batam, 29464 Indonesia
[3]Department of Mechanical Engineering, Faculty of Engineering, Universitas Muhammadiyah Surakarta, Jl. Ahmad Yani, Tromol Pos 1 Pabelan, Surakarta 57162, Indonesia

Abstract. Wellheads and Christmas trees are the main equipment for oil production. They are manufactured in the plant and installed on the casing head to seal the annular space between casing and tubing. They are used to transport oil to pipeline in well field. The loss of wellhead and Christmas tree integrity can result in major accidents, presenting a severe risk to the environment. Therefor a purpose of this study is to investigate of an organization perception for risk management implementation in manufacturing wellhead and Christmas tree equipment. The responses were analysed using SPSS software by using Cronbach's α, mean-value and standard deviation. Data was collected by posting a Google link form, or sent to the email of the companies listed in the Malaysia and Batam, Indonesia. This quantification of the risk management process and a risk identification tool onto the risk management process framework. The model can support and indicate the contribution of an industrial risk manager towards achieving project objectives, as well as making comprehensive decisions regarding analysis of risk management in manufacturing project and operation.

1 Introduction

Oil exploration and production has played a significant part in the financial success of Asia and the Middle East. The main companies supplying wellhead and Christmas tree are FMC Technologies, Aker Solutions, GE Oil and Gas, with market shares in that order. Past accident analysis for offshore oil and gas rigs [1] confirm that accidents are related to the following causes: equipment failure, equipment malfunction (electrical and mechanical), corrosion, and safety system malfunction. In the report, proper risk management was suggested as a way to reduce the number and severity of the accidents. To execute a manufacturing project smoothly, it is important for a project manager or team to use systematic models and tools. These tools can be used to control the project through every stage and process to meet project

* Corresponding author: hamid.abdul9@gmail.com

objectives such as budget, safety plans, and deliverables according to schedule. Key objectives in performing project risk management is to increase the probability of project success by minimizing or eliminating negative risk events and increasing the probability of positive events [2]. To achieve the manufacturing project objectives and meet deliverables, a good planning and risk management model should be implemented before commencing with the manufacturing process, and performing improved material used [3,4]. The model should include all work tasks, engineering design tasks, estimate and cost control sheets, project scheduling and time sheets, the corresponding record sheets in a communication circle with customers, and the internal organizational and vendors/contractors involved for the entire period necessary to complete the project successfully [56,7]. These risks and failures will be anticipated by the engineering side as they perform design calculation and analyses prior to releasing approval of the scenario, meeting client requirements, and providing project deliverables [8]. Recent research on risk management is related to adaptive risk management, which starts out with a conceptual view of the risk and focuses on the consequences and uncertainty involved in the project [9]. Risk model is developed from literature studies and research findings related to organizational behavior.

1.1 Manufacturing of wellhead and Christmas tree equipment

Generally, manufacturing refers to any business that uses an assembly process that involves machines, tools, manual labour, quality control and inspections to convert raw materials into saleable goods [10]. Wellhead and Christmas tree equipment are oil and gas equipment that are produced and fabricated in the plant. After the tender and sales order has been completed, the manufacturing organization can start the process, taking into consideration the technical aspects, lead time, and overall client requirements.

1.2 Risk management in manufacturing project

According to [11], risk is a risk factor that can affect the progress and process of the project. Risk was defined as a function of the likelihood of loss that can arise for an activity, process, and situation [12]. Meanwhile [13] argued that a risk was a sensitive issue or a weakness that companies or organizations did not want to publish in order to maintain market value and keep their competitive edge. Another researcher said that risk was an exposure or a probability of the occurrence of a loss and events [14]. [15] defined risk as an adverse factor that influenced the success of any situation where it is present. The general meaning of the word 'risk' implies a negative attitude towards the environment in question [16]. A risk can be a significant barrier to success [17]. [18] explained how to handle risk factors through three key dimensions: control, elimination and avoidance. In this case, risk cannot be considered a known negative and should be assessed as a likelihood. [19] defines risk as an uncertain event or condition that, if it occurs, has a positive or negative effect on a project's objectives. When put into context, it seems that risk can have a two-dimensional meaning (namely, a negative as well as a positive implication). It is of great interest to be able to separate the meanings of risk.

2 Objectives

The aim of risk management is to develop and establish a risk management framework that includes and combines risks in Safety Health and Environment (SHE) and financial objectives.

3 Method Approach

A risk management method for the study was collected and developed after reviewing available literature. The survey was conducted among the managers, project managers/leaders, engineers, supervisors, and lead technicians to measure their perceptions of the degree of influence of different variable factors on effective risk management implementation. The field study involved observing the manufacturing process of wellhead and Christmas tree equipment at several companies in Batam, and Malaysia. A questionnaire was developed based on existing literature and experts practitioners in manufacturing in order to refine the primary risk management variable factors of company risk management culture.

3.1 Reliability and validity

The internal consistency of the pilot was analysed by using Cronbach's α - the most widely used indicator for internal consistency [20]. A desirable result for Cronbach's α should be greater than 0.60 [21]. SPSS Ver 20 was used to measure and calculate the coefficient of reliability of the instruments, "Cronbach Alpha Formula" – developed by [22] – was used. The questionnaire was reviewed by industrial experts who contributed to the theoretical questions and validated them. The contributions were related to the validity of foundation of the survey content and the dimensions of construct for the risk management culture as a factor dependent on the implementation stage of risk management [23]. The results for the pilot study were analysed by using SPSS. The individual score for each questionnaire construct ranged from 0.705 - 0.826. The reliability test for the working environment construct showed a satisfactory range of 0.705 - 0.826. Based on these tests, all the constructs were found to be reliable, and did not require changes for further analysis

4 Result

This group of questionnaire consists of questions related to company facilities, assessment procedures, activities, and standards. The data was used to understand the needs and behaviours of a company/organization. The strength of association between risk management culture and other variable data were shown in Table 4.1.

Table 4.1. Current research of respondent feedback for risk management (culture vs implementation)

Item	Dimension	Overall mean	F stat	X^2	Level of Sig (p)	Sig (yes or no)
Company Risk Management culture	Culture (n=5)	17.9922	3.448	89.023	.000	yes
	Project knowledge (n=5)	17.3798	2.843	45.403	.001	yes
	Competency (n=5)	17.2791	1.988	42.147	.022	no
	Motivation (n=5)	18.1395	3.683	75.574	.000	yes
Implementation stage of Risk management	Organizational (n=6)	21.4651	3.169	79.992	.000	yes
	Risk assessment (n=5)	17.7287	4.307	85.147	.000	yes
	unlikely events (n=5)	16.8140	2.871	94.628	.001	yes
	Risk utilization (n=7)	26.1783	1.914	60.891	.032	no
	Control Risk (n =8)	28.9225	2.425	84.349	.004	yes

$P < 0.05$ is significant test

Associate analysis of factors that affected risk management implementation was performed. The association between company risk management culture and implementation was measured using the following variables: culture, project team knowledge, competency, motivation. The Phi (φ) test was conducted in this analysis and produced significant results when a Cramers V test was used. The values are reported in Table 4.2.

Table 4.2. Current research of regression analysis

Variable	Phi (φ)	Interpretations
Culture (n=5)	0.406	Moderate
Project team knowledge (n=5)	0.471	Moderate
Competency (n=5)	0.419	Moderate
Motivation (n=5)	0.496	Moderate

N = 216. Less than 0.1 is week. 0.1-0.5 is moderate.Above 0.5 is high

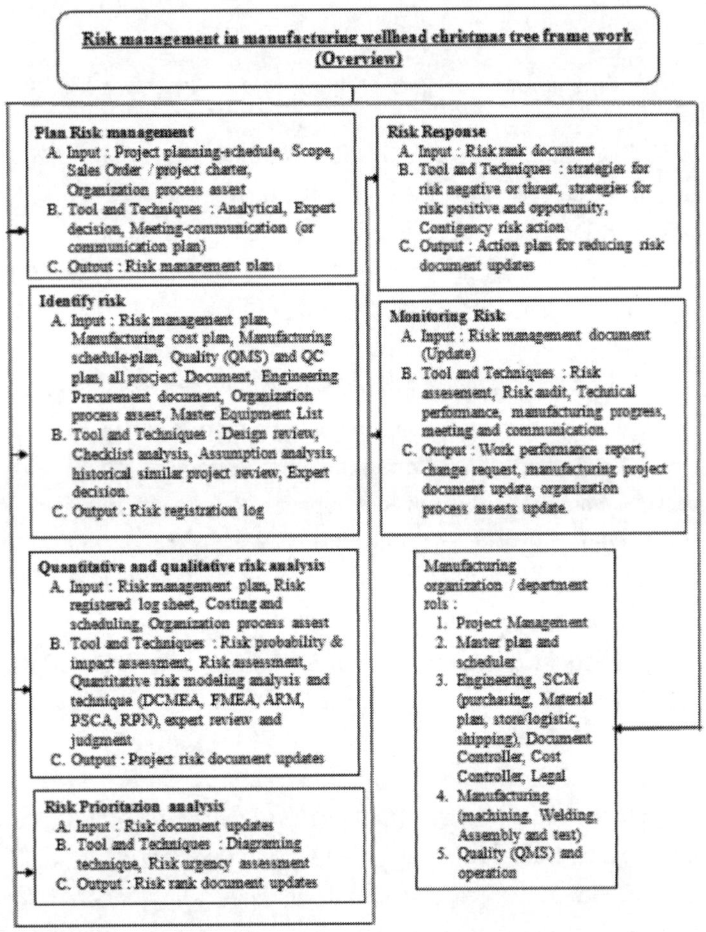

Fig. 4.1. The figure Risk management process in manufacturing wellhead Christmas tree framework (overview)

It is important for static risk and dynamic risk assessment to be implemented during the initial stages in order to detect any risks or unwanted hazard. The risk management framework also considered the input and advice of experts. Reviewers were selected based on their relationships between personnel, work experience requirements, knowledge, positions, and competency.

5 Conclusion

This framework model is emphasized in front of the risk management process to identify risk and proactive risk management processes and used a risk identification tool and technique in practice accordingly. The framework should be involved anyone related to the production of wellhead and Christmas tree equipment such as contractors, suppliers, or sub-contractors for components, services and accessories.

We would like to thank Faculty Engineering Technology and Office for Research, Innovation, Commercialization, and Consultancy Management (ORICC) UThM and grateful for industry partner in Batam island and Malaysia and Universitas Batam Indonesia to significant support on this research project.

References

1. M. Christou, M. Konstantinidou*Safety of offshore oil and gas operations*. J. RC Eur C. Italy , (2012).
2. B. Akatova, E. R. Curran.,. F Assmts Sys RM. *J Modelling Mngmnt*, ***8*(3)** (2013).
3. A.S. Darmawan, W.A. Siswanto, T. Sujitno, R J of App Sc, E & T **5(4)** (2013).
4. A.S. Darmawan, A.D. Anggono, A. hamid, D design Opt.MATEC **154**, 01105 (2018)
5. T. Shimizu, Y. W. Park, P. Hong, P mngrs f R M case f Japan. *An Int J*, **19(4/5)**, (2012)
6. A.Hamid, I. Baba, W. Sani, Prp for R M implmnt phase, MATEC **97** 01067 (2017)
7. A.Hamid, I. Baba, W. Sani, RM frameW in oil f Dev. Projt. IJASEIT, **7 (2),** (2017)
8. F. Barahim, *Hazard Ident, (Hirarc) in a Corrugated Box Manufacturer*. (2010).
9. T. Bjerga, T. Aven,. Adaptive risk mngmnt using new risk persp. **134**, 75–82 (2015).
10. API 6A – Spec. for Wellhead and Christmas Tree Equipment, *552*(3), 443 (2008)
11. N. Chileshe, B.Y. Fianko,. An Evl of R F impact. *J Eng, Dsgn and Tech*, **10(3)**, (2012)
12. P.N. Ranong, W. Phuenngam, Critical S F F Eff. R M Procdr **1–76** (2009)
13. D. Schoenfeld *Organisational Risk Culture,* (Thesis Ph.D). (2013).
14. S. Ameziane*A resilience engineering app.r to safety excellence* (Thesis Ph.D) , (2016).
15. R.Olsson,. RM in a multi-prj envirn. *Int J of Q & R Mngmnt*, **25**(1), (2008)
16. R. Olsson, *Mangng Proj Uncertainty* (Thesis Ph.D open access) (2006).
17. O. Otobo, RM sys to guide B const. proj (Thesis Ph.D) (2016).
18. H. Zhang, M. Yongbo,. Prodt Dev RM in Prod Dev Proc. *Int B & Mngmnt*, **3(1)** (2011)
19. Project Managmnt Institute. *Project Management Body of Knowledge* PMIBOK 2013
20. G.D. Garson,. Testing Statistical Assumptions. *Blue Book Series*, 1–52 (2012)
21. M. Tavakol, R. Dennick, Making S of Cronbach's alpha. *Int J of Med. Edu*, **2**, (2011)
22. G. Groth-Marnat, *Handbook of Psycholgical Assessment (4th Edition)*. (2003).
23. F. Goerlandt, N. Khakzad, G. Reniers,. V & valid of sfty.*Safety Science*, **99**, (2017)

Product Distribution Optimization in Food SMEs with Integer Linear Programming

Syafira Chika Widiyanti, *M.* Dachyar*, and Farizal

Industrial Engineering Department, Universitas Indonesia

Abstract. Efficient product distribution is a key driver of the overall profitability of a company. The high cost of product distribution in food SMEs is caused by the absence of a proper distribution cost calculation to obtain optimal distribution decisions. This research aims to provide SMEs with a product distribution model with the lowest cost. A Heterogenous Fleet Vehicle Routing Problem (HVRP) for SMEs product distribution is modeled in the form of Integer Linear Programming and solved using branch-and-bound algorithm to obtain an optimal route and vehicle assignment with the lowest cost. The result of this research is a model that can be used by SMEs to make daily product distribution decision.

1 Introduction

Small Medium Enterprises (SMEs) is the largest business actor in Indonesia. SMEs contribute about 60% of Gross Domestic Product (GDP) and absorb 96,99% of the total workforce [1]. The competitiveness of Indonesian food SMEs is still low compared to other ASEAN countries. The increase of product price due to the distribution in Indonesia is 15% which is much higher compared to the average 7% increase in ASEAN [2]. Distribution is a major driver of the overall profitability of a company because it directly affects supply chain costs and customer experience. SMEs as the largest business actor in Indonesia requires a proper calculation of optimal distribution costs to improve competitiveness.

2 Literature Review

Logistics is defined as the full process of materials and products movement into, within, and out of a business [3]. Distribution management refers to the physical flow and product storage from the final production point to the customer or end user [4]. Distribution management can be divided into three levels of decision planning: distribution network design, distribution network planning, and transportation planning [5]. Transportation planning involves short-term distribution operations planning and related to delivery planning to different customers. The general decision in transportation planning is to find the shortest route, time, and minimum cost [6].

* Corresponding author: mdachyar@yahoo.com

Vehicle routing problem (VRP) aims to obtain routes for homogeneous vehicles to fulfill the demand of customers. Each customer is visited only once by one vehicle, each vehicle route begins and finishes at the depot [7]. Heterogeneous Fleet Vehicle Routing Problem (HVRP) is a type of VRP in which the vehicle is characterized by different capacities and costs [8]. Previous studies on HVRP aims to minimize the logistics cost that includes the manufacturing capacity and driver work time using branch-and-bound algorithm [9].

Integer linear programming is a mathematical programming problem where a linear objective function is optimized while satisfying a linear constraint and decision variables are integer values. Branch-and-bound (B & B) is a method where a branching and pruning process is used to solve sub-problems in ILP. Upper and lower bound is tightened progressively to the objective value of ILP [10].

3 Problem Formulation

SMEs have different types and number of vehicles, some of them have only one type of vehicles and others have more than one type of vehicles. Graphic G = (V, A) is given, where V = {0,1,...,n} is a set of node n and A is set of arc. Node 1 represents the factory and node set V' = V\{1} represents n customers. Every customer i ∈ V' demands a certain amount of product Qi from the factory (Qo = 0). Heterogeneous vehicles are stationed in the factory and used to deliver the products to customers. Vehicles consist of different vehicle type m, with M = {1,...,k}. For each vehicle type k ∈ M, the vehicle has a capacity of Qk. Each type of vehicle has a fixed cost Fk of vehicle depreciation and driver wage. For each arc(i,j) ∈ A and for each vehicle type k ∈ M, routing cost Cijk is given. Routing cost is calculated by multiplying vehicle variable cost Ck with the distance traveled. The vehicle variable cost consists of fuel cost and vehicle maintenance cost. The route is defined as (R, k), where the route is a circuit in G that contains the factory and k is a type of vehicle that is used in the route. R represents the visiting sequence and a set of customers (including factory) in the route. Route (R, k) is feasible if the total customer demand does not exceed capacity Qk. Distribution cost is the sum of arc sets that build the route and the total cost of vehicles used in the route. HVRP aims to create a feasible set of routes with minimum distribution cost where every customer is visited in a route and a route is assigned for one vehicle k.

4 Mathematical Model

$$\text{Min} \sum_{k \in M} Fk \sum_{j \in V'} Xojk + \sum_{k \in M} \sum_{\substack{(i,j) \in V \\ i \neq j}} cijk.Xijk \tag{1}$$

$$\sum_{k \in M} \sum_{j \in V} Xijk = 1, \forall i \in V' \tag{2}$$

$$\sum_{k \in M} \sum_{i \in V} Xijk = 1, \forall j \in V' \tag{3}$$

$$\sum_{\substack{(i,j) \in V' \\ i \neq j}} qi.Xijk \leq Qk.Xojk, \ \forall k \in M \tag{4}$$

$$\sum_{i \in V} Xipk - \sum_{j \in V} Xpjk = 0, \ \forall p \in V', \forall k \in M \tag{5}$$

$$yik - yjk + Qk.Xijk \leq Qk - qj, \ \forall i,j \in V', i \neq j, \forall k \in M \tag{6}$$

$$qi \leq yik \leq Qk, \forall i \in V', \forall k \in M \tag{71}$$

$$Cijk = Dij.Ck, \ \ \forall i,j \in V, \ i \neq j, \forall k \in M \tag{82}$$

$$yij \geq 0 \quad , \forall i, j \in V,\ i \neq j \tag{93}$$

$$Cijk \geq 0 \quad , \forall i, j \in V,\ i \neq j, \forall k \in M \tag{104}$$

$$Xijk \in \{0,1\},\ \forall i, j \in V, i \neq j, \forall k \in M \tag{11}$$

$$Xojk \in \{0,1\}\ , \forall k \in M \tag{12}$$

This model have an objective to minimize the total distribution cost that consist of vehicle k fixed cost and routing cost. Constraint (2) and (3) ensure that every customer is visited exactly once. Constraint (4) ensure that the total customer demand distributed by vehicle k in one route that does not exceed the capacity of vehicle k that is assigned in the route. Constraint (5) ensure that a vehicle enters and leaves a customer. Constraint (6) and (7) ensure that there is no subroutine formed in the route. Constraint (8) calculates the routing cost based on the distance travelled and vehicle variable cost. Constraint (9) – (12) is the integrality and non-negative restriction towards the variable.

5 Implementation

The mathematical model is modeled with Lingo 11 with a branch-and-bound algorithm to obtain the product distribution decision with the lowest cost for food SMEs. This model is verified because it succeeds to run without error and generated the desired outcome. This model is validated by comparing the outcome of Lingo and manual calculation.

Data needed for the model are customers and factory addresses, customer demands, drivers wages, vehicle maintenance costs, and vehicle characteristics (type, capacity, amount, initial cost, fuel consumption). The addresses are used as an input to generate the coordinates for obtaining distances between locations with Bing Maps embedded in Ms. Excel VBA. In this model, the fixed costs used are drivers wages and vehicle depreciation costs calculated with straight line depreciation. The variable costs used are fuel costs and vehicle maintenance costs.

This model is implemented in 5 food SMEs in Indonesia. VB SME distributes burgers to 6 customers, J distributes biscuits to 7 customers, S and KGG distributes potato chips to 6 customers each, and MK distributes noodles to 14 customers. Vehicle information in each SMEs are presented in Table 1.

Table 1. Vehicle information in each SMEs

SME	VB		J	S	MK	KGG
Type	Motorcycle	Pick-up	Pick-up	Motorcycle	Motorcycle	Motorcycle
Amount	1	1	1	2	3	2
Capacity	250	2100	2500	50	90	50
Fixed cost (Rp)	131,393	226,712	196,233	114,110	140,088	128,196
Variable cost (Rp)	124	387	402	124	122	124

6 Results and Discussion

Food SMEs in this research differs in product, vehicle characteristics, and customer. The product distribution in these SMEs has been optimized and the total distance, route, vehicle assignment with the lowest distribution cost is generated from the model. The distribution decision result for each SMEs are presented in Table 2. The route and vehicle assignment result for MK SME is shown in Figure 1.

Market fluctuation will affect food SMEs daily product distribution activities. Scenarios with a different number of customers and demands are implemented in each SMEs. The scenarios are: (1) if the number of customers increased by 50%, (2) if the number of customers decreased by 50%, (3) if demand increased by 50%, (4) if demand decreased by 50%, (5) if the number of customers and demand increased by 50%, (6) if the number of customers and demand decreased by 50%, (7) if the number of customers increased by 50% and demand decreased by 50%, and (8) if the number of customers decreased by 50% and demand increased by 50%. The results showed that a change in customer amount and demand would change the product distribution decision with the lowest cost.

Table 2. Model implementation result for each SMEs

SMEs	Distance (km)	Distribution Cost (Rp)	Vehicle used	
			Type	Amount
VB	169.91	Rp 387,539	Motorcycle	1
			Pick-up	1
J	195.67	Rp 302,435	Pick-up	1
S	64.69	Rp 236,241	Motorcycle	2
MK	81.28	Rp 430,342	Motorcycle	3
KGG	144.21	Rp 274,274	Motorcycle	2

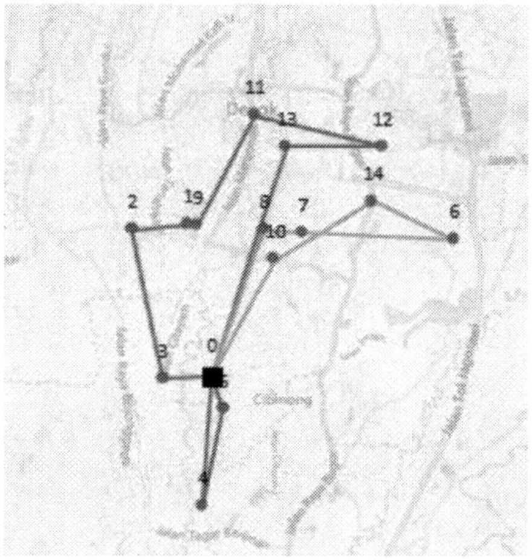

Fig. 1. Visualization of MK SME distribution decisions

7 Conclusion

In this research, food SMEs product distribution is formulated as a heterogeneous fleet vehicle routing problem in the form of integer linear programming. Product distribution optimization is implemented in 5 food SMEs in Indonesia and distribution decisions regarding the routing and vehicle assignment with the lowest cost were obtained. A distribution cost calculation tool is developed and could be used by food SMEs for daily product distribution decisions in the future.

This research is funded by Universitas Indonesia – PITTA 2018

References

1. Kementerian UKM dan Koperasi, *Sandingan Data UMKM 2012-2013* (2013)
2. Investor Daily Indonesia, (2012). [Online]. Available: http://id.beritasatu.com/home/industri-makanan-minuman-indonesia-di-peringkat-50/46326
3. M. Dachyar, G. Novita, ARPN J. Eng. Appl. Sci., **11**, 4539 – 4546 (2016)
4. C. P. Rushton A, *The Handbook of Logistics and Distribution Management,* (2006)
5. R. F. Akkerman, OR Spectrum, **32**, 863-904 (2010)
6. J. Gu, M. Goetschalckx, L. McGinnis, Eur. J. Oper. Res. 177, **1**, 1-21 (2007)
7. C. Koc, & G. Laporte, Comput. Oper. Res., **91**, 79-91 (2018)
8. R. Baldacci, M. Battarra, D. Vigo, *The Vehicle Routing Problem: Latest Advances and New Challenges,* 1-25 (2008)
9. J. Oesterle, T. Bauernhansl, *Procedia CIRP,* **41**, 573-578 (2016)
10. D. Chen, R. G. Batson, Y. Dang, *Applied Integer Programming: Modeling and Solution* (2010)

Patients' Loyalty Improvement in Public Hospital

M. Dachyar[*], *Farizal*, and *Ika Prisila Minar*

Industrial Engineering Department, Universitas Indonesia

Abstract. The purpose of this research is to improve outpatient loyalty in using National Health Insurance in a public hospital. Structural Equation Modeling (SEM) method is used to analyze factors that influence patient loyalty. The discriminant analysis method is used to know the priority for improvement. An assessment of 5 experts was performed to obtain latent variables and indicators. The questionnaire was given to 400 patients. The SEM model is concluded, satisfaction and trust are two strong factors that affect the loyalty of patients.

1 Introduction

The increasing number of the Indonesian population, with a 1.19% growth rate, indicates that the need for health and health facilities will also increase. However, the condition of health services, especially in public state hospitals, is still disappointing. The needs for proper healthcare are seen as potentially beneficial to private hospitals. Moreover, the implementation of National Health Insurance since 2014 is considered as a high volume business and very promising for private hospitals. It can be seen from the number of private hospitals that cover National Health Insurance continues to increase. Private hospitals in the future are predicted to be able to control 67% ownership of clinics and specialized hospital's non-maternity, laboratory clinics and check-up clinics [1]. The entry of the private sector in the hospital industry has become a matter of concern. If this continues, it is possible that the public state hospital will lose their patients.

The patient's demand for health services is getting higher, not only wanting their illness to be cured, but also wanting an excellent overall service. Public hospital as a hospital managed by the government should also improve itself in improving the quality of service to be able to satisfy the needs and expectations of patients. These improvements aim to be able to retain patients of public hospitals and acquire new potential patients. The purpose of this research is to improve outpatient loyalty in public hospital that using National Health Insurance.

2 Literature Review

Service quality is defined as a post-consumption evaluation comparing expectations with performance perceptions [2, 3, 4]. Measurement of service quality in public hospitals has a positive impact on patients, such as short patient waiting times, decreased hospital infections,

[*] Corresponding author: mdachyar@yahoo.com

decreased medical errors [5]. Perceived value is defined as a tradeoff between the quality or benefits gained in a product or service against the sacrifice provided by paying for the product or service [6, 7]. In a public hospital, perceived value is the patient's perception of the difference or incompatibility between total benefit and total sacrifice. Therefore, if the total sacrifice is higher, the perceived value of hospital services will be low, whereas small sacrifices with high benefits will lead to high patient satisfaction [8].

Trust represents the reliability, integrity, intensity, and attitude of the hospital felt by the patient [9, 10]. Trust becomes one of the factors affecting patient loyalty to hospitals [10, 11, 8]. Patient satisfaction is considered increasingly important and seen as a vital component of hospital services [12]. Patient satisfaction will lead to patient loyalty [13], as patients tend to re-visit [14] and give word-of-mouth [15]. In addition, patient satisfaction will also enhance the reputation and image of the hospital [14].

There are four types of behaviors that indicate customer loyalty [16], i.e. making a regular repetitive purchase, buying inter-product lines and services, referring to others, and cannot be influenced by competitors to move. Loyalty represents the trustworthiness of the customers [17]. On the other hand, disloyal patients will very easily move to other health services and spread the negative news to others [18].

2.1 Structural Equation Modeling (SEM)

Structural Equation Modeling (SEM) is a multivariate analysis approach used to test and estimate the causal relationship between variables, even when the relationship is hypothetical or not directly observable. SEM is considered to have advantages over other methods, such as multiple regression [19]. The SEM method consists of five main stages [20], namely the stages of specification, identification, estimation, model test, and re-specification.

3 Research Methodology

3.1 Development of Initial Research Model

At the specification stage, a literature study was conducted to find out the latent variables that influence patient loyalty and its indicators. Then, latent variables and indicators are assessed by five experts. From the results of the assessment of five experts who then averaged using the geometric mean, four latent variables affecting loyalty were obtained, which are service quality, perceived value, satisfaction, and trust.

From the results of the expert assessment, there are eight initial hypothesizes to be tested in this research, namely:
H1: Service quality affects the perceived value
H2: Quality of service affects the patient's trust
H3: Quality of service affects patient satisfaction
H4: Service quality affects patient loyalty
H5: Perceived value affects patient satisfaction
H6: Patient confidence affects patient satisfaction
H7: Patient confidence affects patient loyalty
H8: Patient satisfaction affects patient's trust
H9: Patient satisfaction affects patient loyalty

Then, indicators of each latent variable, 18 indicators for service quality, 4 indicators for trust, 3 indicators for patient satisfaction, 5 indicators for patient loyalty were obtained based on the expert's assessment.

3.2 Development of Questionnaire

The questionnaire used in this study is divided into two main parts, wherein the first part is a question related to the identity of the patient and the second part is statements that state the level of agreement on the latent variables. The questionnaire consisted of 32 indicator variables. A 5-point Likert scale was used to measure the level of respondent's approval of questions related to the latent variables used in the study, ranging from strongly disagree to strongly agree. A pilot test was carried out to 40 patients to check the validity and reliability of the questionnaire and to ensure that the use of words and sentences in the questionnaire was accurate and easy to understand.

3.3 Sample and Data Collection

The population of this study was patients from Pasar Rebo Public State Hospital located in Jakarta. Samples from this study were outpatients using BPJS and had visited the same doctor at least three times. Patient selection is done randomly. To be able to meet the adequacy of data, at least five times from the number of indicators, which is 150 respondents, conducted data collection to 400 patients. Respondents consisted of 295 women (74%) and 105 men (26%).

4 Results and Discussion

4.1 Measurement Model

Previously, measurements of the degree of freedom and normality of data were performed. A total of 32 questions used (variables studied) were divided into five latent variables, each latent variable has a good level of validity with a standard loading factor that ranges from 0.539 to 0.832 and a good level of reliability with CR that ranges from 0.79 to 0.94. A construct model has good reliability if its construct reliability (CR) is more than 0.7 and its variance extracted (VE) value is more than 0.5.

4.2 Structural Model

Previous research recommends reporting some model fit indicators, which represent at least three categories, absolute fit, model comparison, and parsimonious fit [20]. The goodness of fit test that is considered in absolute fit indices is Root Mean Square Error Approximation (RMSEA) and Standardized Root Mean Square Residual (SRMSR). The Incremental fit measure can be estimated from the NFI (Normed Fit Index) and CFI (Comparative Fit Index) The parameters that can be used for parsimonious are CFI and RMSEA. The results of the model goodness of fit test and the standard of each parameter [21] is acceptable structural model that can be seen in Table 1.

Table 1. The goodness of Fit Indices of Model

GOF Criteria	Fit Criteria	Result	
RMSEA	< 0.08	0.070	Fit
SRMR	< 0.08	0.015	Fit
CFI	> 0.90	0.901	Fit
NFI	> 0.80	0.857	Fit
CMIN/DF	< 3	2.928	Fit

4.3 Causal Relationship

Once an acceptable structural model is formed, it is necessary to evaluate the statistical significance of SEM. By looking at the value of the p-value of the regression weight calculation, it can be seen whether the relationship between latent variables is significant or not, between exogenous variables and endogenous variables. If p-value <0.05, it can be said there is a significant relationship between latent variables. In Table 2., the results show the calculation of the model regression weight.

Table 2. Summary of Hypothesis Testing Results

			Standardized Regression Weight	P	
Perceived Value	←	Service Quality	0.751	***	Significant
Trust	←	Service Quality	0.111	0.309	Not Significant
Satisfaction	←	Perceived Value	0.696	***	Significant
Satisfaction	←	Service Quality	0.524	***	Significant
Loyalty	←	Trust	0.403	***	Significant
Loyalty	←	Satisfaction	0.599	***	Significant
Loyalty	←	Service Quality	-0.053	0.502	Not Significant
Satisfaction	←	Trust	-0.185	0.337	Not Significant
Trust	←	Satisfaction	0.835	***	Significant

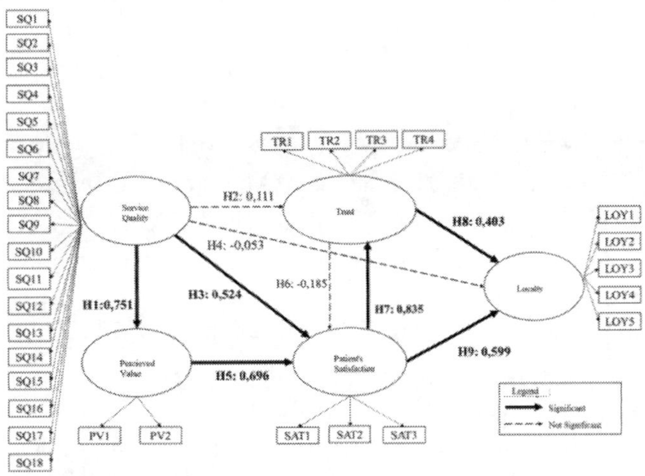

Fig. 1. The Estimated Structural Model

From the calculation above, it is known that the ρ-value marked with three stars is significant because ρ-value is less than 0.001 or in other words to the strong relationship between the two variables and the hypothesis accepted. Loyalty model can be seen in Fig.1. Out of the 9 hypothesizes testing conducted, it is known that 6 hypothesizes are accepted on the model. Service quality and percieved value greatly affect patiens' satisfaction. patients' satisfaction greatly affects trust. Trust and patients' satisfaction greatly affect loyalty of patients.

Based on Fig. 1, it can be seen that satisfaction and trust are factors that have a direct relationship to patient loyalty in hospitals. Satisfaction does have a greater impact on loyalty than trust. Satisfaction becomes the factor that connects service quality [22] and perceived value [8] to loyalty. To improve satisfaction, the public hospital needs to improve the indicator of perceived value, trust, and quality service.

5. Conclusion

This research obtained 5 latent variables and 32 indicator variables in patient loyalty model in public hospital. Factors that have the most important effect on patient loyalty is the service quality. Satisfaction and trust are two strong factors that affect the loyalty of patients. To gain the patient's trust, management can focus on several factors such as reliability, integrity, intensity, and attitude of doctors. To improve patient satisfaction, the public hospital needs to improve their effort in perceived value, trust, and quality service.

This research funded by Universitas Indonesia – PITTA 2017

References

1. Ernst &Young, (2015).
2. Parasuraman, V. A. Zeithaml, and L. L. Berry, J. of Mkt. **49**, 41-50 (1985).
3. M. Dachyar and L. Hananto, Mgmt. and Tech. K., Svc., T. & Hosp. (2014).
4. Grönroos, Int. J. of Ops & Prod. Mgmt., 9-19 (1988).
5. S.-J. Yoon, D.-C. Choi and J.-W. Park, The Svc. Ind. J., 371-388 (2007).
6. V. A. Zeithaml, L. L. Berry, and A. Parasuraman, J. of Mkt, 35-48 (1988).
7. K. B. Monroe, *Making Profitable Decisions* (1990).
8. M. A. Moliner, J. of Sci. Mgmt., 76-97 (2009).
9. M. J. Sanchez-Franco, A. F. V. Ramos and F. A. M. Velicia, Info. & Mgmt, 196-202 (2009).
10. R. M. Morgan and S. D. Hunt, J. of Mkt., 20-38 (1994).
11. J. Singh and D. Sirdeshmukh, J. of the Ac. of Mkt. Sci., 150-167 (2000).
12. R. N. Jandavath and A. Byram, Int. J. of Pharm. and H. C. Mkt. (2016).
13. Anbori, S. N. Ghani, H. Yadav, A. M. Daher and T. T. Su, Int. J. for Q. in H. C., 310–315 (2010).
14. S. M. Lee, D. Lee, and C.-Y. Kang, The Svc. Ind. J., 17-36 (2012).
15. R. E. Herzlinger, *Harvard Business Review*, 57 (2006).
16. D. L. Stum and A. Thiry, T. & Dev. J., 34-36 (1991).
17. M. Dachyar and Fatkhurrohman, W. Ac. of Sci., Engr. and Tech., 845 - 849 (2011).
18. C.-W. Chang, T.-H. Tseng and A. G. Woodside, J. of Svcs. Mkt., 91-103 (2013).
19. C. B. Astrachan, V. K. Patel, and G. Wanzenried, J. of F. Bus. Str., 116-128 (2014).
20. M. S. Khine, *Application of Structural Equation Modeling in Educational Research and Practice* (2013).
21. D. Hooper, J. Coughlan, and M. Mullen, The Elect. J. of Bus. Res. Mtd, 53-60 (2009).
22. M. G. Mbuthia and E. Thaddaeus, Op. Acc. Lib. J. **2** (2015).

MATEC Web of Conferences **248**, 03016 (2018) https://doi.org/10.1051/matecconf/201824803016
ESTIC 2018

Optimization of Production Planning Using Goal Programming Approach at Chocolate Factory

Noviyarsi.[1,*], *Lestari* Setiawati[1] , and *Arhamna* Arisya[1]

[1]Industrial Engineering Department, Bung Hatta University, Padang, Indonesia

Abstract. Company X is a mini chocolate factory in West Sumatera Province. The Company produces three types of product, that is paralin chocolate, bar chocolate 75 gr and bar chocolate 20 gr. Today, production planning only based on customer demand without any analysis an optimized amount of production to maximize profit. The purposed of this research was the optimization of production planning to maximize company profit. Goal programming approach was used in this research. There were four goals that need to optimize, that are, production volume, profit, processing time and raw material. Horizon plan was set up for six periods. LINDO software was used to compute an optimized result of goal programming. The final result shows an average of production amount for six periods, that were paralin chocolate about 65,95 kg, bar chocolate 75 gr about 883 bars and bar chocolate 20 gr about 3725 bars with maximum profit about Rp. 17.806.782,-

1 Introduction

Production planning is related to the planning of the type and quantity of products to be produced by the company. Production planning aims to achieve production and labor stability against demand fluctuations. Company X is a small industry engaged in the processing of chocolate. There are 3 products produced are paralin chocolate, 75 gram chocolate bar and 20 gram chocolate bar. At this time, the determination of production quantities and types of products to be produced only based on demand. The problem faced by the company was cannot fulfill customer demand. The company had never done production planning to predict the availability of raw materials, labor and time to meet consumer demand. It does not affect the optimal level of production that can maximize profits. This is due to the availability of cocoa stock which is still dependent on the cocoa harvest season, the imbalance in the capacity of the cocoa processing machine and the productivity of the employee working time.

Goal programming is one mathematical model that can be used in optimizing production planning with multiple goals. Goal Programming is one of the mathematical models that is viewed in accordance with multi-purpose problem solving because through its deviation variable, goal programming automatically captures information about the relative achievement of the objectives [1]. The only difference lies in the presence of a pair of deviation variables that appear on the purpose function and constraint functions [2]. This

* Corresponding author: lestarisetiawati@bunghatta.ac.id

model requires various inputs (inputs) from existing production systems in the factory to support the decision to be generated. Therefore goal programming is suitable to solve production planning problems at the chocolate factory. Based on this, the purpose of this research was to optimize the production to produce maximum profit.

2 Reseacrh Methodology

2.1 Data Collection

The data collected in the form of primary data and secondary data. Primary data is obtained by doing self-observation while secondary data obtained from company document. The data required in this studied were production data, production cost data, processing time data and demand data.

2.2 Aggregation

Aggregation had been done to facilitate forecasting production. This is because the products produced by the factory used the same raw materials, production processes, machinery, and production equipment. The resulting chocolate product is distinguished only on the basis of the mold and the weight of the product. The aggregate calculation of chocolate products had been done by combining the product based on the unit of measure or weight of each product.

2.3 Forecasting

Forecasting was used to predict amount of demand in the future. Forecasting had been done using *Win QSB* software for six periods.

2.4 Disaggregation

The disaggregation calculation is performed to determine the production forecasting result of each chocolate product. The disaggregation calculation is done by percentage approach by multiplying the forecast result by the proportion of each product.

2.4 Goal Programming

In preparing a goal programming model must first determine the decision variables of the optimization model, then formulate the function constraints and objective functions to be achieved. In the goal programming model there is a deviation variable in the constraint function. These variables serve to accommodate the deviation of the completion of the objectives to be achieved. In the process of processing the model, the number of devisional variables will be minimized in the objective function [3].

2.4.1 Decision Variables

X_{it} : Amount of production product-i at period-t
i : type of product. $i = 1,2,3$
t : production period. $i = 1,2,3,4,5,6$

2.4.2 Constraints

a. Production volume
b. Profit
c. Processing time capacity
d. Raw material availability

2.4.3 Goal Programming Formulation

Min $Z = (d_1^+ + d_1^- + d_2^+ + d_2^- + \dots + dn^+ + d_n^-)$

3 Result and Conclusion

3.1 Aggregation

Table 1 shows result of aggregation from three types of chocolate product.

Table 1. Product Aggregation

Période	Produc			Agrégat			Sum Agrégat (gram)	Sum Agrégat (kg)
	Pralin (kg)	Bar 75 gr (unit)	bar 20 gr (unit)	Collât Pralin (gram)	Collât Batang 75 gr (gram)	Collât Batang 20 gr (gram)		
1	81	582	2182	81000	43650	43640	168290	169
2	91	657	2466	91000	49275	49320	189595	190
3	91	560	2101	91000	42000	42020	175020	176
4	84	607	2277	84000	45525	45540	175020	176
5	104	752	2819	104000	56400	56380	216780	217
6	110	789	2961	110000	59175	59220	228395	229
Total	562	3948	14804	561000	296025	296120	1153245	1157

3.2 Forecasting Result

Table 2 shows forecasting result for 6 period for each product. Forecasting had been done using *Win QSB* software.

Table 2. Forecasting Result for Product

Periods	Praline	Bar 75 gram	Bar 20 gram
1	111,15	799	2995
2	116,06	838	3146
3	131,39	809	3034
4	126,48	914	3429
5	131,68	953	3570
6	137,44	986	3700

3.2 Goal Programming Formulation

Min $Z = (d_1^+ + d_1^- + d_2^+ + d_2^- + d_3^+ + d_3^- + d_4^+ + d_4^- + d_5^+ + d_5^- + d_6^+ + d_6^- + d_7^+ + d_7^- + d_8^+ +$
$d_8^- + d_9^+ + d_9^- + d_{10}^+ + d_{10}^- + d_{11}^+ + d_{11}^- + d_{12}^+ + d_{12}^- + d_{13}^+ + d_{13}^- + d_{14}^+ + d_{14}^- + d_{15}^+ + d_{15}^- +$

$d_{16}^+ + d_{16}^- + d_{17}^+ + d_{17}^- + d_{18}^+ + d_{18}^-) + (d_{19}^- + d_{20}^- + d_{21}^- + d_{22}^- + d_{23}^- + d_{24}^-) + (d_{25}^+ + d_{25}^- + d_{26}^+ + d_{26}^- + d_{27}^+ + d_{27}^- + d_{28}^+ + d_{28}^- + d_{29}^+ + d_{29}^- + d_{30}^+ + d_{30}^- + d_{31}^+ + d_{31}^- + d_{32}^+ + d_{32}^- + d_{33}^+ + d_{33}^- + d_{34}^+ + d_{34}^- + d_{35}^+ + d_{35}^- + d_{36}^+ + d_{36}^-) + (d_{37}^+ + d_{37}^- + d_{38}^+ + d_{38}^- + d_{39}^+ + d_{39}^- + d_{40}^+ + d_{40}^- + d_{41}^+ + d_{41}^- + d_{42}^+ + d_{42}^- + d_{43}^+ + d_{43}^- + d_{44}^+ + d_{44}^- + d_{45}^+ + d_{45}^- + d_{46}^+ + d_{46}^- + d_{47}^+ + d_{47}^- + d_{48}^+ + d_{48}^- + d_{49}^+ + d_{49}^- + d_{50}^+ + d_{50}^- + d_{51}^+ + d_{51}^- + d_{52}^+ + d_{52}^- + d_{53}^+ + d_{53}^- + d_{54}^+ + d_{54}^- + d_{55}^+ + d_{55}^- + d_{56}^+ + d_{56}^- + d_{57}^+ + d_{57}^- + d_{58}^+ + d_{58}^- + d_{59}^+ + d_{59}^- + d_{60}^+ + d_{60}^- + d_{61}^+ + d_{61}^- + d_{62}^+ + d_{62}^- + d_{63}^+ + d_{63}^- + d_{64}^+ + d_{64}^- + d_{65}^+ + d_{65}^- + d_{66}^+ + d_{66}^-)$

Constraint :

1. Demand $$\sum_{i=1}^{t}\sum_{j=1}^{n}(d_1^- - d_1^+ = D_{ij})$$
2. Profit $$\sum_{i=1}^{t}(X_i + d_1^- - d_1^+ = P_i)$$
3. Processing time $$\sum_{i=1}^{t}(X_i + d_1^- - d_1^+ = T_i)$$
4. Cocoa $$\sum_{i=1}^{t}\sum_{j=1}^{n}(X_{ij} + d_1^- - d_1^+ = C_j)$$
5. Sugar $$\sum_{i=1}^{t}\sum_{j=1}^{n}(X_{ij} + d_1^- - d_1^+ = S_j)$$
6. Milk $$\sum_{i=1}^{t}\sum_{j=1}^{n}(X_{ij} + d_1^- - d_1^+ = M_j)$$
7. Lestin $$\sum_{i=1}^{t}\sum_{j=1}^{n}(X_{ij} + d_1^- - d_1^+ = L_j)$$

Where :

i	: time period
j	: number of product
X	: Production Plan
d_1^-	: deviation variable if performance below production target
d_1^-	: deviaton variable if performance above production target
D_{ij}	: Demand product j at period i
P_i	: Total profit at period i
T_i	: Total processing time at period i

C_i : Total Cocoa at period i

S_i : Total Sugar at period i

M_i : Total Milk at period i

L_i : Total Lestin at period i

The solution of the goal programming model shows that for the optimum fulfillment of chocolate demand of chocolate in periods 4, 5 and 6. In brown stems 75gr per optimum period and chocolate stems 20gr per period is not optimal, the average optimization of paralin chocolate production is 65 , 93 kg, 75 kg stem chocolate as much as 883 stems and 20 gr bar chocolate as much as 3725 stems. The optimal advantage that can be achieved from the goal programming model solution is an average of Rp. 17,806,782 per period. For the utilization rate of time consuming process will be optimal with average optimization of 230.3 hours per period. For the consumption level of brown fat, brown paste, sugar, milk, lecithin and soda as a whole will be below the level of availability and above the availability level or there is excess stock and shortage of raw material stock. From the average goal programming model optimization solution for the use of brown fat raw material as much as 55,80 kg, chocolate pasta 51,66 kg, sugar 51,66 kg, milk 45,47 kg, lecithin 0,91 kg and soda as much as 0.14 kg.

References

1. Charles, *Optimization Modelling With Lingo*, Lindo System, Inc, Chicago, (2006)
2. Siswanto, A, *Goal Programming dengan Menggunakan LINDO*, Gramedia, Jakarta, (2007)
3. Anis, M. Nandiroh, S. dan Utami, A.D, Optimasi Perencanaan Produksi Dengan Metode Goal Programming, *Jurnal Ilmiah Teknik Industri.* Vol.5, No.3: page 133-143 (2007)

Quality Improvement of Industrial Products Bungo Mayang to Customer Satisfaction with The Kano Model Approach

Yesmizarti Muchtiar[1,*], *Dessi* Mufti[1], *Vil* Yosri[1]

[1]Industrial Engineering Bung Hatta University, Padang, Indonesia

Abstract. The abundance of the home industry appeared in the Batusangkar requires home industry Bungo Mayang should retain the factors that affect the quality of the product so that it can continue to exist and remains sought after by the public. This research aims to identify the attribute customer needs to improve product quality 'kue karambia' Bungo Mayang and classifying the attributes customer needs based on model of Kano. The instrument used was a questionnaire functional and dysfunctional questionnaires grouped into attributes of the Kano must-be, one dimensional, attractive, indifferent, quesionable result, and reverse. From the results of the calculation model Kano note that attributes have the highest grade IE product quality is good (O = 66), the attractive packaging (O = 61), availability of product information (O = 56), and compliance rates (O = 43).

1 Introduction

Among many traditional cakes produced by Bungo Mayang Industry, a home industry originated from Batusangkar, 'kue karambia' is one of the products which are very less in demand. This is due to the packaging which is only using plain plastic and there is only one original coconut flavor. In order to keep exist in the market; it is of vital importance that product should fits the customer needs. Satisfied customer will be loyal to the produced product.

Customer satisfaction is a determining factor for the success of a business. Hence, it is important for the businessman to keep a continuation on the improvement of the product to win the customer. The businessman has to always understand the needs and the willing of the customer. One of the methods to get to know the level of customer satisfaction is through Kano model. Kano model is a diagram that shares to the customer the satisfaction of the product. It is also a theory to reveal the satisfaction influencing factors of a product or a service, such as taste which is suitable with the customer needs. final printed area will be 130 x 210 mm. Do not add any page numbers.

* Corresponding author: yesmizartimuchtiar@bunghatta.ac.id

2 Literature Review

2.1 Customer Satisfaction

Today marketing concept is one of the management orientations which assume that "customer is the king". Customer is placed in the main position to be served (Adinata, 2013). This marketing concept is implemented to enable the company to provide a more effective and efficient satisfaction to the customer compared to the competitor.
Tjiptono (2002) stated that mostly customer refer to many factors and dimensions in evaluating their satisfaction toward the product, service or certain company.

2.2 Kano Concept

KANO model developed by Prof. Noriaki Kano (1984) is a model which aimed to categorized attributes of the product or service based on its ability to satisfy customer needs.

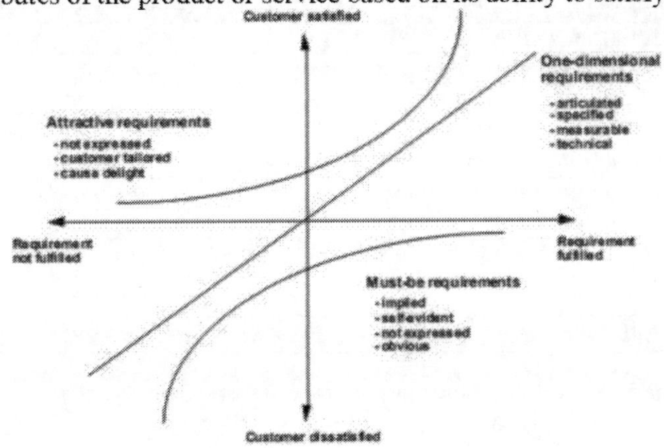

Fig.1. Kano Diagram

Table 1. KANO Method Evaluation Table

Customer Needs ↓ →		Dysfunctional Questions (negative)				
		1. Like	2. Must-be	3. Neutral	4. May	5. Dislike
Functional Question (Positive)	1. Like	Q	A	A	A	O
	2. Must-be	R	I	I	I	M
	3. Neutral	R	I	I	I	M
	4. May	R	I	I	I	M
	5. Dislike	R	R	R	R	Q

This table allowing conclusion drawing on whether the needs of customers can be categorized into one of the following:
a. A = Attractive
b. M = Must-be
c. O = One-Dimensional
d. R = Reverse
e. Q = Questionable
f. I = Indifferent

2.3 Kano Stages

This research was resulted from the interview and filled questionnaire in the form of customer needs characteristic toward quality performed by Bungo Mayang Industry.

3 Result and Discussion

3.1 Kano Model Method

The determination of Kano category can be obtained from the questionnaire distribution. The product and service concepts, which consist of Core Product, Actual Product, and Extended Product, produce result mapping as can be seen on table 2.

Table 2. Kano Attribute Based on the Product and Service Concepts

No	Attribute	Kano Concept
1	Attractive packaging	*Core Product*
2	The availability of product information	*Extended Product*
3	The availability of color variation	*Core Product*
4	Product size suitability	*Core Product*
5	The availability of product flavor variation	*Core Product*
6	Famous brand	*Actual Product*
7	The uniqueness of a product	*Actual Product*
8	Good quality product	*Actual Product*
9	Price suitability	*Actual Product*
10	Attractive Logo	*Core Product*

The functional and dysfunctional questionnaires were employed to conduct calculation whichis based on KANO evaluation table. Attribute categorization is implemented to see the influence of each attribute toward the customer satisfaction.

- *One Dimensional* (O)

It is a very important attribute categorization to put into priority by Bungo Mayang Industry. This is due to the level of satisfaction is in linear correlation with the attribute performance. Factors including in this category can be seen on table 3.

Table 3. One Dimensional Factors

No	Attribute	*Grade*
1	Attractive packaging	O
2	The availability of product information	O
3	Good quality product	O
4	Price suitability	O

- *Indifferent* (I)

It is not a crucial attribute categorization for the customer. Therefore, its availability will not influence the improvement and decline of the customer satisfaction level.
Factors including in this category can be seen on table 4.

Table 4. Indifferent Factors

No	Attribute	*Grade*
1	The availability of color variation	I
2	Product size suitability	I
3	The availability of product color variation	I
4	Famous brand	I
5	The uniqueness of a product	I
6	Attractive Logo	I

3.2 Kano Method Assessment

This stage implemented the calculation of better and worse values. Better value indicates the improvement level of customer satisfaction on the availability of A&O feature. Meanwhile worse indicates the decline of the customer satisfaction on the unavailability of O&M feature. The following is the formulation to calculate better and worse:

$$Better = \frac{A + O}{A + O + M + I} \quad \ldots\ldots\ldots\ldots\ldots\ldots\ldots\ldots\ldots (1)$$

$$Worse = \frac{A + M}{A + O + M + I} \quad \ldots\ldots\ldots\ldots\ldots\ldots\ldots\ldots\ldots (2)$$

Table 5. Better and Worse Values Recapitulation

Variable	A	M	O	I	*Better*	*Worse*
1	17	14	61	9	0,77	0,74
2	10	18	56	15	0,66	0,74
3	26	10	15	45	0,42	0,26
4	17	17	26	35	0,45	0,45
5	14	12	10	51	0,27	0,25
6	20	12	10	52	0,31	0,23
7	18	15	7	57	0,25	0,22
8	4	15	66	14	0,70	0,81
9	8	20	43	25	0,53	0,60
10	27	10	24	35	0,53	0,35

Considering the customer expectation toward the product of Bungo Mayang Industry, the attributes that need to keep improved are as follow:

1. Attractive packaging
2. The availability of product information
3. Good quality product
4. Price suitability

4 Conclusion

Based on research conducted, it can be concluded that:

1. The calculation result of KANO model reveals respondent answer assessment with KANO diagram, it is known that the biggest interpretation only exist at O and I categories, as follow:
2. Based on the calculation of better and worse values, it is known that attributes which influence the customer satisfaction improvement level are as follow:

- Attractive packaging, where customer satisfaction level reaches 77 %.
- Good quality product, where customer satisfaction level reaches 70 % only if Bungo Mayang Industry can fulfill the requirement.
- The availability of product information, where customer satisfaction level reaches 66%.
- Price suitability, where customer satisfaction level reaches 53% only if the requirement can be fulfilled.

References

1. Nofirza and Indrayani, K. Aplikasi Metode Kano dalam Analisis Indikator Kualitas Pelayanan di Rumah Sakit Arifin Ahmad Pekanbaru. Jurnal Sains, Teknologi, dan Industri. **9(1). 1-8**.(2011)
2. Noriaki, Kano. Attractive Quality and Must-be Quality. The Journal of the Japanese Society for Quality Control. (1984)
3. Rangkuti, Fendy. Measuring Customer Satisfaction. Jakarta. PT. Gramedia Pustaka Utama (2000)
4. Tjiptono, Fendi. Prinsip-Prinsip Total Quality Service. Yogyakarta. Andi Offset (1997)
5. Tjiptono, Fendi. Service Quality & Satisfaction. Yogyakarta. Andi Offset (2008)

Scale Up Production Indonesian Liquid Propolis From Raw Propolis and Wild Beehive Using Bubbling Vacuum Evaporator

Muhamad Sahlan[1,2*], *Andhika* Akhmariadi[1], *Diah Kartika* Pratami[3], *Heri* Hermansyah[1], and *Anondho* Wijanarko[1]

[1]Department of Chemical Engineering, Faculty of Engineering, Universitas Indonesia, Depok, 16425, West java, Indonesia
[2]Research Center of Biomedical Engineering, Faculty of Engineering, Universitas Indonesia, Depok, 16425, West java, Indonesia
[3]Department of Pharmacognosy and Phytochemistry, Faculty of Pharmacy, Pancasila University. South Jakarta, 12640, DKI Jakarta, Indonesia

Abstract. The process of production propolis on a laboratory scale using a rotary evaporator with capacity one liter of ethanol extract propolis include propylene glycol as solvent medium. Ethanol is evaporated at a temperature of 65°C and reduced water content at 80°C. For the quantity of products can be increased by increasing the capacity of the ethanol extract of propolis and the medium while solvent evaporation to 12 liters and modify the rotary system with a bubbling system (bubbling vacuum evaporator). Then the solvent medium used was replaced with glycerol. The result of 8.02 kg of raw propolis produced 10.85 liters with the average of total flavonoid content 914.8 mg/l, while for 8 kg of raw material forest beehive produced 7.65 liters with the average of total flavonoid content 307.2 mg/l. With the capacity production average reaches one liter per day. The resulting product by organoleptically sweet taste, no smell and distinctive of saliva and the color of the product are brown.

1 Introduction

Products derived from bees include honey, pollen, royal jelly, propolis, and beeswax. One of the bee products that have high economic value and health benefits are high for propolis. Propolis or bee glue, a substance produced by bees and serve to patch up and sterilize the hive [1]. The content of bioactive propolis different in each region depends on the location where the bees live [2,3]. Contained bioactive flavonoids in propolis dominated, with levels nearly 50%. Flavonoids are organic compounds that function as an antibacterial and anticancer, ferulic acid serves as an antibiotic while functioning as an antiviral terpenoid [4-6] .

From the results of field surveys in the fact that propolis is obtained for the majority of these imports on the market, while Indonesia has the potential of propolis that is not less

* Corresponding author: sahlan@che.ui.ac.id

great. Among beekeepers were still a few who take advantage of using propolis trap [7- 8]. Propolis obtained by extraction of raw propolis, or beehives, which are used for the extraction solvent is ethanol solution; the result is called the extraction of ethanol extract of propolis (EEP) [9]. Propolis wax contained in the ethanol extract of propolis are separated by the addition of water [10,11]. After that extract plus solvent medium for the vacuum distillation process is then performed using a vacuum evaporator bubbling to separate the alcohol and reduce the amount of water in the extract. Separation process in the production of propolis ethanol solvent Indonesia using a rotary evaporator before the capacity of one liter with propylene glycol solvent media. Production capacity by using the tool to only one liter per week and similar products such as propolis produced in the market (propylene glycol base). In the present study was carried out modifications to the tool rotary evaporator, by enlarging the capacity of the extract and replace the rotary system to bubble system. Besides the solvent media used is changed from the propylene glycol to glycerol. Thus propolis produced sweet and odorless typical saliva.

2 Material and Methods

2.1 Material

Raw propolis obtained from bee farms in Jakarta, while the honeycomb obtained from the Sumatran jungle. Auxiliary raw materials such as 96% ethanol, glycerol, and distilled water obtained from the supplier of industrial chemicals.

2.2 Propolis Extraction

The propolis extraction from beehive using Sahlan et al. methods [12]. A total of two kilograms of raw materials are weighed, then crushed materials first become a much smaller size. Ethanol plus 96% as much as 10 liters and stirred using an overhead stirrer at 150 rpm for eight hours. Extract and then allowed to stand overnight, then filtered using a Buchner funnel which has been coated filter paper (pore diameter, ten μm), performed with the aid of vacuum filtration of the water jet pump. The filtrate is called ethanol extract propolis (EEP) 96%, while the cake is called resin.

2.3 Separation Wax Propolis

The EEP 96% diluted by the addition of distilled water until the concentration of ethanol to 70%, so that propolis wax deposits formed in a solution of 70% EEP. Before precipitate filtered propolis wax, wax deposition solution of propolis extract and advance in aging at 50 ° C for 30 minutes using a water bath, and then be continued stored in a freezer at a temperature of -4 ° C for overnight. After precipitating of propolis wax filtered with a Buchner funnel. The filtrate pH (EEP 70%) was raised to 6.4 by the addition of 5% Na_2CO_3.

2.4 Distillation Solvent using Bubbling Vacuum Evaporator

EEP as much as 10 liters of 70% plus two liters of glycerol into the bubbling Vacuum Evaporator, is distilled at a temperature of 65 ° C until ethanol stops dripping on the bottom of the condenser, then the temperature was increased to 80 ° C until the product becomes viscous.

3 Result and Discussion

3.1 Propolis extraction

Propolis raw material extraction is done by maceration method, which is one method for the extraction of ingredients that can not stand the heat, that is by soaking the material with a specific solvent and in a certain period. Solvents used are ethanol 96% to extract more robust use of dense concentration. The resin produced from raw material forest honeycomb can be processed into beeswax by heating above 100°C until the wax melted and filtered to remove impurities. A resin derived from the raw propolis cannot be recycled as resin from beehives forest, but it is very likely to be other processed as Scrubb [13,14].

3.2 Separation wax propolis

After the filtration process, a solution of EEP combined with distilled water until the concentration of ethanol to 70%. The purpose of this dilution is to separate the wax propolis contained in raw materials. From the results of scanning the degree of separation of wax propolis, the optimal concentration for the separation of wax propolis ethanol concentration was 70% [10]. When added to water extract solution became turbid due to precipitate formed propolis wax. EEP solution then aging at 50 °C for 30 minutes, then continued to be stored in the freezer overnight to achieve complete separation of wax. Propolis wax precipitate which was originally floated and separated into fused after the process. A solution of ethanol extract of propolis that has been free of wax propolis increased the pH to 6.4 by addition of Na_2CO_3 5% of its original pH of about 5. Indonesia pH of water in an average of about 6.4 below pH 7, to obtain a high solubility in water then the pH should be adjusted. Media solvent used is glycerol so that the organoleptic taste sweeter products, such as the brown color of the original extract of propolis and odorless saliva [12].

3.3 Distillation solvent

Distillation process takes 24 hours, at this stage of the separation of the ethanol extract and a reduction in the amount of water added. In the first stage, the temperature is set at 65 ° C, to separate the ethanol found in EEP, so the bioactive propolis ethanol moves to the glycerol. Ethanol separation process carried out for 16 hours or until ethanol stops dripping from the bottom of the condenser. Recycle the ethanol concentration reached only 77%, to be used for extraction is necessary to re-distillation until the concentration reaches 96%. The second stage the temperature was increased to 80 °C, the objective is to reduce water content in the product. The process of reduction of water content takes up to 8 hours.

3.4 Bubbling vacuum evaporator

Bubbling Vacuum Evaporator is distillation equipment used for the separation of ethanol from the extract and a reduction in moisture content is the result of modification of the rotary evaporator and a capacity of 12 liters of solution, which consists of 10 liters of EEP solution and 2 liters of solvent media. Where previously the process of separation of ethanol and reduction of water content using a rotary evaporator done. Ethanol is successful in the recycle less than half of the EEP solution that goes into the tool, as well as a longer distillation time. The production capacity of the rotary evaporator was only one liter per week. Scheme of bubbling vacuum evaporator presented in Figure 1.

3.5 Total flavonoid content (TFC) of propolis from raw propolis

Flavonoids are one of the parameters determining the quality because almost half of the bioactive compounds contained in propolis are flavonoid [6,12,15]. TFC of propolis cattle averaged about 939 mg / l in a solution of 70% EEP and 914 mg / l in the product. From the data, the trend can be concluded that many TFC is successfully extracted, reflected in a solution of 70% EEP. So if we want to increase the levels of TFC in the product, we can reduce the ratio of glycerol solvent media, only a consequence of the amount of product produced less.

The TFC in EEP of 96% and 70% EEP decreased dramatically up to a fifth, due to propolis wax in a solution of 96% EEP flavonoids participate as a measurable, while the TFC in the EEP 70% and the product tends to be stable. Departing from previous experience, to obtain greater flavonoid content in the product, then the ratio should be reduced glycerol solvent media. In the batch 3 and batch 4 of propolis Riau forest amount of solvent glycerol media reduced by half, the result is total flavonoid content in the product increased to two-fold [16,17].

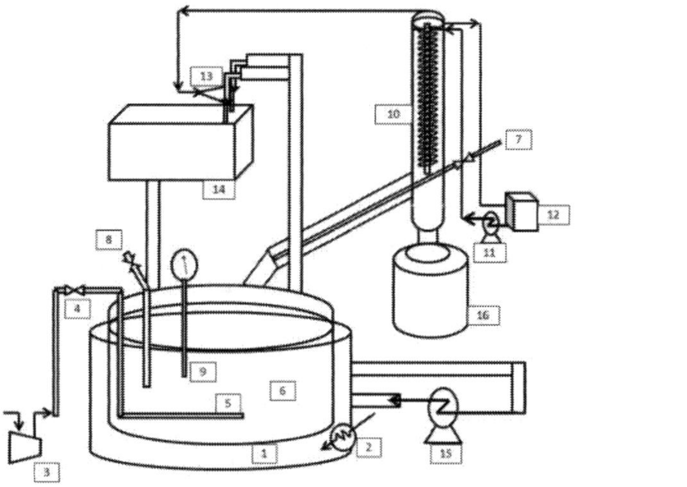

Equipment:
(1) Water Bath,
(2) Thermocouple,
(3) Compressor,
(4) Valve Bubbling,
(5) Bubbler,
(6) Distillation Vessel,
(7) Inlet Extract,
(8) Outlet Product,
(9) Thermometer,
(10) Coil Condensor,
(11) Condensor Pump,
(12) Cooler Sirculation,
(13) Water Jet,
(14) Water Jet Sirculation,
(15) Water Jet Pump,
(16) Distillation Vessel.

Fig. 1. Bubbling Vacum Evaporator Scheme.

3.7. Mass balance of ethanol

Ethanol mass balance results can be seen in Figure 2 below :

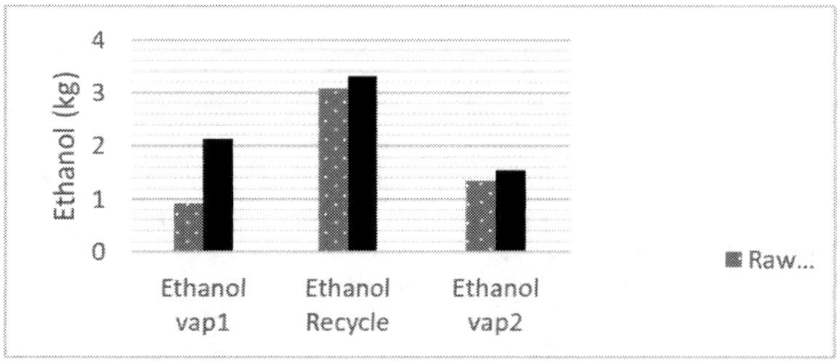

Fig. 2. Mass Balance of Ethanol from Raw Propolis and wild beehive

Mass balance calculations only focus on components that are considered important, such as flavonoids, ethanol, and water. From mass balance calculations when ethanol is added at the beginning of the process shrink by more than 50% during the process. This loss occurs during the process of ethanol extraction for 8 hours, the filtration process, the process of aging at 50 °C for 30 minutes until the distillation process, there is a small portion of ethanol is sucked into the water jet system. So that the fore in order the production process more effective and efficient again is still needed a lot of development. As an example, the process of extraction in order ethanol not too much who evaporate preferably inside it reactor.

4 Conclusion

Production propolis Indonesia from 8.02 kg propolis raw propolis can obtain 10.85 liters and total flavonoid content of an average of 914.8 mg / l and 8.04 kg for propolis from wild bee hives in the woods as much as 7.65 liters total flavonoid content of an average of 307.2 mg / l. Ethanol is lost during the process of achieving an average 50% of ethanol were added to the process. The production process a batch of propolis Indonesia as a whole process takes 24 hours or three working days with the volume of products reach more than three liters when averaged over a production capacity of one liter per day.

This work is supported by Grant of Indexed International Publication of Student Final Project (Publikasi International Terindeks Untuk Tugas Akhir Mahasiswa / PITTA) 2018 funded by DRPM Universitas Indonesia (No. 2486/UN2.R3.1/HKP.05.00/2018).

References

1. M. Simone-Finstrom, R. S. Borba, M. Wilson, and M. Spivak, *Insects* **8**, 2 (2017)
2. M. Simone-Finstrom and M. Spivak, *Apidologie* **41**, 3 (2010)
3. V. Bankova, S. De Castro, and M. Marcucci, *Apidologie* **31**, 1 (2000)
4. B. Trusheva, M. Popova, E. B. Koendhori, I. Tsvetkova, C. Naydenski, and V. Bankova, *Nat. Prod. Res* **25**, 6 (2011)
5. W. Syamsudin, Simajuntak, P, Heffen,WL. *Am. J. Biochem. Biotechnol* **5**, 4 (2009)
6. Syamsudin, *ASIAN J. Chem* **22**, 1 (2010)
7. J. M. Sforcin and V. Bankova, *J. Ethnopharmacol.* **133**, 2 (2011)
8. Mahani, B. Nurhadi, E. Subroto, and M. Herudiyanto, in *Proceeding University Malaysia Terengganu Annual Sciences* (2011)
9. D. K. Pratami, A. Mun, A. Sundowo, and M. Sahlan, *Pharmacogn. J.* **10**, 1 (2018)
10. M. Sahlan and T. Supardi, *Int. J. Pharma Bio Sci.* **4**, 1 (2013)
11. M. Sahlan, D. Dienayati, D. Hamdi, S. Zahra, H. Hermansyah, and M. Chulasiri, *Makara J. Technol.*, **21**, 1 (2017)
12. D. Christina *et al.*, in *AIP Conference Proceedings* **1933, (**2018)
13. M. Sahlan and E. Rohmatin, *Int. J. Pharm. Pharm. Sci.*, **9**, 5 (2017)
14. A. G. Mahadewi *et al.*, *AIP Conf. Proc.*, **1933**, 20005, Feb (2018)
15. M. Funakoshi-Tago *et al.*, *Int. Immunopharmacol.*, (2015)
16. V. Bankova, *J. Ethnopharmacol.*, **100**, 1–2 (2005)
17. C.-C. Chang, M.-H. Yang, and H.-M. Wen, *J. Food Drug Anal.*, **10**, 3 (2002)

Chitin and Chitosan from Green Shell (*Perna Viridis*): Utilization Fisheries Wastes from Traditional Market in Jakarta

Flora Elvistia Firdaus[1*], *Indah* Purnamasari[2], *Pandu* Gunatama[3]

[1,2,3] Chemical Engineering, Jayabaya University, Jl Pulomas Kav 23 Jakarta 13210, Indonesia

Abstract. Many advantages of burden wastes which adversely impact to the environment in the form of solid waste. Green shellfish wastes are used for resources of chitosan. The aim of this works is to identify the effectiveness of chitosan application into 2 kinds of moist foods; wet noodles and meatballs. The concentration of chitosan are 0%, 0.5%, 1%, 1.5%, and 2%. The samples are immersed for 0,15, 30, 45, and 60 minutes. The study has found the water content of chitin is 8.31% and chitosan is 6.83%. The degree of deacetylation of chitin is 37.81% and chitosan is 82%. The best concentration of chitosan for wet noodle is 1.5% -2% with the immersion of 45 -60 minutes, and for meatballs is 1.5%-2% with the immersion of 15 minutes. The organoleptic test, on day-3 noodle and meatball, is still in good condition. If no chitosan added the noodles sample is day-1 is in fair condition, while meatball should be consumed directly because in day-1 is becoming chewy with a sour taste, and on day-2 are overgrown with mushrooms.

1 Introduction

Indonesia is referred to as a country with the largest archipelago. Fisheries is a major source of marine resources for food, industry, and other reliable sources of income. Besides the benefits, it also generates wastes. Green shellfish is the other type of wastes generated from the fisheries sector. The global annual estimates the shellfish processing discarded is around 5.118 x 106 metric ton [1]. The abandoned of the shell is severe with the limit process of wastes which creates damage to the environment, it needs proper management [2]. Exploring its value as an effort of reducing wastes has increased its economic outcome. The mollusk containing chitin can be processed to a broad range of industrial application [3]. The non-toxic and easily degradable of chitin in nature has encouraged the modification of chitin with the aim of optimizing the usefulness as well as expanding the field of chitin application [4]. Chitosan is one of the derivatives of chitin, it is the product of deacetylation. It can be produced in the form of powders, pastes, films, and fibers [5]. The bio-degradable property of the chitosan [6] 6so is certain safely to be consumed.
The property of it's in acidic media provides an antibacterial activity which makes it an ideal polymer for food science [7]. The presence of hydroxyl and amino groups along the chain of

* Corresponding author: flora_elvistia@yahoo.com

polymers in chitosan are effectively adsorbed the cations of organic substances such as proteins and fats. Based on this facts, the chitosan can perform as food preservatives which can inhibit fermentation, acidification or other forms of damages by damaging cell membranes of microorganisms [8].

Chitin and chitosan of Green Shellfish

Green shellfish (*Perna viridis*) lives in tropical seas, especially in coastal waters. The common name of green shellfish in Indonesia is kaung-kaung, kedaung and etc. Chitin of green shellfish is a linear polymer [9]. Chitin is an amorphous or crystalline solid, white, and biodegradable [10], it is soluble in acidic solutions and then form chitosan [9].

Chitosan is a heteropolymer of N-acetyl-D-glucosamine and D-glucosamine, which can be obtained from chitin through partial deacetylation process. Commercial chitosan has deacetylation degree of 2% to 40% [11]. The more acetyl groups lost from the chitosan biopolymer, the stronger the interaction between ion and hydrogen bonding from chitosan [12-13]. The capacity to adsorb chitin and chitosan increases with the increase of free amino group [14]. The most influential factor of chitosan as a food preservative is the number of amines (NH_2) groups.

2 Experimental

The experimental flow diagram for chitosan fabrication is drawn in Figure 1.

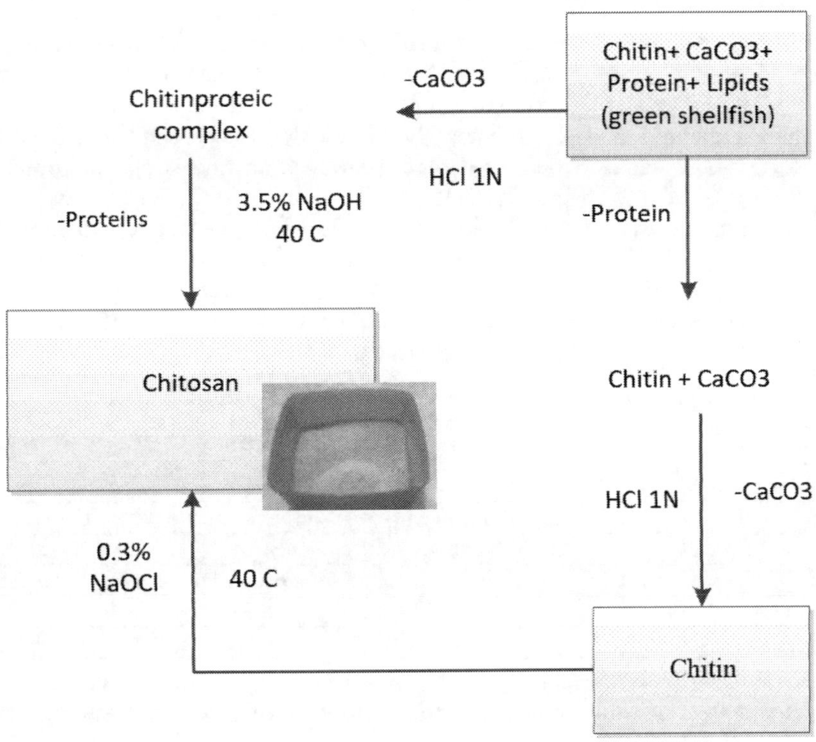

Fig1. Flow Diagram of Chitosan

2.1 Characterization of Chitosan

1.Organoleptic test. Testing includes shape, color, and odor. It performed by the appearance of shell powder, chitin, and chitosan.
2.Moisture Content. AOAC method. Weigh the empty porcelain. Weigh a sample of 1-3 grams in a porcelain. Put in the oven at 100-105^{0}C for 1-2 hours. Cooled for approximately 30 minutes and weighed. Heat again, and cooled until the weight is constant.
3. Measuring the Degree of Deacetylation (DD) by FTIR. The degree of deacetylation is the percentage of acetyl group removed during the chitin deproteination process. The chitin is treated by adding 50% NaOH which causes the hydrolysis of the acetyl group of the chitin acetamide group.

2.2 Immersion of noodles and meatballs into chitosan solution.

The chitosan solution was prepared by blending chitosan powder with acetic acid 2%. The desired chitosan concentration are 0.5%; 1%; 1.5%; and 2% (v/v). Soak the wet noodles and meatballs into chitosan solutions. Immersion is done at the desired time 15; 30; 45; and 60 minutes. Drain the wet noodles and meatballs. Let it stand for room temperature. Conducted the organoleptic test in 0,1,2, and 3 days.

3 Results and Discussion

The crushed green shell is treated by 1N HCl to removed the $CaCO_3$ salt, and continued the process to proteins and lipids removal by 3.5% NaOH, according to Remya the best concentration of NaOH is 5% [15]. The degree of deacetylation is address to differentiate between chitin and chitosan. The depigmentation into chitin has caused to forming the white chitosan powder. Chitosan is a safe ingredient to be used as an inhibitor of microbial activity. This is due to the positively charged amino groups interacting to microbes, so weaken active microbes. The limited solubility of chitosan in many solvents but perfectly soluble in acetic acid.The percentage of acetyl group removed during the extraction process by adding NaOH which causes hydrolysis of an acetyl group from acetamide group. The degree of deacetylation is determined by IR spectrum. The degree of acetylation of chitin is 37.81% which is lower than chitosan 82%, as summarizes in Table 1.

Table 1. Physical characteristics of raw materials and products

Parameter	Shellfish	Chitin	Chitosan
Form	Rough powder	Fine powder	Fine powder
Color	Greyish White	Brown	Pale white
Odor	Fishy	Little Fishy	No smell
Water (%)		8.31	6.83
Degree Deacetylation		37.81	82

The water content of chitosan is 6.83% this is possible of concurrent water molecules as well the nature as hygroscopic. The Standard of water content in chitosan is in the range of 2-10% [6]. In this study, the application of chitosan has been carried out on foods such as wet noodles and meatballs. The indicator used to know that chitosan works effectively in extending food life is from odors. The food stale because preservatives are not working properly. Some initial occurrence is from the odor, texture, and possibly of mushroom growth which is the sign of beginning of decomposition.

In wet noodles, 15 minutes of food immersion is still fair good until day 2. Overall the best concentration is in the range of 1.5-2% with immersion for 45-60 minutes, as can be seen in Figure 2.

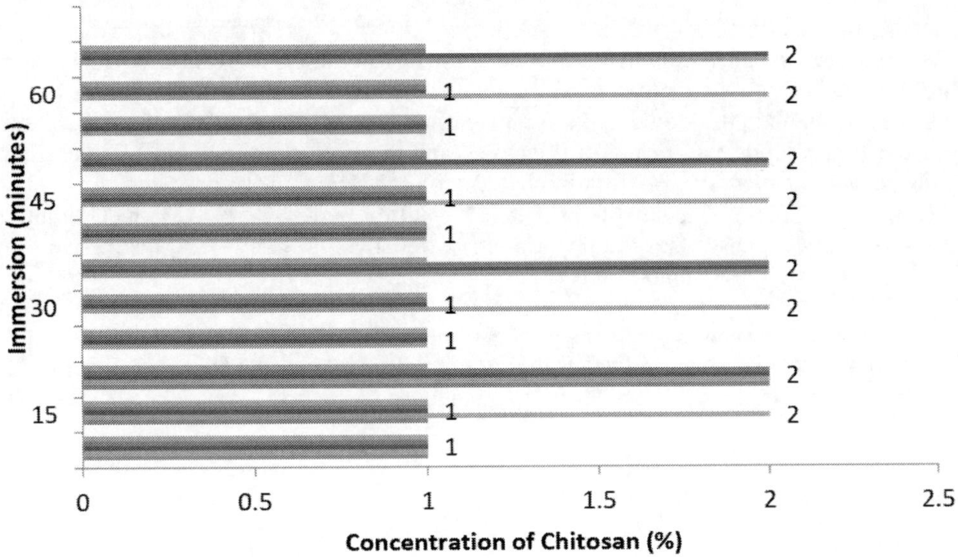

Fig 2. The concentration of chitosan to shelflife of noodle
1: smell flour 2: smell sour

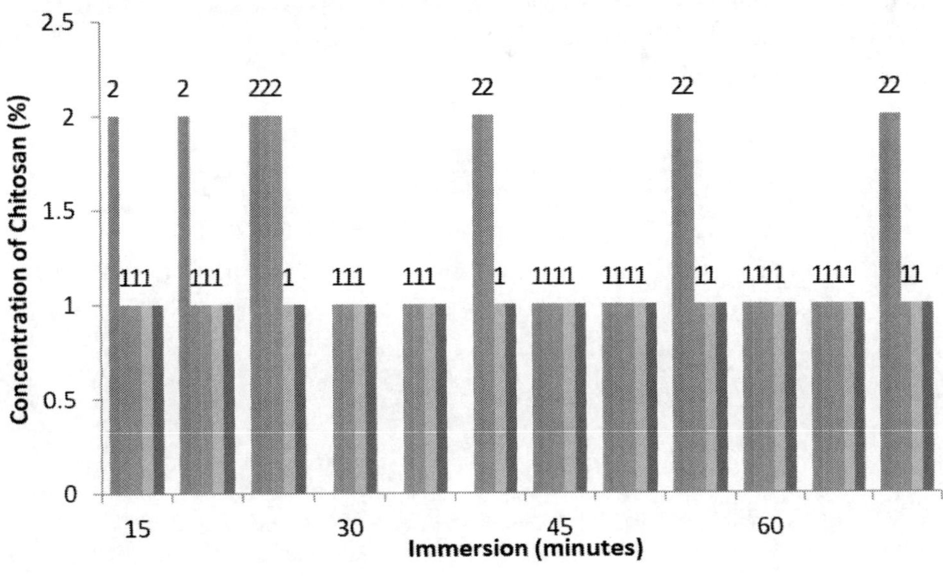

Fig 3. The concentration of Chitosan to Shelflife of meatball
1: smell typical meat 2: smell sour

The sour aroma of meatball will smell on average on the day-3. Meatballs on the immersion day 0.5% - 1% have not effectively preserved and shelflife during storage. Until day-2 the sample still has in the physical characteristics; the chewy and distinctive smell of meat and with no mushroom. The best concentration is 1.5% - 2% in 15 minutes, as can be seen in Figure 3.

4 Conclusions

This study can be concluded :
1. From resource of green shellfish, the degree of deacetylation for Chitin is 37.81% and Chitosan is 82%.
2. The best immersion of noodle in chitosan solution 1.5 % - 2% with 45 – 60 immersion. . From the organoleptic parameters all samples are in good condition until 3 days, with physical characteristics; smells of flour, chewy and not yet encountered moldy.
3. The best immersion of meatballs in chitosan solution best is 1.5% - 2% in 15 minutes. Until day-3 the sample still has the physical characteristics; the chewy and distinctive smell of meat and the sample with no moldy.

The author gratefully acknowledges the support of the Indonesia Ministry of Research and Technology, Department of Higher Education as funding the research

References

1. D,Knorr, Food Technol **38**, 85 (1984)
2. V,Ferraro *et al,* Food Research International, **43**, 221 (2010)
3.JV, Ornum, INFOFISH Internat, **6**, 48 (1992),
4. NK,Hong and SP, Meyers, J. Aquatic Food Prod. Technol, **4**, 27 (1995)
5. PR, Austin *et. al*, Chitin: new facets of research. **212**: 749 (1981)
6. B, Krajewska, Separation and purification Tech., **41**, 305, (2009)
7.H, Liu *et. al,* Internat J of Food Microbiology. **95**, 147 (2004)
8. K, Jolanta *et.al.*, Marine. Drugs, **8**, 1567(2010)
9.CKS, Pillai *et al*, Progress in Polym Sci, **34**, 641(2009)
10. DE, Montassir *et al*, **19**, 342 (2014)
11.PK, Dutta, *et al,* J of Scientific & Industrial Research, **63**, 20 (2004)
12. J,Yang and ZX, Tang *et.al.*, Brazilian J of Chem Eng, **28**, 353 (2011)
13. PA, Sudha *et.al*, Industrial App of Marine Biopolym, Ch 1, Taylor & Francis(2017)
14. J,Synowiecki, NA,Al-Khateeb, J Critical Rev in Food Sci. and Nutr, **43**, 145 (2003)
15. S.Remya, J Food Sci Technol., **53**, 685, 2016

The Effectiveness Biosorption of Durian (*Durio zibhetinus*) Rind Pectin on Handling Liquid Waste Containing Heavy Metal (Pb II)

Flora Elvistia Firdaus [1, *], *Amida* Redella [2], *Sintani* Nursabila[3]

[1,2,3] Chemical.Eng, Jayabaya University, Jl Pulomas Selatan Kav 23 Jakarta 13210 Indonesia

Abstract. Heavy metal is one of the most substances occur in water pollutants. It is harmful to humans and other living things. The biosorption of pectin from durian rind is used as a heavy metal binder. However, the type of pectin presented in organic waste is generally HMP (High Methoxyl Pectin), which previously should be demethylation. Durian rind used in this study are originated from Bogor West Java. This study aims to determine the effectiveness of pectin biosorbent in reducing the concentration of heavy metal (Pb II). Pectin extraction was carried out at 80 °C within a time variation of 1; 5; and 6 hours. The results showed that the most optimum pectin is with 6 hours of extraction time where the methoxyl content is 3.46%; weight equivalent 3860 % galacturonate content; the degree of esterification 20.29%; and lead (Pb II) uptake 97%. While durian seeds do not contain pectin.

1 Introduction

The mining industry in Indonesia like any other country produces heavy metals which sufficiently to the fulfillment of infra structure. The worrying consequences are led to metal contamination on surface water if managed inappropriately. The health problems and other environmental degradation can be impacted to the quality of life. Efforts of handling heavy metal pollution can actually be done by using chemical processes which expensive and tends to cause new problems [1], the process produces high toxic [2]. Another alternative is using biological materials namely biosorption, with low cost, high efficiency, minimization of mud formation, and ease of regeneration process [3]. Pectin is an ionic plant cell [4] can be used as biosorbent [5]. The active group present in pectin will bind metals [6-7] and forms complex compounds [8].

Pectin is a polymer of 1-4 linked D-galacturonic acid where widely used in industries [9], can be used as biosorbent because of many active groups which are generally High Methoxyl Pectin (HMP) type. It must be demethylated to LMP to be applied to heavy metals. Citrus pectin absorbs metal Pb (II) to almost 90% in the amount of pectin 0.1 g/L and Pb concentration of 0.1 mm [10]. The purpose of this study is to determine the effectiveness of pectin biosorbent in reducing the concentration of heavy metal (Pb II).

* Corresponding author: flora_elvistia@yahoo.com

2 Experimental

2.1 Sample Preparation

The durian rind and seed are collected from local fruit stall in Bogor origin, West Java, Indonesia. The rind was washed and dried with oven for 3 hours. The fine durian rind and seed samples were added aquadest with the ratio 1:10 (gr/mL) and HCL 2 N until it reached pH 2. The samples were heated in an oven at 80 °C for 1; 5; and 6 hours. Strain and wash with 96% ethanol solution. Put in the oven with temperature 40°C until the sample is dry.

Pectin is dissolved in aquadest until it reaches a concentration of 1.5%. Add 3N NaOH solution to show pH 10. Then incubated with 55 °C for 1 hour. Cooled, then add HCL 3N solution to reach pH 3 and allowed to stay overnight. It was precipitated with 96% ethanol. Filter the pectin residue and washed with acetone. Dried in the oven. Blend until smooth and sifted with Mesh 60.

2.2 Qualitative Identification

The pectin yield is the amount of pectin in grams resulting from durian rind and its seed extraction, at various treatment times of extraction (1, 5, and 6 hours).

$$\text{Yield} = \frac{Pectin\ (mgr)}{\text{Weight of dry sample (gram)}}$$

The equivalent weight was determined to 0.5 grams of pectin samples which inserted into 250 mL Erlenmeyer, add 2 mL of 96% ethanol, dissolve it with 1 gr of NaCl solution in 100 mL aquadest, then add 6 drops of Phenolphthalein indicator. Perform titration with a standard 0.1 N NaOH, and calculates the equivalent weight using the formula:

$$\text{Equivalent Weight} = \frac{Pectin\ (mgr)}{\text{V NaOH x N NaOH}} \quad (1)$$

The neutral solution of the equivalent weight determination plus 25 ml of 0.2 N NaOH solution. Then added 25 mL of 0.2 N HCl solution and 5 dropped of phenolphthalein, titrated with 0.1 N NaOH solution until it turns pink.

$$\text{Methoxyl Content (\%)} = \frac{mL\ NaOH\ x\ N\ NaOH\ x\ 31}{\text{Pectin (mgram)}} x100 \quad (2)$$

The Galacturonate levels are calculated from the milliequivalent (mEq) of NaOH obtained from the determination of equivalent weight and methoxyl content. The value 176 is obtained from the lowest equivalent weight of pectic acid. The calculation using the formula:

$$\text{Galacturonate Content (\%)} = \frac{(mEq\ Wt+mEq\ methoxyl)x\ 176}{\text{Pectin (mgram)}} x100 \quad (3)$$

The degree of esterification is calculated from the methoxyl and galacturonate acid levels obtained.

$$\text{Degree of Esterification (\%)} = \frac{176\ x\ \%\ methoxyl}{\text{31 x galacturonate}} x100 \quad (4)$$

2.3 Capacity the capacity Lead (Pb II).

Pectin samples from various treatments of 1 g were fed into a measuring flask each contain of 50 mL $Pb(NO_3)_2$ 10 mmol /L solution. The solution was centrifuged at 3000 rpm for 5 min. The supernatant was taken and the concentration of lead (Pb II) measured by Atomic Absorption Spectrum (AAS) at 283.3 nm wavelength [5].

3 Results and Discussion

3.1 Qualitative Identification of Durian Rind Pectin and Durian Seed.

The pectin of modified durian rind by treatment of different extraction times has yielded a positive response. The durian seed does not form anything during the qualitative test, it can be concluded it does not contain pectin. The hypothesis to this result, durian seed pectin powder is only a sap of durian seed and is not a pectin so no further research on durian seed pectin.

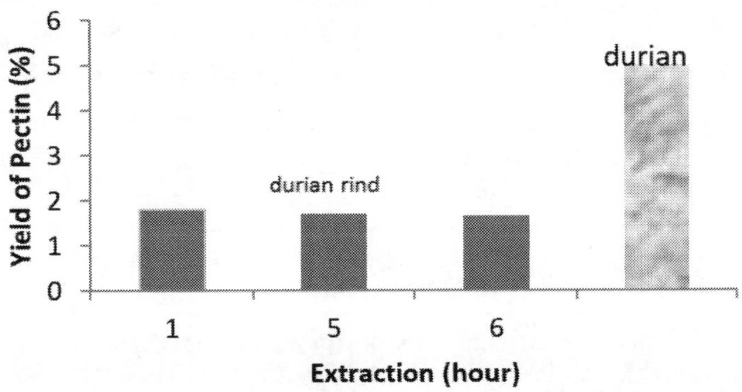

Fig 1. The Yield of Pectin in an average of Modified Durian rind and Durian Seed

The pectin yield is the amount of modified pectin in grams resulting from rind extraction at various times of extraction; 1; 5; and 6 hours and the durian seed the result then compared to dry powder. The yield of modified durian rind pectin with an hour extraction time shows the value of 1.80%; the yield of 5 hour extraction time has shown the value of 1.59%, and the yield of 6-hour of extraction is 1.49%. In the modified durian rind pectin the yield value decreases as the time of extraction take longer. The durian seeds show the value of 5%, this could be other substances than pectin that took part in the solid as seen in Fig 1.

The equivalent weight is the content of free un-esterified galacturonate acid consist of methyl ester group which are not esterified. The lower the pectin level causes the lower equivalent weight [11]. The highest equivalent weight in average is 5022 gr/eq within 1 hour of extraction. The equivalent weight of 5 hours and 6 hours of extraction of durian rind are respectively 4179 gr/ eq and 3860 gr/eq. This means a modified durian rind pectin within 6 hour extraction time has the most pectin compared to extracted for 1 or 5 hours.

3.2 Methoxyl content

Pectin is comprised into two groups, namely low methoxyl pectin (LMP) and high methoxyl pectin (HMP) [12]. The lowest methoxyl content was in modified pectin with an hour extraction of 3.09%. While the highest is achieved in 6 hours of extraction time, is 3.46%. It can be concluded that modified pectin produced pectin with low methoxyl content as illustrated in fig 2. Galacturonate content produced in this study was 84.11% for 1-hour extraction, 91.25% for 5 hours extraction, and 96.78% for 6 hours extraction. The minimum permissible galacturonate level according to International Pectin Producers' Association

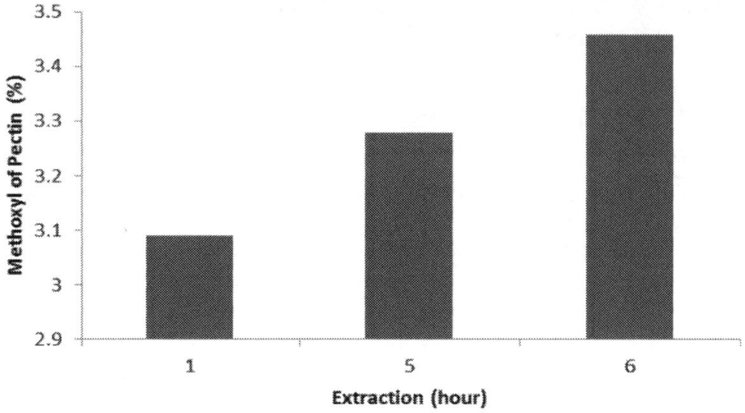

Fig 2. The Methoxyl content in an average of Modified Durian Rind Pectin

(IPPA) is 65%, so the resulting pectin has met the requirements of pectin quality. The lower the methoxyl pectin level, the low gel forming [13]. The degree of esterification of modified durian skin pectin for 1 hour is 20.83%, in 5 hours of extraction is 20.38%, the lowest is 6 hours of extraction time is 20.29 %. These are classified as low methoxyl pectin (LMP).

3.3 The capacity for Pb (II) Uptake

Pectin modification causes esterification to decrease, so increase the activity of the adsorption because of the lower the degree of pectin esterification the more active pectin

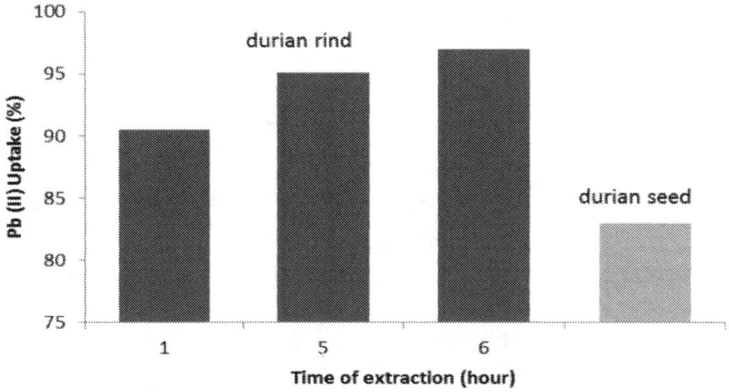

Fig 3. The Pb (II) uptake Capacity of Modified Pectin rind

group [5]. In durian rind pectin, for the time of extraction 1; 5; and 6 hours are respectively 90.46%; 95.1%; and 97%. The highest value is with 6 hours of extraction. So it is concluded that the longer the extraction time, the higher the absorption of Pb (II). Durian seeds do not show any response to various qualitative tests conducted, it is concluded that durian seeds do not contain pectin. Its absorbency shows a high enough value which is 83% as illustrated in Fig 3.

4 Conclusions

Durian waste can be used as raw material for making pectin. Modified durian skin pectin has a brownish yellow color. Optimum condition of pectin extraction results when viewed from the characteristics and absorption of lead metal was with 6 hours of extraction time. With a yield of 1.49%; weight of equivalent 3860.7 gr/eq; 3.46% methoxyl content; levels of 96.78% galacturonate acid; degree of esterification 20.29%; and 97 % of lead (II) uptake. While durian seed, throughout various of qualitative tests, was conducted does not give any responses, so it is concluded durian seeds do not contain pectin.

The author gratefully acknowledges the support of the Indonesia Ministry of Research and Technology, Department of Higher Education as funding the research.

References

1. B.Volesky., Z.R Holan, Biotech. Prog. **11**, 235 (1995)
2. AG, Prasad, and M.A.,Abdullah, J of Appl Sci in Env Sanitation, **4**, 273 (2009)
3. SE,Bailey, Olin, TJ, Bricka, and Adrian, Water Research, **33**, 2469 (1999)
4. CMGC, Renard, MJ, *Crépeau*, JF, Thibault,Carb. Research **275**, 155 (1995)
5. WW, Abbas, FMA, Liong, ME, Azhar., Internat Food Research J., **15**, 363 (2008) Kartel, Kupchik, Veisof, Chemosphere **38**, 2591 (1999)
6. N, Ahalya, TV, Ramachandra and RD, Kanamadi, Research J. of Chem. and Env., **7**,72 (2003)
7. LA, Kupchik, NT, Kartel, ES, Bogdanov, OV, Bogdanova and MP, Kupchik, Russian J. of appl. chem, **79**, 457 (2005).
8. M, Rodriguez, M, Zalba, and A, Debbault, J.Colloid Polym, **285**, 119 (2006)
9. A, Balaria, Ankit, and S,Schiewer, Separation and Purification Tech., **63**, 579 (2008)
10. APC, Saratha *Devi* & VM, Devi, Internat. J. of Eng. Research & Tech.,**3**, 1925 (2014)
11. BM, Yapo and KL, Koffi, Foods, **3**: 1, (2014)
12. BR, Sharma, L, Naresh, and NC, Dhuldhoya, Times Food Processing Journal, **23**, 44 (2006)

The Production of Bioethanol from Rimau Gerga Lebong (Rgl) Orange Waste as an Alternative Energy

Vike Darliyasi[1], *Kurnia* Herlina Dewi[2*], Budiyanto[2]

[1)] Student of Agricultural Technology, Faculty of Agriculture, University of Bengkulu
[2)] Department of Agricultural Technology, Faculty of Agriculture, University of Bengkulu, Indonesia

Abstract. Bioethanol from Rimau Gerga Lebong (RGL) orange waste is one of the solution to overcome fuel oil problem. The aim of this research is to get the type of microorganisms and fermentation time that produce the highest ethanol from RGL orange waste. The research method used was Randomized Block Design (RBD) of two factors, namely type of microorganisms (*Trichoderma viride*, *Saccharomyces cerevisiae*, and *Trichoderma viride* + *Saccharomyces cerevisiae*) and fermentation time (3 days, 5 days, and 7 days. Within the three type of microorganisms with variations of fermentation time showed that the pH was able to carry out the fermentation process smoothly. The highest total dissolved solids were in the type of *Trichoderma viride* 3 days and 5 days, and the type of mix of microorganisms on the 3rd day. The highest ethanol content is in the type of *Sachharomyces cerevisiae* for 7 days. ANOVA result showed that the interaction between two treatments on the total dissolved solids experienced significant differences, so it continue with the DMRT test at a significant level of 0.5%. However, it is different from the results of ANOVA on ethanol content which showed that there were significant differences between ethanol content and types of microorganisms, but there was no significant difference on fermentation time

1 Introduction

Orange is one of the fruits which containing cellulose. Orange juice contains cellulose around 0,52 gram per 50 ml and cell wall of orange fruit contain cellulose of 25-35% [1], orange peel contains cellulose of 5.36% [2]. [3] Reported that orange fruit yield ethanol levels of 9.75% on acid hydrolysis. Whilst, [4] utilizing orange peel with enzymatic hydrolysis resulted in the highest ethanol content of 40.298%.

According to the Central Bureau of Statistics, the production of orange fruit in the province of Bengkulu reached 10,132 tons. Rimau Gerga Lebong (RGL) orange is originally orange species, which developed in Bengkulu Province. This orange becomes the pre-eminent commodity of Lebong Regency [5]. RGL orange have competitive advantages, namely the colour of yellow-orange fruit, fruitful throughout the year, large fruit size, high

* Corresponding author: nia_unib@yahoo.com

fruit juice and also has a good market potential as the demand for orange RGL reached 100 tons/month, and contains 11.4 grams of carbohydrates [6].

About 30% of fruit fall due to various factors is one of shortcoming of RGL orange, and it makes farmers often suffer considerable losses [7]. The fallen fruit directly throw away as a waste, even though the content of the cellulose can be processed into bioethanol, whether by means of (1) hydrolyzing cellulose into glucose (reducing sugar) using acid or enzyme or (2) reshuffling glucose (reducing sugar) to ethanol. The enzymatic breakdown of cellulose using cellulose enzyme, while *Saccharomyces cerevisiae* is widely used for ethanol fermentation

This study aims to obtain the type of microorganisms and fermentation time that can produce the highest ethanol from bioethanol manufacture of Rimau Gerga Lebong (RGL) orange waste.

2 Research Methods

The materials used are whole fruit of RGL orange waste, aquades, Saccharomyces cerevisiae and Trichoderma viride.

Randomized Block Design (RBD) used with 2 factorial i.e. microorganism type and fermentation time. Microorganism type consists of Trichoderma viride, Saccharomyces cerevisiae and mixture of Trichoderma viride + Saccharomyces cerevisiae respectively. Meanwhile, the fermentation time selected 3, 5 and 7 days respectively. Each treatment repeated three times (27 experimental units).

2.1 Preparation of Raw Material

RGL orange waste directly blended without sterilization process. The raw material is prepared by using NaOH until it reaches pH 5 for the fermentation, which Trichoderma viride used as the microbial, and pH 3.92 for Saccharomyces cereviseae. Then stirred for homogeneous and put it into container fermenters.

500 ml of RGL orange waste is then fermented. Fermentation carried out in three types, namely, using 10 ml Trichoderma viride, 4 grams Saccharomyces cerevisiae and mixed of 10 ml Trichoderma viride + 4 grams Saccharomyces cerevisiae. Enzymatic hydrolysis carried out first for the mixed fermentation process, using Trichoderma viride for 3 days. Then incorporate 4 grams of Saccharomyces cerevisiae and conducted the fermentation process for 3, 5 and 7 days. The next step was distillation, aims to separate ethanol from other products. The distillation would be conducted by simple distillation with 78°C–90°C for 5 hours.

2.2 Data analysis

The data obtained plotted to found out the relationship of treatment observation parameters. Data analysis using ANOVA (Analysis of Variance) was used to determine the effects of treatment on observation result, if the ANOVA result was significantly different, and then continued with DMRT (Duncan Multiple Range Test) test at 0.5% significant level.

3 Results And Discussion

3.1 Changing pH on the Fermentation Process

pH is one of the important factor that can affect in the growth of microorganisms and final product on the fermentation process. All the types of microorganisms which used in the fermentation process shows that the longer the fermentation the more the pH decreased, as shown in figure 1. The occurrence of pH decline is caused by many factors, such as the presence of byproducts produced resulting in a decrease pH. Each microbes can work to produce something that suits their respective roles, so it needs the optimal pH, as in the fermentation with *Trichoderma viride* that can produce rough cellulose enzymes

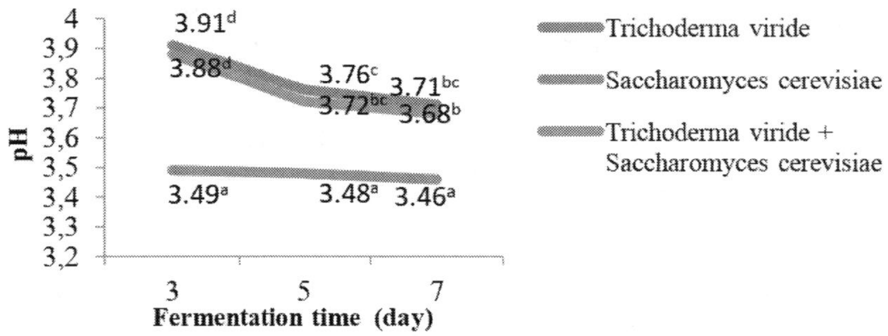

Fig. 1. Graph of pH on the Fermentation of Bioethanol RGL Orange Waste

3.2 The Changing of Glucose Levels

Bioethanol is ethanol derived from ingredients that contain carbohydrates or glucose, so the role of glucose itself is essential to produce high levels of ethanol. Figure 2 showed that the highest glucose levels are in the type of Trichoderma viride microorganisms in which this type of microbe can produce cellulose enzymes that function to convert cellulose to glucose. The process of changing cellulose to glucose using this cellulose enzyme namely enzymatic hydrolysis. On the 3rd and 5th days the highest glucose level, which was 7.05 brix, decreased on the 7th day by 6.03 brix. The tendency to increase glucose levels during enzymatic hydrolysis is suspected, because at the beginning of enzymatic hydrolysis the synergistic cellulose activity between endoglucanase, cellobiohydrolase, and beta-glucosidase. Endoglucanase hydrolyzes the 1,4 binding randomly and acts on the amorphous portion of cellulose fibers. Furthermore, cellobiohydrolase hydrolyzes the ends of cellulose chains to produce cellobiose, where they are hydrolyzed by β-glucosidase to glucose.

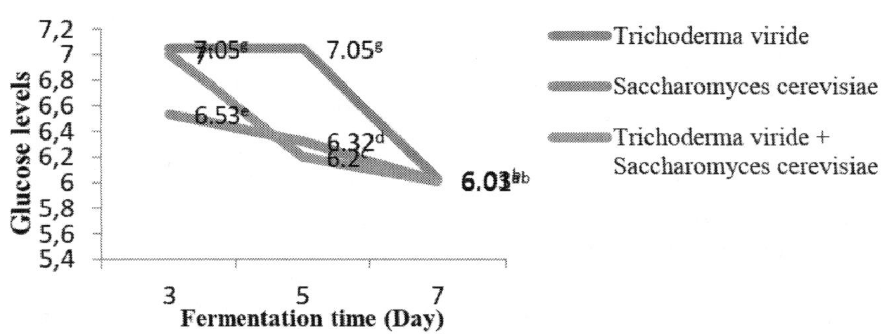

Fig. 2. Graph of Glucose levels of Bioethanol RGL Orange Waste

Fermentation using *Saccharomyces cerevisiae* microbes produce the lowest glucose levels than others. This is because *Saccharomyces cerevisiae* can convert glucose into ethanol, so that the glucose content contained in the raw material is directly converted to ethanol, which will cause the glucose level to decrease as it has been converted to ethanol.

Similarly, mixed fermentation using a mixture of enzymatic hydrolysis results for 3 days followed by the addition of *Saccharomyces cerevisiae*.
ANOVA analysis proof that type and length of fermentation were significance to glucose level so that tested continued to DMRT (Duncan's multiple range test) at level 5%

3.3 The Changing of Ethanol Levels

Ethanol content obtained from the distillation by using a simple distillation tool and measured using a 25 ml pycnometer. The result then converted by using alcohol metric table. Result of analysis of average ethanol level to treatment factor shown in figure 3.

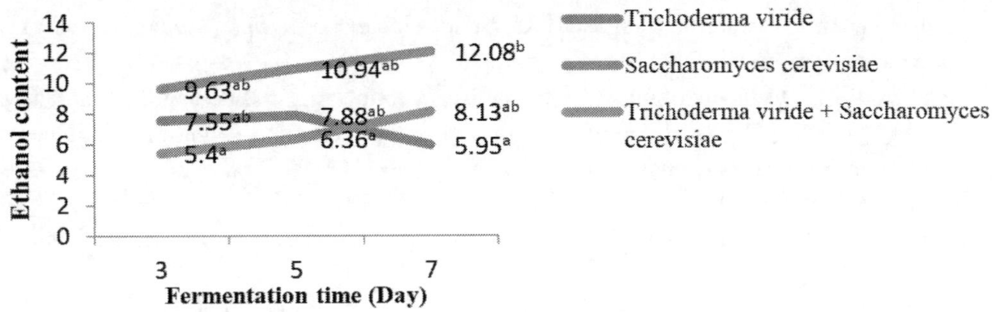

Fig. 4. Graph of Ethanol content of Bioethanol RGL Orange Waste

Based on the figure 4, the type of microorganisms that produce the highest ethanol is the *Saccharomyces cerevisiae* microorganism, which increases day by day. This is because yeast works well in converting glucose to ethanol. In this process, glucose is not directly converted to ethanol. The yeast will convert glucose and fructose into pyruvic acid through the reaction stage of the Embden-Meyerhof-Parnas pathway, which will then proceed, with the process of decarboxylation of pyruvic acid into acetaldehyde. Acetaldehyde then undergoes dehydrogenation process into ethanol compounds.

However, this result inversely proportional to the fermentation using Trichoderma viride, which is up and down. The occurrence of a decrease on the 7th day is due to the resulting glucose levels is also small which will lead to low levels of ethanol. So, the role of Trichoderma viride microbe is not to produce ethanol, but convert cellulose into glucose, but when distilled and measured ethanol content, this type of microbe can produce ethanol as well. This is allegedly due to the presence of natural microbes that can directly convert raw materials into ethanol. This also proven by spontaneous fermentation experiments on RGL orange garden waste, i.e. without any microbial mixture, the results obtained in the form of spontaneous fermentation can produce ethanol, even the ethanol content is small.

In contrast to the fermentation type of mixtures, in which can produce the highest levels of ethanol compared to others, because the working microbes have also been maximized, such as Trichoderma viride, which can convert cellulose into glucose, and *Saccharomyces cerevisiae*, which can convert glucose into ethanol. Although the results are less than the type of fermentation using Saccharomyces cerevisiae alone, but this type of fermentation increases day by day.

ANOVA test showed that fermentation type significantly different to ethanol content, but there is no different with fermentation time.

4 Conclusion

The results concluded that:

1. The microorganisms that produce the highest ethanol content was of microorganisms that use Saccharomyces cerevisiae
2. The fermentation time to produce the best ethanol content was 7 days, but there is no effect on the ANOVA.

References

1. Kadarisman, D., Sunarmani dan M. Arintawati.. Buletin Penelitian Ilmu dan Teknologi Pangan. Bogor. **IV**(1): 61-68. (2004)
2. Ichwani, R.. Pemanfaatan Limbah Kulit Jeruk Keprok (Reticulata Blanco syn) Sebagai Bahan Penguat Nanokertas Selulosa Bakteri dari Air Kelapa. Skripsi. Medan: Universitas Sumatera Utara. (2013)
3. Andini, C., E. Mahajoeno dan R. Setyaningsih. Production of Bioethanol From Citrus Fruit (Citrus Sp) Waste By Acid Hydrolysis and Fermentation Using *Saccharomyces cerevisiae*. Seminar Nasional X Pendidikan Biologi FKIP UNIS. Surakarta : Universitas Sebelas Maret. (2013).
4. Megawati dan R. Ciptasari.. Prosiding SNST ke-6. Semarang : Universitas Negeri Semarang. (2015)
5. Rambe, M dan S. Suryani. Menelisik Inovasi Gerga Bersama Bu Yeni. Badan Penelitian dan Pengembangan Pertanian Bengkulu. Majalah Sinar Tani Edisi 19 - 25 Agustus No.3619 Tahun XLV (2015)
6. Pancawarman, R. Permintaan Jeruk Gerga Lebih 100 Ton Perbulan. (http://www.antarabengkulu.com/berita/5129/lebong-kembangkan-jeruk-gerga . (2017).
7. Sugandi, D. Inovasi Teknologi Jeruk Gerga Harga Mati. Badan Penelitian dan Pengembangan Pertanian Bengkulu. Majalah Sinar Tani Edisi 19 - 25 Agustus 2015 No.3619 Tahun XLV. (2015).

Non-Ionizing Radiation (NIR) Exposure Map Development Using GIS for Gong Badak, Terengganu, Malaysia

S.N. Hazmin[1,2] , *A.R.S.N.* Dianah[1], *M.K.A.* Kamarudin[3], *R.* Umar[3,*], and *H.N.* Syafiqah[3],

[1]School of Fundamental Science, Universiti Malaysia Terengganu, 21030, Kuala Nerus, Terengganu, Malaysia
[2]Pusat Islam Sultan Mahmud, Universiti Malaysia Terengganu, 21030, Kuala Nerus, Terengganu, Malaysia
[3]East Coast Environmental Research Institute, Universiti Sultan Zainal Abidin, Gong Badak Campus, 21300, Kuala Nerus, Terengganu, Malaysia

Abstract. Public Non-Ionizing Radiation (NIR) exposure should be monitored by the government since it has proven harmful for human. Even though some studies reported that low frequency spectrum is safe, the other says that long and repetitive exposure may affect health. Thus, this study is conducted to measure the ambient NIR exposure value in the selected Terengganu area, Malaysia. The values were then used to develop a spatial model of exposure to see the radiation exposure pattern. From the findings, we found that the value was higher nearby base stations which are normally erected around residential area where the telecommunication network user is high. Thus, this study is crucial to determine the level of exposure for future plan. This findings can be used by the government to control the radio-frequency transmission and active spectrum service provider in order to limit the radiation exposure for public health concern.

1 Introduction

Non-Ionizing Radiation (NIR) has been proven may cause adverse health effect by numerous studies [1], Some finding shows that most of people who highly exposed to radio frequency has higher chance to diagnose to have brain tumor and affect the reproductive system other than mild pain such as headaches, insomnia, dizziness and dermatitis [2]. From the study done by Urbinello et al. (2014), they found the arithmetic mean values for mobile phone base station exposure at Basel, Brussels and Amsterdam were far below the international reference level proposed by ICNIRP (International Commision on Non-Ionizing Radiation Protection)[3]. They further their study on the temporal trends of RF-EMF exposure levels in different microenvironment of three European cities (Basel (Switzerland), Ghent and Brussel (Belgium)) and found that an increase of RF-EMF exposure levels has been observed between April 2011 and March 2012. They suggested a continuous monitoring is needed to identify high exposure areas and to anticipate critical development of RF-EMF exposure at public places [4].

In this study, the focus will be on the measurement of ambient NIR exposure value in the vicinity of Gong Badak which is located in the district of Kuala Nerus, Terenganu state of Malaysia. This research findings will provide the map of NIR exposure for future monitoring actions. This is essential to make sure the recommended public exposure value by International Radiation Protection Association (IRPA) is complied.

The stakeholders such as telecommunication service provider, internet service provider, radio & TV broadcasting and other active spectrum user may also benefited from the map to manage the location of their base stations and transmitter for better coverage. The Ministry of Health will be assisted with the management of exposure value limit which is crucial to provide healthy environment to the community. This is due to the fact that overdose non-ionizing radiations are proved may cause long term effect to human body & systems, and also bad for health. The most important thing, this research outcome will propose the Malaysia government, a policy on electromagnetic radiation. This will control the radio-frequency transmission and active spectrum service provider in order to limit the radiation exposure for public health concern.

2 Methodology

The step-by-step methodology is as illustrated in Figure 1 below.

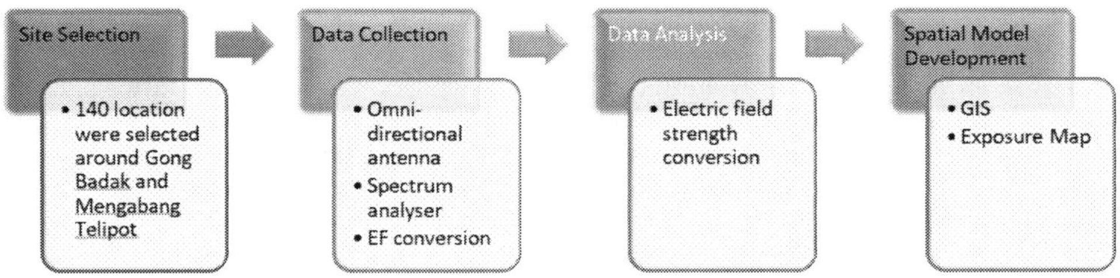

Fig. 1. Flow of the study

2.1 Site selection

140 location has been selected around Gong Badak. The data collection sites is determined based on accessibility and the presence of base station tower (BST).

2.2 Data collection

In this stage, measurements were conducted using the fabricated antenna specifically for this study, connected to spectrum analyser at selected sites. The signal were recorded for 10 minutes at each selected point roughly from 10.00 am until 4.00 pm. Based on the ICNIRP postulate, the temporal average value of exposure over any 6 minutes interval must be used for the comparison with the reference levels suggested [5]. In this study, frequencies below 3GHz were focussed since daily human activities on spectrum use lies between this range as permitted and allocated by Malaysian Communications and Multimedia Commission (MCMC).

2.3 Data analysis

The raw data was in the form of signal power level value which are in the unit of decibels relative to one millivolt (*dBmV*). These data were converted to electric field (EF) strength value where the unit is volt per meter (*V/m*). For each site, the maximum exposure value and the average value were obtained. This values were compared with the international reference level proposed by ICNIRP and then used for the exposure value mapping in the next stage.

2.4 Spatial model development

The map of ambient exposure level of Gong Badak were developed based on maximum value. The spatial model (map) were developed using ARCGIS software. Interpolation method was used in this study. It is a mathematical technique that enables estimating the values of a curve at any position between known points. Inverse Distance Weighting (IDW) techniques were employed to interpolate the datasets. IDW implements an algorithm that effectively deals with highly variable data during interpolation. This technique computes the value at each assigned grid node by considering the surrounding data points which are between the user-defined radiuses. The most influence factor contribute to the exposure level were determined.

3 Result and Discussion

From the measurement, 70 points of ambient NIR exposure value in term of EF strength were collected around Gong Badak. The EF strength at ten points were collected for each sites as shown in Table 1. The shaded cell indicate the highest value recorded for each site.

Table 1. Electric field strength value for each site.

Sites	Kubang Badak	UMT	Gong Badak	Kg. Jati	UniSZA1	UniSZA2	Kg. Pak Tuyu
Population	High	High	Moderate	Moderate	Low	Moderate	Low
Point 1	5.06	3.54	11.06	0.02	4.09	10.97	3.84
Point 2	10.91	2.34	11.09	3.02	5.73	8.73	8.76
Point 3	37.82	11.42	12.77	14.63	5.24	14.63	4.05
Point 4	9.47	34.74	11.23	12.69	8.33	7.15	9.46
Point 5	6.57	12.04	22.35	5.08	9.79	4.36	11.04
Point 6	2.03	12.12	6.50	5.23	10.14	7.89	0.22
Point 7	3.30	2.31	3.87	10.09	11.13	0.08	0.96
Point 8	12.12	2.03	7.65	13.46	10.45	1.37	3.21
Point 9	0.64	6.57	7.63	0.08	5.48	4.30	1.19
Point 10	34.74	8.33	13.68	15.04	5.26	3.07	2.79

The maximum value obtained from the table were then used to develop a spatial model using interpolation method in ARCGIS. This map illustrates the pattern of exposure at surrounding of the study area as shown in Figure 2.

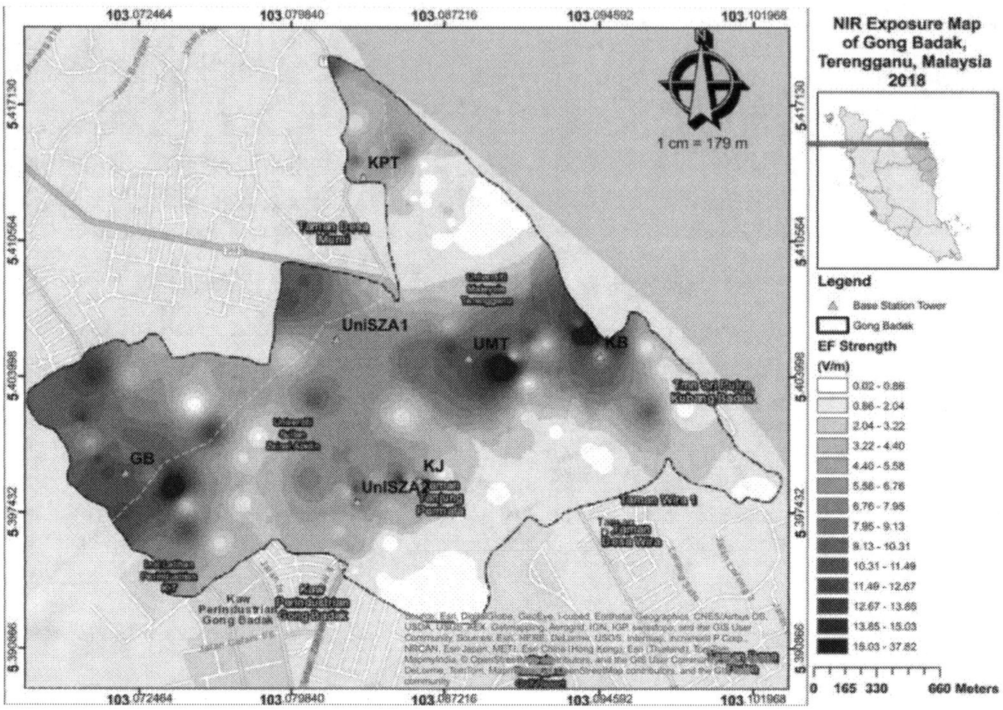

Fig. 2. Spatial model of NIR exposure value at Gong Badak

From the spatial model, we can see the darker area which indicates the higher exposure value were mostly located nearby BST (small triangle). The farther the distance from BST, the lighter the colour, which means the lower the exposure value. However, Kampung Pak Tuyu (KPT) recorded lower exposure value compared with other sites even there is BST nearby. This may because of less resident living in the neighbourhood, contribute to the less data traffic in the area. RF exposure is proved varies on a population basis, where high populated area shows an increasing trend compared to low dense area [6].

Base on Table 1, Kubang Badak has the highest exposure with 37.82 V/m, followed by UMT with EF strength of 34.74 V/m. Kubang Badak is close to UMT hostel where the mobile phone user are high. This is the reason why Kubang Badak recorded high exposure value compared to other sites. Moreover, the number of BST nearby will increase the propagation of multipath, thus, easier the interaction between them to occur. This interaction can be divided into two: constructive interferences and destructive interferences [7-8]. The constructive interference occurs when two signal with the same wavelength that propagates aligned with each other and resulted the larger amplitude. Meanwhile, the destructive interference occurs when the two signal were not perfectly aligned or shifted then may resulted the smaller amplitude or no signal at all [9-10]. Therefore, can affect the quality of communication and then, it reduce the exposure [11-12]. Further study on NIR exposure at other location on population basis should be done to investigate the trend and the effect on human for health monitoring purposes.

4 Conclusion

NIR exposure value in term of EF strength were measured around Gong Badak, Terengganu, Malaysia to visualize the exposure pattern in the area of concern. From the findings, it can be concluded that the highest EF strength is recorded at Kubang Badak and

UMT where the data traffic is high due to high population in the neighbourhood where there is located UMT's student hostel. Since NIR exposure has proven can cause adverse health effect, the study on this kind of exposure is crucial for future health concern.

This study is made possible by the usage of grant FRGS/1/2015/SG02/UNISZA/02/1, TPM/68006/2016/79 and UMT 68006/INSENTIF/60. Our sincere gratitude for all assistance and help from Assoc. Prof Dr. Ahmad Nazri b Dagang and Dr. Hajar Ja'far. The authors also thank to Universiti Malaysia Terengganu, Universiti Sultan Zainal Abidin and Universiti Teknologi MARA for the facilities provided.

References

1. Fatin Syaza Syariff Khan, Roslan Umar, Nor Hazmin Sabri, Mohd Khairul Amri Kamarudin, Azmi Hassan, Hafizan Juahir. Journal of Fundamental and Applied Sciences, 9(2S), 308-316. (2017)
2. Dianah review Siti Nur Dianah Abdul Rahman, Nor Hazmin Sabri, Roslan Umar, Mohd Khairul Amri Kamarudin, Ahmad Nazri Dagang. Journal of Fundamental and Applied Sciences, 9(2S), 182-198. (2017)
3. Damiano Urbinello, Wout Joseph, Anke Huss, Leen Verloock, Johan Beekhuizen, Roel Vermeulen, Luc Martens and Martin Roosl. 2014. Environment International 68. pp49-54. (2014)
4. Damiano Urbinello, Wout Joseph, Leen Verloock, Luc Martens and Martin Roosl. Environmental Research 134, pp134-142 (2014)
5. ICNIRP, International Commission on Non-Ionizing Radiation Protection 1998 Guidelines for limiting exposure to time varying electric, magnetic, and electromagnetic fields (up to 300 GHz), Health Physics, 74 (4) pp 494–522.(1998)
6. Jean-François Viel, Elisabeth Cardis, Monika Moissonnier, René de Seze, Martine Hours 2009 Environment International 35 pp 1150–1154 (2009)
7. Rappaport T S. Wireless Communications *Principles and Practice* 2nd ed. (Singapore: Pearson Education, Inc.) (2002)
8. Haykin S and Moher M Modern Wireless Communications (Singapore: Pearson Education, Inc.) (2002)
9. Mark J W and Zhuang W. Wireless Communications and Networking (New Delhi: PHI) (2005)
10. Feher K Wireless Digital Communications *Modulation and Spread Spectrum Applications* (Upper Saddle River, NJ: Prentice Hall) (1995)
11. Blake R Wireless Communications Technology (Delmar, Singapore: Thomson Asia Pvt Ltd) (2004)
12. Agarwal D P and Zeng Q A Introduction to Wireless and Mobile Systems (Nelson, India: Thomson Learning) (2007)

MATEC Web of Conferences **248**, 05002 (2018) https://doi.org/10.1051/matecconf/201824805002
ESTIC 2018

The Benefit of Wind Distribution Analysis for Coastal Construction Design in East Java Province

Surya Hermawan[1,*], *Joko* Purnomo[2], and *Daniel* Tjandra[3]

[1,2,3] Civil Engineering Department, Petra Christian University, Siwalankerto 121-131 Surabaya, Indonesia

Abstract. To create a safe and durable coastal structures design, a comprehensive data including minimum, maximum, and significant sea wave's height are needed. Normally, to obtain such massive data, engineers need to elaborate specialized consultant to carry out direct measurement on site, which typically consumes enormous financial resources and involves exceptionally rare experts. In such case, engineers can use a set of 10 to 20 years' wind data as a substitution. The suggested duration is expected to ensure that the set of data has sufficient information, which includes extreme wind speed and its fluctuation along the respective years. Therefore, it is common to adopt available data at the nearest wind stations, such as airports and harbors, to estimate the wind features on site. This paper presents a normative wind speed distribution model based on 16 years' wind data at a location obtained from National Oceanic and Atmospheric U.S. Department of Commerce (NOAA). The outcomes of the research show that the maximum sea wave height prediction in Sidoarjo Regency East Java with respect to analytical and graphical method are 0,465 m and 0,840 m, respectively. The results are considered satisfactorily reliable to be used as primary data to design coastal structures.

1 Introduction

To create a safe and durable coastal structures design, a comprehensive data of sea waves is needed. However, particularly in Indonesia, such data is rarely being well recorded because it consumes enormous time to collect and spends expensive cost. As an alternative method, wind data consisting wind speed, direction, and duration as well as fetch length can be used to estimate sea wave's height.

It is common for local engineers to address for natural phenomenon—which cannot be easily predicted, due to ineffective and inefficient design of coastal structures in Indonesia. Such problem can be avoided if reliable design data exists. Currently in Indonesia, a well-calibrated data of sea waves' height is very limited. It, if available, usually covers only one-month or one-year period at max. This also extends similar problems when identical data for new sites are needed.

This paper presents a normative wind speed distribution model based on 16 years' wind data at a locations of research study: Sidoarjo Regency in the surrounding area near Surabaya.

* Corresponding author: shermawan@petra.ac.id

The data are obtained from National Oceanic and Atmospheric U.S. Department of Commerce (NOAA) [1] to fill the gap between design and construction phase. The data will include a one to six hours-based wind speed measurement from 2000-2006 [2–4]. The data will be rigorously analyzed to further be used as a normative data to estimate sea waves height for designing a safe and durable coastal structures in East Java, Indonesia [5, 6].

2 Study literature

2.1 Fundamental theory of sea wave formation

Sea waves can be mainly generated by five factors, i.e. wind, moon-sun gravitational force interactions, eruption of volcanos, earthquakes, and any physical disturbance in the surrounding area [7]. Wind flowing over the ocean causes normal and tangential stresses, through which wind energy is dissipated to ocean surface. As a result, water particles move in the same direction as wind flow.

2.2 Wind speed

The term wind can be generally defined as a moving mass of air in relatively horizontal direction. This movement is triggered by the potential difference at two locations. Particularly in Indonesia, sea waves are highly affected by two distinct seasonal wind direction, i.e. wet monsoon and dry monsoon. The wet monsoon causes the wind blowing from the southwest between May and September, and conversely the dry monsoon brings the wind blowing from northwest between October and April [7].

2.3 Wind direction

Indonesia is highly affected by two distinct seasonal wind direction caused by the potential difference between Asia continent and Australia continent, i.e. wet and dry monsoon. The wet monsoon causes the wind blowing from the southwest between May and September, and conversely the dry monsoon brings the wind blowing from northwest between October and April [8].

2.4 Wind duration

Wind duration dictates the amount of dissipated energy from wind flow to sea surface, and thus plays a key role to the formation of sea waves. Furthermore, as the duration controls the dissipated energy, the longer the wind blows the height of formed sea wave is also higher. Besides, wind direction also affects the period of sea waves, which is also an important factor to consider in designing coastal structures.

2.5 Fetch length

Fetch length is the distance travelled by air particles during wind blows. In case of sea wave's formation, fetch is defined and limited by land, i.e. continents, islands, and mountains, in the surrounding area. Fetch length can be evaluated using empirical equations.

2.6 Methods of sea wave forecasting

Numerous method to estimate sea wave height are widely available, such as Sverdrup Munk and Bretschneider (further *SMB method*) method and Darby Shire Method (further *DSM method*). Both method uses forecasting to estimate the sea wave height based on past meteorological data.

For forecasting the wave height, a typical conversion chart has been provided by CERC to estimate the deep sea wave height [9]. Moreover, for coastal wave height estimation, another chart has also been generated [10]. In order to obtain reliable results, a set of ten years' consecutive data is needed.

3 Methodology

To achieve reliable results, primary and secondary data was considered. Primary data was obtained by direct measurement at the studied location, while secondary data was obtained from NOAA website [1]. The later data was in one to six-hours based wind speed record for 16 consecutive years' measurement. Meanwhile, primary data was retrieved to ensure that the secondary data can be representatively used for forecasting and determining correction factors when necessary. Furthermore, the data was analyzed to construct wind rose to obtain dominant wind direction. After the wind direction was confirmed, design wind duration was selected.

After the wind parameters, i.e. speed, direction, and duration, and fetch length were obtained, forecasting method was performed via both graphical and analytical method. Forecasting by graphical method was done using the existing graph in Shore Protection Manual (SPM) [9]. To use the graph, first and foremost, the wind data needs to be converted to wind stress factor (U_A). As all graph's parameters, i.e. fetch length, wind stress factor, as well as wind direction, are known, the wave height forecasting via graphical method can be done. Meanwhile, the analytical method was done by applying shallow water equations.

In order to provide reliable data, direct measurement as primary data was carried out. Figure 1a shows the research personnel as well as surveying helper, which includes local officers, army, and local people. Furthermore, Figure 1b shows weather station type AW 003, which was used to monitor real time wind data on site.

Fig. 1. Field Survey and Data Measurement

Figure 1c shows the process of harvesting the minute-based real time wind data from the weather station probe. The data was further transferred into laptop for advance processing. On the other hand, Figure 1d shows the inspection through the river to identify the key location to the coastal flood problem. In final stage, the data obtained from different spot were combined to produce comprehensive analysis and be used as parameters to decide proper type of coastal flooding structure design.

4 Results and Discussion

The outcomes of this research is presented by referencing to the accompanying drawings. The identification of research site, at where the maximum sea waves' height will be predicted, is based on precise full latitude, longitude, and altitude position. Determination of the location can be based on coordinate projection system and World Geodetic System (WGS) 1984-based Cartesian system. For the particular research, projection system is better to be used because it provides higher accuracy as well as presented in standard unit (SI), i.e. meters (m).

In identifying the accurate location of the research site, it was important to calibrate the position of the synoptic data recorder station from the anemometer [1,5,7,8,9,10,11]. The synoptic data was periodically sent to National Oceanic and Atmospheric Administration (NOAA) from the station. After the location was confirmed, the number of usable stations for collecting data at the research site was displayed. The available stations were displayed alphabetically, which represented the relative distance to the research site such that representative stations can be selected. The analysis results in effective fetch length of 125.4 km. Furthermore, based on analysis of 16 years' wind data from 2000 until 2016, analysis based on graphical method (SMB method) and analytical method (DSM method) were performed. It is found that the maximum wind speed is 34.96 knot in 2009.

Table 1 shows the prediction of sea waves' height via graphical and analytical method. The results clearly show that the maximum wave's height is 0.840 m and 0.475 m by graphical and analytical method, respectively. The results appropriate the previous result [11], which revealed that the sea waves' height at Panciran, East Java, were between 3.2 and 82.9 cm. It is also notable that the prediction result obtained via DSM method is more accurate than the SMB method.

Table 1. Prediction of sea waves' height

Year	Maximum wind velocity						Wave's height prediction (m)	
	knot	U (m/s)	RL	U_w (m/s)	U_A (m/s)	U_A (km/h)	Graphical	Analytical
2000	12.336	6.346	1.300	8.249	9.516	34.258	0.290	0.305
2001	12.850	6.610	1.320	8.725	10.196	36.704	0.320	0.316
2002	30.326	15.600	1.000	15.600	20.835	75.006	0.740	0.453
2003	12.850	6.610	1.300	8.593	10.006	36.022	0.310	0.313
2004	35.466	18.244	0.900	16.419	22.189	79.882	0.840	0.468
2005	34.438	17.715	0.950	16.829	22.873	82.341	0.840	0.475
2006	20.560	10.576	1.150	12.162	15.340	55.225	0.500	0.389
2007	14.906	7.668	1.250	9.585	11.444	41.200	0.360	0.335
2008	16.448	8.461	1.220	10.322	12.537	45.134	0.450	0.351
2009	34.952	17.979	0.895	16.091	21.646	77.924	0.840	0.462
2010	11.882	6.112	1.320	8.068	9.259	33.334	0.280	0.301
2011	12.850	6.610	1.300	8.593	10.006	36.022	0.310	0.313
2012	14.392	7.403	1.280	9.476	11.285	40.627	0.340	0.333
2013	28.784	14.806	1.050	15.547	20.748	74.693	0.730	0.452
2014	12.850	6.610	1.300	5.593	10.006	36.022	0.310	0.313
2015	23.644	12.162	1.100	13.379	17.248	62.094	0.600	0.412
2016	25.186	12.956	0.950	12.308	15.566	56.039	0.520	0.391

5 Conclusion

The lack of sea wave height measurement data in a remote area / research location is identified as one of the main problems for the development of the maritime industry for many years in Indonesia. Then, it is need to have an accurate and effective method for analyzing and determining maximum sea wave height in every territorial waters of Indonesia. The scientist or experts typically carry out the measurements of field data of 2 weeks at research sites / projects which are then used as the basis for planning. Meanwhile, to obtain maximum sea wave height must be based on data in a long time, minimum 10 years so that can be used as guidance design criteria of the coastal construction.

Two methods to predict sea waves' height, namely SMB and DSM method, was performed to estimate the sea waves' height at the research site. Both method was used to predict the sea waves' height based on 16 years' synoptic wind data. This method is a new technique which accommodate an accurate procedure to provide realistic and high quality prediction results. It is found that the results obtained from both methods fit the results from previous research. The results also show that the maximum waves' height are 0.840 m and 0.475 m for graphical and analytical method, respectively. Thus, the results can be used to design a safe coastal structures. In particular, for this research area, in the future the results can be used to decide the best adaptation strategy against coastal flood, which consider another physical hydrodynamic condition, including sea level rise along with topographic and hydrometric data.

The authors would like to express appreciation for the support of the sponsors of Petra Christian University project number: 07/HB-Penel/LPPM-UKP/I/2018 and Directorate General of Higher Education Indonesia and Kopertis VIII, project number: 002/SP2H/LT/K7/KM2017.

References

1. National Oceanic And Atmospheric Administration (NOOA) retrieved from: https://www.ncdc.noaa.gov/data-access/model-data/model-datasets/reanalysis-1-reanalysis-2, (2017)
2. S. Hermawan, H. Gunawan, JTEC Spec. Ed., (2017)
3. S. Hermawan, Civil Eng. Dim., 18, 2 (2016)
4. S. Hermawan, Adv. Env. Bio., 9, 7 S2 (2015)
5. H. Edy, I.H.N. Yuwono, D. HE, *Model distribusi kecepatan angin*, (Universitas Gadjah Mada, 2004)
6. J.J. Hidayat, M. Yusuf, E. Indrayanti, J. Ocean., 2, 3 (2013)
7. B. Triatmodjo, *Pelabuhan* (Beta Offset, Yogyakarta, 1993)
8. B. Triatmodjo, *Teknik pantai* (Beta Offset, Yogyakarta, 1996)
9. CERC, USA Eng. Waterways Exp. Sta., 2, (1984)
10. M. Darbyshire, L. Draper, J. Eng., 195, (1963)
11. D.N. Sugianto, Ind. J. Marine Sci., 15, 3 (2012)

E-census Implementation: A Case study in Naikoten II, Kupang, Indonesia

Lily Puspa Dewi[1,*], *Adi* Wibowo[1] , and *Ngakan* M.A. Immanuel[1]

[1]Informatics Department, Petra Christian University, Surabaya, Indonesia

Abstract. Growth in human population around the world affects all people through its impact on the economy and environment. Therefore, the population in a region should always be recorded. Government as a public decision maker has a big responsibility in the welfare of the community. The government needs information about the people lives to determine what policies should be taken. Government needs census data accurately to assist in making decisions. The government of Naikoten II wants to bring this village become digital village using technologies in their public services. In Naikoten II, the population data collection process (census) still uses paper form using Canvaser method which is filled in by the census officer by visiting and interviewing residents directly. It has a big risk of data loss. The main objective of this research was to provide a viable solution that allows surveyors to conduct the census efficiently. This research also provides a web based application to help data analyst to manage the census data. These application have been designed and implemented. The testing has been conducted within the small group of stakeholder that volunteered to use this early version of the application for a specified amount of time to provide feedback for possible changes and improvements.

1 Introduction

The population of a region is constantly changing, sometimes in a certain period, the changing of population is high, stable, and even low. Population can grow rapidly if public awareness of life quality is lacking, the rising welfare and improvement of the health facilities. Moreover, nowadays, rate of population growth is high as the outcome from a gap between the birth rate and the death rate. A simple explanation about this gap is the birth rate is greater than the death rate, which led to rapid population growth around the world. On the other hand the reason that made population grow stable is because the other factors that influenced the high and fluctuating death rates, such as disease outbreak, malnutrition, natural disaster, plague, war and the limitation of technology at that time [1].

In general, the reason for the growth in the population of Indonesia is more or less the same as the reason of the population growth in other countries, which is an improvement in the distribution of food, improving public health and the conquest of disease. These three reasons are the main factor that affect why the population growth in Indonesia can be very high. In addition, there are other factors as well contribute to the growth of the Indonesian

* Corresponding author: lily.puspa.dewi@gmail.com

population; these factors are religion and culture [2]. The growth in human population around the world affects all people through its impact on the economy and environment [3]. Therefore, the population in a region should always be recorded.

In order to support good governance, population policies need to be formulated nationally and thorough that is poured into the population programs that integrated, directed and sustainable. To strengthen the formulation of policies in the field of population complete and detailed data are required through the procurement of the Population Census. Government considers this issue by conducting population census [4]. Based on Government Regulation, the definition of Population Census is collection, processing, presentation, assessment, and data analysis, number and nature of socioeconomic and socio-cultural population of residents who at the time of the Population Census are residing or located in the geographical region of the Republic of Indonesia [5, 6]

Government as a public decision maker has a big responsibility in the welfare of the community. The government needs information about the people lives to determine what policies should be taken. Government needs census data accurately to assist in making decisions.

Population census conducted by the Statistics Indonesia (Badan Pusat Statistik/BPS) has been causing problems. A long span of ten years is a problem for agencies that use census data from Statistics Indonesia. Within 10 years, population census data have been obsolete and many changes that caused death, birth, migration of the population. Census officers are still using paper and the data recording process is still done manually thereby increasing the risk of data loss [7].

Naikoten II is one of the urban villages in Kota Raja (district), Kupang, East Nusa Tenggara, Indonesia. The government of Naikoten II wants to bring this village become digital village using technologies in their public services [8]. In Naikoten II, the population data collection process (census) still uses paper form using Canvaser method which is filled in by the census officer by visiting and interviewing residents or respondents directly. It has a big risk of data loss. In this research, a mobile based for population census application is needed to assist the government, especially the Kupang (in this case Naikoten II as study case), in conducting the census in the city level government, so that the government get information on the population in the region quickly and accurately. e-Census system to minimize the cost of census-taking and maximize efficiency of available resources.

2 Literature Review

2.1 Population census

Population census is an activity carried out by taking the population data in a region based on the population number of gender, age, ratio and density [9].

Population census is a concept of social geography which is one of the oldest and most extensive statistical activities undertaken by governments throughout the former duo oriented for military and taxation estimates. The census was also developed to gather information on housing, manufacturing, agriculture, mining and the business world.

There are two common methods of conducting the population census, i.e [10]:

1. Householder method. In this method, filling in a list of questions about population data is left to the residents or respondents, so that the population is given a list of questions to fill in and will be taken back some time later. Such a method can only be done in areas where the education level of the population is relatively high, because they are able to understand and answer any questions submitted to them.

2. Canvaser method. In the Canvaser method, the questionnaire is filled in by the census officer by visiting and interviewing residents or respondents directly. The census officer asks the questions according to the list and the residents who were replied verbally in accordance with the actual circumstances.

Based on the resident status of the population, the census can be distinguished into the de facto census and the de jure census. Below is the population census difference between the de facto census and the de jure census method [10].

1. De facto census. In the De Facto method, data collection is conducted by counting people as residents who are present physically to be at the time the census is taken. Only those that are present physically are seen and counted.
2. De jure census. In the De jure method, population records are performed by officers only for residents who are officially registered of a specific area. It does not matter whether the person is present or not.

The last population census in Indonesia was done in 2010. In the implementation, the population census used two stages, namely complete enumeration and sample enumeration. The de jure and de facto approach was applied to cover everyone in the enumeration area. Those who had permanent identity card were approached with de jure approach. They were recorded in accordance with their formal residence; whereas those who did not have permanent identity card approached with a de facto approach and recorded where they are. The result of the census was 237,556,363 people, consisting of 119,507,580 men and 118,048,783 women [8].

2.2 Naikoten II, Kupang

Naikoten II is one of urban village in Kupang, East Nusa Tenggara, Indonesia. The government of this village has program to create smart village using technology for its public service. People can use this service digitally, without being limited by space and time. Kupang is the capital city of East Nusa Tenggara (province of Indonesia). This municipality is the largest city on Timor Island located on the coast of Kupang Bay, northwest of the island of Timor. As the largest city in the province of East Nusa Tenggara, Kupang is inhabited with various ethnic groups. Significant ethnic groups in Kupang are Timor, Rote, Sabu, Tionghoa, Flores and a small number of migrants from Bugis and Java. The total area of Kupang City is 180.27 km^2 with a population of about 450,360 people [8]. This area is divided into 6 districts and 51 sub-districts.

2.3 Bootstrap

Bootstrap is an open source toolkit for developing with HTML, CSS, and JS. [11]. Bootstrap was developed by Mark Otto and Jacob Thornton at Twitter, and released as an open source product in August 2011 on GitHub. Bootstrap has become a de facto standard for web design. Bootstrap is a front-end framework of HTML, CSS, and JavaScript to create templates based on typography, forms, buttons, tables, navigation, modals, image carousels and many other optional JavaScript plugins, as well as sidabootstrap aims to create responsive and mobile-first website design [12].

3 E-census: system design

From the analysis of business process and problem analysis in Naikoten II, Kupang, it can be concluded that Naikoten II, Kupang needs e-census as real population data collection.

The e-Census system was developed under objectives related with survey environment, low cost, high efficiency, improvement of data quality, and shortening data release time. To meet the objectives, the e-Census system was organized into three entities including Surveyor, Data compiler and Residents.
The main features of this application are:

1. Login
2. Input the surveyor
3. Data management such as surveyor, district, city, province and census period.
4. Questionnaire management
5. Census result.

This research build two application, which are web based application and mobile application (Android version). The mobile application is used by surveyor meanwhile the web based version for editing and coding questionnaires, handling statistical computations, and reporting the census result.

4 Result and discussion

Residents list conducted to retrieve data from the existing data in Statistics Indonesia before doing the population census. Surveyor should login to the application to access the features. The main menu for surveyor as presented in Fig. 1.

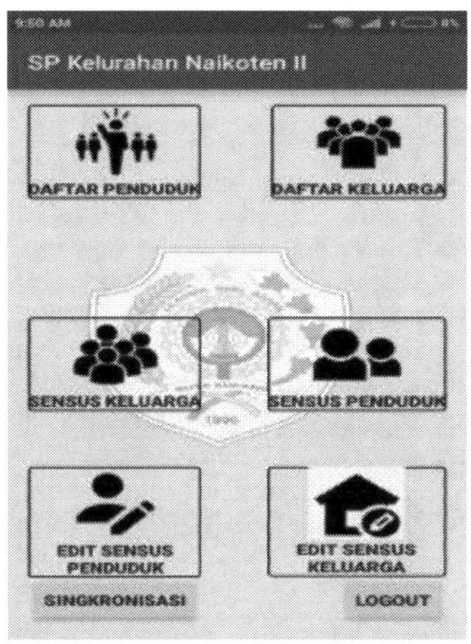

Fig 1. The main features of the application

Surveyor can choose the menu and the page will appear. If the resident is not in the list, the surveyor can create new data to input the resident data. Once the resident data is enrolled, the surveyor conducts the population census by clicking on the population census menu. After completing the data, the surveyor can save the census results by clicking the save button. The data will be saved in database in the server.

If the internet connection lost when the surveyor performs the census and input the data, the answer will be accommodated into the local database and will appear notice as in Fig. 2. Surveyor can return to the main menu and click the sync button when there is internet connection.

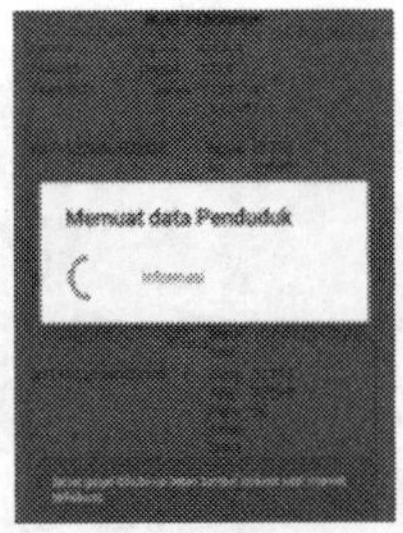

Fig 2. The warning message when the internet connection is lost

5 Conclusions

The main objective of this research was to provide a viable solution that allows surveyors to conduct the census efficiently. This research also provides a web based application to help data analyst to manage the census data. The testing has been conducted within the small group of stakeholder that volunteered to use this early version of the application for a specified amount of time to provide feedback for possible changes and improvements. The questionnaire of their experience using the application has been distributed to these respondents. The result of the questionnaire for Android interface, 80% of respondents stated good. For instructions in the application 86% of respondents stated very well. User friendly level on the application 88% of respondents stated good. For application performance 92% of respondents stated very well. For the whole application 90% of respondents stated very well.

References

1. M. P. Todaro, *Economic development 12th ed,* (Trans-Atlantic Publications, 2014)
2. I.K.W.T.Putera, *Population Growth and Economic Growth in Indonesia,* (Tilburg University, 2011)
3. G. Dang, L. Sui Pheng, *Infrastructure Investments in Developing Economies: The Case of Vietnam*, (Springer, Singapore 2015).
4. X. Gao, J. Lee, E-government services and social media adoption: Experience of small local governments in Nebraska state, Gov. Inf. Quarterly, **34**, 4, 627-634 (2017).
5. Indonesia Government Regulation (Peraturan Pemerintah), *Implementation of population census (Pelaksanaan sensus penduduk) No.21 1979*, (Indonesia Government, 2 Juli 1979).
6. Indonesia Government Regulation (Peraturan Pemerintah), *Economic census (Sensus ekonomi) No.29 1985,* (Indonesia Government, 1985)
7. Statictics Indonesia, Statictics Indonesia retrieved April from: https://www.bps.go.id/menu/1/informasi-umum.html#masterMenuTab1 (2018)
8. Statictics Indonesia, *Kabupaten Kupang dalam angka 2017*, (BPS, Jakarta, 2018).
9. Fathoni, A, Sensus penduduk : pengertian dan jenis-jenis, Retrieved: Oktober 25, from: http://www.zonasiswa.com/2014/10/pengertian-dan-jenis-jenis-sensus.html (2017)
10. Tukiran, Sensus Penduduk di Indonesia, J Kependudukan dan Kebijakan, **11**, 1, (2000).
11. F. Hussain, *Responsive Web Design by Example: Embrace responsive design with HTML5, CSS3, JavaScript, jQuery and Bootstrap 4* (Packt Publishing, 2017)
12. M. Lambert, *Learning Bootstrap 4 - Second Edition* (Packt Publishing, 2016)

Understanding the Behavioral Intention to Use a University Web-Portal

Ezra Aditia[1,*], I Nengah Tela[1] , *Nursyam Saleh*[1] , *Desi Ilona*[2] , and Zaitul[3]

[1]Civil Engineering and Planning Faculty, Universitas Bung Hatta, Padang 25133, Indonesia
[2]Economics Faculty, Universitas Putra Indonesia YPTK, Padang 25221, Indonesia
[3]Economics Faculty, Universitas Bung Hatta, Padang 25133,Indonesia

Abstract. Behavioral intention to use the information technology is a predictor of behavioral to use it. There are several studies investigating the effect of factors from Technology Acceptance Model (TAM) on attitude toward use and its effect on behavioral intention to use which based on Theory of Plan Behavior (TPB). However, there is a limited studies emphasizing on academic portal in a university. Therefore, this study investigates the effect of perceived ease of use (PEU) and perceived usefulness(PU) on attitude toward use (ATU) and Behavioral intention to use (BIU). Further, this study also aims to determine the relationship between attitude toward use (ATU) and behavioral intention to use (BIU). Using 123 registered students in four faculties, smart path least square (PLS) is applied to analysis the data. The findings show that PEU has a positive significant relationship with BIU and ATU. In addition, ATU also has a significant positive effect on BIU. However, PU do not have a significant relationship with BIU and ATU. This study has practical and theoretical implication and it discuss in detail in paper.

1 Background of Study

A modern higher education system is needed for a modern society. It focuses on knowledge management, lifelong learning and information technology. University as center of knowledge should have a good management practice. One of management aspects that must be covered by university is technology and information (TI), such as internet. Internet technology play an important role to support the learning process in a University. *Universitas Bung Hatta (*UBH*)*, one of the biggest private university in Indonesia, has developed a management information system, including academic information system (academic portal; portal.bunghatta.ac.id). Analog with library portal, academic portal is a customized, personalized and integrated information service to aggregate all kinds of academic resources and services through a single access and management point for users[1]. In academic portal, there are several information needed by university stakeholders. This portal can be assessed by students, lecturers, administration staffs, and managers (Rector, vice rector, dean, head department etc.).To access the portal, University also provide students and other stakeholders with Wi-Fi. However, students always complain with academic portal. IT department report

* Corresponding author: adi@bunghatta.ac.id

that there is low rate of using this portal to get academic and non-academic resources from university. In addition, there is no conclusive answer why students have low rate of using the university portal.

Theoretically, there are several theories why it happens. Technology Acceptance Model (TAM) is one of theories why acceptance or use of technology vary among individual [2]. Ease of use and perceived usefulness are factors affecting the acceptance of technology.[3]conclude that TAM is useful model to describe the behavior of technology use in Canada. In addition, [4]conclude that social influence and cognitive instrumental process have a significant effect on use acceptance in US. [5]study of behavioral intention to use the hotel internet application. The determinants of intention are an ease of use, usefulness, credibility and subjective norm in US. [6] examine the relationship between e-satisfaction and e-retention in e-learning context of United Emirate Arab (UEA). [7] confirm that the explanatory power of TAM in adoption of social network in Portugal.[8]found that attitude toward usage and subjective norm significantly influence the behavioral attention to use the technology. [9]investigate the effect of perceived privacy and government support on the intention to use internet banking services in Indonesia.[10]study an internet adoption and use among men and women among Indonesia's students. Finally, [11]study an internet cafe users in Indonesia.

However, less attention has been paid this issue in higher education institution, such University. [12]state that Education institution is a potential key to fostering greater internet use and large body of research has addressed the psychological variables that influence technology use or non-use, such as academic portal. In addition, the developing countries have benefited from the internet or IT in general and it, however, has receiveda little research attention [12]. According to[2], there are several determinants of the attitude toward use of Technology : ease of use and perceived usefulness. Therefore, this study investigates the effect of Perceived-Ease of Use (PEU) and Perceived Usefulness (PU) on Attitude Toward Use (ATU) of academic portal (see figure 1). Further, it also examine the relationship between attitude toward use (ATU) and Behavioral Intention to Use(BIU) an academic portal(see figure 1) that suggested by Theory of Plan Behavior[13]. This study differs from study of [14]where they study based on TAM, this study use two theory: TAM and TPB. Therefore, this studies combine two theories: Technology Acceptance Model (TAM) and Theory of Plan Behavior (TPB) in context of university students. This study contribute to the theory of TAM [2], [4] and TPB[13]. Therefore, this studies have practical implication in the sense that University can use the findings for IT related decision to improve student's performance.

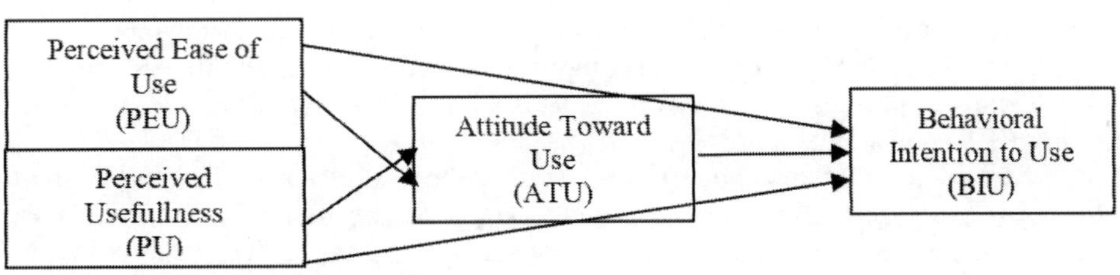

Fig. 1. Conceptual Model.

[12]argue that perceived usefulness refers to the expectation that the technology will enhance one's job performance and perceived ease of use defined as belief that using the technology will be free of efforts. Thus, [8]defined the attitude toward use as the way individuals respond to and are disposed towards an object. Behavioral intention to use is defined as strength of one's intention to perform a specified behavior[10].

2 Methodology

UBH has seven faculties with more than 8700 registered in 2018. This study uses the students in four faculties in UBH, i.e. Economics, Education, Law and Humanities. Convenient sampling method is applied in this study. A structured questionnaire was designed and implemented in the context of university portal. All construct measurement was adapted from[7]. Each item in questionnaire is constructed as follows: (i) Three items for BIU, (ii) Three items for ATU, (iii) seven items for PU and (iv) five items for PEU. Five-point Likert scale was used to measure the respondents' perception on the survey question. Data was analyzed using smart-pls. There are two analyses: measurement and structural model. There are four criteria's for measurement model [15] : (i) indicator reliability (using outer loadings, > 0.70), (ii) internal consistency (using composite reliability and Cronbach alpha,> 0.70) (iii) convergent validity (using AVE, > 0.50), and (iv) discriminant validity (using Fornell-Lacker criterion, greater than its cross loading). In addition, there are two criteria for structural model [15]: (i) predictive relevance (using Q-square, > 0), and (ii) Predictive power (using R-square). The objective of research is answered by assessing the significance of structural path coefficients: (i) 1.65 for 10%, (ii) 1.96 for 5% and (iii) 2.58 for 1%.

3 Result and discussion

Students in four faculties are object of studies. 150 questionnaires were distributed and the researcher managed to collect 135(return rate of 90%). During screening, 12 incomplete questionnaires was excluded from analysis. Demographic analysis of respondent from gender, female students are 70 students or 56.90% and the rest is male student (43.10%). According to respondent's age, student with age of 18 to 20 years is 55 students or 44.72%, followed by 60 or 48.78% are student with age of 21 to 23 years. Further, 8 students (6.50%) are greater than 23 years old. However, majority of respondents are from faculty of economics (54.50%), followed by education (29.30%), law (10.60%) and humanity (5.70%).

3.1 Analysis of measurement model

The result of measurement model is shown in Table 1. [16]state that the validity of measurement model depends on the indicator loadings, Cronbach alpha (CA), construct reliability (CR), average variance extracted (AVE) and discriminant validity. The result show that indicator loading for all items are greater than 0.7 and it can conclude that all items surpass the cut off level [17]. In addition, result of reliability test is also achieved the criteria because the value of Cronbach's alpha and composite reliability are greater than the threshold level of 0.7. further, the result of convergent validity which applied the average variance extracted (AVE) also indicate more than 0.5.

The last criteria are discriminant validity. This kind of validity is assessed by using the Fornell-Lacker analysis. This value is created through square root of the AVE for each construct. Decision whether construct is valid or not by comparing this value with other constructs. All constructs have a satisfied discriminant validity because all value of square root of AVE each constructs is higher. For example, square root of BIU's AVE is 0.707601, which is greater than correlation between BIU & ATU (0.653447), BIU& PEU (0.663595), and BIU & PU (0.469327).

Table 1. Measurement Model.

Construct	Item	Loadings	CA	CR	AVE
BIU	biu_1	0.766297	0.714121	0.839807	0.636255
	biu_2	0.827066			
	biu_3	0.798446			
ATU	atu_1	0.798663	0.782333	0.873484	0.697436
	atu_2	0.879254			
	atu_3	0.825443			
PEU	peu_1	0.824804	0.859955	0.899821	0.899821
	peu_2	0.715683			
	peu_3	0.862417			
	peu_4	0.78361			
	peu_5	0.816111			
PU	pu_1	0.764462	0.910676	0.929165	0.929165
	pu_2	0.810103			
	pu_3	0.773517			
	pu_4	0.852532			
	pu_5	0.882891			
	pu_6	0.748115			
	pu_7	0.81503			

3.2 Analysis of Structural model

The result of structural model is demonstrated in Table 4. The overall explanatory power (R square) show that the model accounts for 42.69% and 40.92% of BIU and ATU respectively. In addition, second assessment is Q square. Q square is 0.27 for BIU and 0.25 for ATU respectively. These value is more than zero, which satisfy the model's predictive validity [16]. Based on the path coefficient and t value, three constructs have a significant relationship: (i) perceived ease of use (PEU) and attitude toward use (ATU), (ii) Perceived ease of use (PEU) and BIU, and (iii) attitude toward use (ATU) and BIU.

Table 3. Structural Model Analysis.

Endogenous Construct	R-Squared	Q-Squared	Relation	Path Coefficient	T Value
BIU	0.426993	0.270757	ATU -> BIU	0.379359	3.707242***
ATU	0.409186	0.254627	PEU -> BIU	0.428289	3.038777***
			PEU -> ATU	0.638142	4.854615***
			PU -> BIU	0.001302	0.011789
			PU -> ATU	0.001055	0.007794

Notes: *** indicate a significant at 1%

PEU has a significantly positive relationship with ATU of academic portal. It means that the higher the PEU, the better ATU of academic portal. This finding is consistent with result of Pinho & Soares (2011). In addition, there is also a positive significant relationship between PEU and BIU the academic portal. This finding is aligned with evidence of [5]. Further, the effect of ATU on BIU is positive significant. The better the ATU of academic portal, the higher BIU. This result is supported by [8]. However, the effect of PU on ATU and BIU are not significant and this findings are not consistent with finding of [7], [8]. In brief, these findings confirm that TAM is partially explain the technology acceptance among university students. Thus, it also confirms that the relationship between attitude toward use and behavioral intention to use is underpinned by Theory of Plan Behavior.

4 Conclusion

The important of university portal, such as academic portal, has been documented by several researchers. This study investigates the effect of Perceived Usefulness (PU) and Perceived-Ease of Use (PEU) on Attitude Toward Use (ATU) and the relationship between ATU and Behavioral Intention to Use (BIU). Students from four faculties are as research object. The findings show that PEU has a significant relationship with ATU and BIU. In addition, ATU also has a positive effect on BIU. This study has a theoretical and practical implications. This study contributes the TAM and TPB. These findings also could be used by university manager to design, revise, and develop the academic portal. This study has limitation in the sense that there is a limited data (only 123 students) and do not investigate the role of ATU as mediating variable. there is some avenue for future researchers who interested in this topic. They can add more respondents and testing the mediating role of ATU between PEU & PU and BIU.

References

1. M. N. Masrek, A. Jamaludin, and S. A. Mukhtar, . *Libr. Rev.*, vol. **59**, no. 3, pp. 198–212 (2010)
2. F. D. Davis, *Int. J. Man-Machine Stud.*, vol. **38**, pp. 475–487, (1993)
3. P. Legris, J. Ingham, and P. Collerette, *Inf. Manag.*, vol. **40**, pp. 191–204 (2003)
4. V. Venkatesh and F. D. Davis, *Manage. Sci.*, vol. **46**, no. 2, pp. 186–204 (2000)
5. J. S. Kim, *Int. J. Contemp. Hosp. Manag.*, vol. **28**, no. 8, pp. 1535–1553 (2016)
6. M. A. Al-hawari and S. Mouakket, *Educ. Bus. Soc. Contemp. Middle East. Issues*, vol. **3**, no. 4, pp. 299–314 (2010)
7. J. C. M. R. Pinho and A. M. Soares, *J. Res. Interact. Mark.*, vol. **5**, no. 2/3, pp. 116–129 (2011)
8. T. Teo and C. B. Lee, *Campus-Wide Inf. Syst.*, vol. **27**, no. 2, pp. 60–67 (2010)
9. A. Susanto, H. Lee, H. Zo, and A. P. Ciganek, *Inf. Dev.*, vol. **29**, no. 4, pp. 309–322 (2012)
10. F. Wahid, *Electron. J. Inf. Syst. Dev. Ctries.*, vol. **32**, no. 6, pp. 1–8 (2007)
11. B. Furuholt, S. Kristiansen, and F. Wahid, *Electron. J. Inf. Syst. Dev. Ctries.*, vol. **22**, no. 3, pp. 1–16 (2005)
12. M. Fusilier and S. Durlabhji, *Campus-Wide Inf. Syst.*, vol. **22**, no. 4, pp. 233–246 (2005)
13. I. Ajzen, *Organ. Behav. Hum. Decis. Proccess*, vol. **50**, pp. 179–211 (2013)
14. T. Handayani, Kusrini, and A. Sunyoto, *Angkasa*, vol. **5**, no. 1, pp. 63–74 (2013)
15. V. E. Vinzi, W. W. Chin, J. Henseler, and H. Wang, *Handbook of Partial Least Square: Concepts, Methods and Applications*. Berlin, German: Springer (2010)
16. J. F. Hair, G. T. M. Hult, C. M. Ringle, and M. Sarstedt, *A Primer on Partial Least Squares Structural Equation Modeling (PLS-SEM)*. Thousand Oaks: Sage (2013)
17. J. F. Hair, C. M. Ringle, and M. Sarstedt, *J. Mark. Theory Pract.*, vol. **19**, no. 2, p. 2011 (2011)

MATEC Web of Conferences **248**, 05005 (2018) https://doi.org/10.1051/matecconf/201824805005
ESTIC 2018

Identification of Implementation Program of Electricity Competency

Ija Darmana[1,*], *Nizwardi* Jalinus[2], *Ganefri*[3] and *Sukardi*[4]

[1] Faculty of Industrial Technology, Universitas Bung Hatta Padang, West Sumatera, Indonesia.
[1] PhD Students (S3), Faculty of Engineering, Universitas Negeri Padang, West Sumatera, Indonesia.
[2,3,4] Faculty of Engineering, Universitas Negeri Padang, West Sumatera, Indonesia.

Abstract. This study aims to identify the implementation of technical personnel competency testing program. This study applies the Kirkpatrick model at four levels, namely: 1) the level of satisfaction of participants in the implementation of the program; 2) the level of understanding of the subject by the participants; 3) changes in the professional behavior of the participants in the proficiency tests after their return to work; 4) the impact of changes in the professional behavior of the participants in the skills tests. The identification of assessment instruments for each level is obtained from bibliographic studies relevant to the program. The results of the study concluded that the application of the Kirkpatrick model in this study was as follows: 1) short-term skills testing program; 2) Kirkpatrick model widely used to evaluate program training programs

1 Introduction

The ASEAN Economic Community calls for the abolition of rules that previously prevented the recruitment of foreign workers. Restrictions, particularly in the professional power engineering sector, are encouraged to be removed. MEA offers foreign workers opportunities to fill jobs and professions in Indonesia. A study by the World Labor Organization predicted that many companies would find their employees less qualified because of a lack of professional training.

Winterton skills [1, 2] are needed and expected in the workplace and refer to the knowledge, skills and attitudes needed for workers to work. Recognition of professional skills is obtained after conducting skills tests conducted by training institutions.

According to Sedarmayanti [3], competence is a fundamental characteristic of someone who has a direct influence or can predict good performance. Skill according to Sedarmayanti in his research "Testing competence rather than intelligence" is: skills and knowledge.

The phenomenon that occurs, participants in proficiency tests that run skill associations or institutions in West Sumatra, are relatively few, which will affect the mental attitude and professional ethics of the technical staff to deal with the competition on the free market. Participants in the skills tests held by APEI West Sumatra on December 1, 2016 in Padang,

* Corresponding author: ija.darmana@bunghatta.ac.id

twenty-two people. On the basis of the phenomenon mentioned above, "the implementation of the proficiency test must be identified", ie whether a program of implementation of the skills tests requires an evaluation model to evaluate the program

2 Literature Review

Khandker [4], this evaluation approach can be carried out using quantitative methods (such as the collection of survey data or simulations) before or after the implementation of the program. This assessment predicts the impact of the program using the data before and after the implementation of the program. Although each evaluation model still has limitations, choosing the right model will have a direct impact on the quality of the information produced during the evaluation.

The researchers conducted a study of several research results in the form of journals and dissertations, including: Hoover [5] conducted a thesis study titled Evidence-based Interventions: Early Examination and Continuing Education.

Homklin [6] conducted a thesis entitled: Skills Certification System Training: The Case of the Automotive Industry in Thailand. Most research and training practices in Thailand focus on level one (response) and level two (learning) with the evaluation of the Kirkpatrick model. Preston [7] conducted a thesis study entitled: Perception of Leadership in Results and Evaluation of Return on Investment Training. This study applies a five-level evaluation of the return-on-investment model, level five. The purpose of this study was to assess how effectively the program has increased the value of organizations in the six largest Catholic health organizations in the United States. accurately reflect the expectations of organizational leaders.

Chang [8] conducted a thesis entitled: An Empirical Study of Kirkatrick's Evaluation Model in the Hospitality Industry. This study examines the Kirkpatrick training evaluation model [9] by evaluating sales training programs conducted in hotel organizations in Florida. The results of his research show that: the results of the analysis largely support this research hypothesis. There is a significant increase in the fact that booking agents completing the training deepen their knowledge of the content and skills required.

Farjad [10] in the journal Procedia-Social and Behavioral Sciences 46 (2012) 2837-2841, entitled: Effectiveness of the evaluation of university education by Kirkpatrick Model (case study: Islamshahr University). The results of the study indicate that training efficiency is below the standard level. The results of the effectiveness evaluation show that the reaction, learning, behavior and organizational level need to be improved.

Donald [11] from the perspective of health information management. Spring 2011, vol. 8 Number 2, p1-5. 5p, Using the Kirkatrick Evaluation Framework to Determine the Effectiveness of Health Information Management Programs and Programs. Researchers used the Kirkpatrick assessment framework to present models that can be used by health information instructors. According to the review literature, the identification of competency tests is necessary to prepare a competency test evaluation model.

3 Methodology

Moldovan [12], Procedia Technology 22, entitled Results Model Evaluation Training. Kirkpatrick evaluates the improvement of the model by linking the level of evaluation to the correct objectives. By following the PDCA cycle. The level of objectives includes results, performance, knowledge and motivation related to the level of evaluation of reactions, learning outcomes, and behavior.

4 Results and Discussions

Level 1, at this level, a person must be able to perform tasks / routine work based on the understanding of work procedures / instructions under the supervision of the direct supervisor.
Level 2, at this level, a person must be able to perform routine tasks / work based on the application of procedures / instructions and the execution of tasks and work requiring: 1) the ability to solve problems.
Level 3, at this level, a person must be able to perform routine tasks / tasks based on procedures / instructions and perform tasks and tasks requiring the existence of: 1) Ability to analyze problems, 2) Ability to solve problems 4) Ability to coordinate and supervise subordinates. The pooling of similar skills is grouped into certain positions to carry out multi-level, fully trained and educated tasks. Clustering is called skill-based professional qualification.
Level 4, at this level, a person must be able to perform tasks / jobs that are able to use a number of procedures, in different, ordinary or unusual contexts, and with extended scope and demanding reasoning skills technical.
Identification results obtained:

1. Research instruments are needed. It is a tool used by researchers to collect research data by measuring them. This is a written directive in the form of a list of questions prepared to obtain information from respondents.
2. Questionnaire Instrument, the questionnaire is used in the needs analysis as a tool to determine the need to develop models for the assessment of the skills of the technical staff of the commercial electricity construction services. The respondents to this questionnaire are the users, namely the Institute or the Electricity Association and the Director.
3. Validation form, this instrument is used as a tool to determine the validation of the technical engineering competency test evaluation model of the electricity generation service activity that has already been designed to the need for a proficiency test.
4. The evaluation instructor seeks to know the validity of the evaluation model of the proficiency test of the technical staff of electrical engineering services by experts. Before conducting the research, the researchers compiled instruments based on indicators of implementation of skills tests at the BNSP and professional associations.
5. The scale used to compile the validation instrument in this study is the Liker scale. The validation form of the Electrical Engineering Technical Competency Test Evaluation Template contains evaluation elements that include elements of the electrical engineering services engineering competency assessment model and a questionnaire.

5 Conclusion

The identification of training participants 'assessments should be based on trainees' ability to achieve the objectives. Step 1, for feedback on training, how much training material was designed, and how well the trainer performed well. Step 2, learning includes written tests, oral exams and practical tests. In Step 3, the attitude / behavior includes the work of the employees at work to see if the employee correctly applies his knowledge and skills. In Step 4, the impact includes the effect of the training on the organization on the impact of the training. Training is one of the most comprehensive methods to increase individual productivity and improve job performance in the work environment.

References

1. Jonathan Winterton, *Typology of knowledge, skills and competences: clarification of the concept and prototype*. Italy, (2006)
2. Jonathan Winterton, *What Is Competence?*, Human Resource Development International, Vol. 8, No. **1**, 27–46, (2005)
3. Sedarmayanti, *Human Resources and Work Productivity*, Mandar Maju, Bandung, (2009)
4. Shahidur R. Khandker, Gayatri B. Koolwal and Hussain A. Samad, *Handbook on Impact Evaluation, Quantitative Methods and Practices*, Washington DC 20433, (2010)
5. N. Hoover, *Evidence Based Intervention: An Examination of Pre-Service Training, Profesional Development, And Classroom Implementation*. Texas A&M University-Corpus Christi Corpus, (2013)
6. Tassanee HOMKLIN, *Training Effectiveness of Skill Certification System : The Case of Automotive Industry in Thailand*. Hiroshima University, (2014)
7. Kevin F. Preston, *Leadership Perceptions Of Results And Return On Investment Training Evaluations*. Colorado State University, (2010)
8. Ya-Hui Elegance Chang, *An Empirical Study of Kirkpatrick's Evaluation Model in the Hospitality Industry*. Florida International University, (2010)
9. Donald L. Kirkpatrick And James D. Kirkpatrick, Third Edition. *Evaluating Programs The Four Levels*. Westchester Book Group, (2006)
10. Shahrooz Farjad, Procedia-Social and Behavioral Sciences **46** (2012) 2837–2841. *The Evaluation Effectiveness of training courses in University by Kirkpatrick Model (case study: Islamshahr university)* (2012)
11. Rouse, Donald, *Employing Kirkpatrick's Evaluation Framework to Determine the Effectiveness of Health Information Management Courses and Programs*. Perspectives in Health Information Management. Spring2011, Vol. **8** Issue 2, p1-5. 5p, (2011)
12. Liviu Moldovan, *Training Outcome Evaluation Model*. Procedia Technology **22** 1184–1190, (2016)

Antecedents of Intention to Use E-Learning

Khairudin[1], *Susi* Herawati[1], *Desi* Ilona[2] and *Zaitul*[3*]

[1]Faculty of Teacher Trainning and Education, Universitas Bung Hatta, Padang, Indonesia
[2]Faculty of Economics, Universitas Putra Indonesia YPTK, Padang, Indonesia
[3]Faculty of Economics, Universitas Bung Hatta, Padang, Indonesia

Abstract. This study investigates the effect of Attitude toward e-learning, Facility condition, and Personalization on intention to use e-learning. theory of plan behaviour is used to understand the antecedents of intention to use e-learning. Using academic staffs as a research object and SEM-PLS, we conclude that Attitude toward e-learning and Personalization have a positive relationship with intention to use e-learning. However, Facility condition has a negative effect on intention to use e-learning. This study partially contributes to theory of plan behaviour. This finding also could be used by university to formulate e-learning concept by considering the significant antecedents in this study.

1 Background of Study

Higher Education Institution (HEI) has obtained remarkable opportunities indebted to the development of Information and Technology [1]. Reference [2] argue that there are two important ways by which we study and get information through Information technology (IT). First, IT enable us to conduct things more flexible, more efficiently, faster, and with substantial access for all. Next reason, Information technology permits us to carry out many of things that we cannot do know, or to do them in ways that are notably dissimilar. In fact, Information Technology creates an experience of learning in outside of curricula, classrooms, text-based formats to which we are customized [2]. In addition, it has been critical to assimilate information technology into e-learning in the era of knowledge economy. Reference [3- 4] state that there is increasing a dependency on the use of e-learning to ease learning and development in the workplace. The web-based application and internet technologies development in education serves unknown chances to aid lecturers effectively conveys the learning materials to students and teach every one, anytime and any places [5]. In fact, e-learning is worthwhile method because it has a low variable cost, flexible learning content, free of temporal and spatial constraints, and cost effectiveness [3]. [5] state that e-learning is more pliable way for learning, and it is specifically convenient for persons with high self-control that allow them to acquire a knowledge at isolated places based on their own necessity.

E-learning requires the use of electronic media, such as the mobile phone, CD-ROM, internet, DVD, videotapes, television, etc. [6]. Universitas Bung Hatta is one of private universities located in west Sumatra, Indonesia. The IT Centre has developed the e-learning

* Corresponding author email: zaitul@bunghatta.ac.id

system decade ago. However, it was not used optimally and usage rate was very low. A year ago, the IT Centre has been revised the system and use html-platform. No study has been done to see whether academic staffs are willing to use this system. However, the study of e-learning has been done by many researchers in different location and context, such as [1– 9].

From literature review of previous researches, there is a gap in literature, especially intention to use the e-learning system in private university in Indonesia. Therefore, the study in this subject would enrich the literature of Technology acceptance [10] and behaviour intention [11]. This study uses the attitude, facilitating condition, and personalization as antecedent of an intention to use the e-learning. These variables are selected due to inconclusive of previous studies findings. Therefore, this study aims to investigate the effect of these three variables on intention to use e-learning. The article is organized as follow. First session is background of study. Second session about methods and proceed to result and discussion. Final session is conclusion and recommendation.

2 Method

This study use an academic staff registered in Universitas Bung Hatta. Academic staffs are distributed a questioner using google form and share through WhatsApp (WA). For those using WA and registered as Gmail or google mail are expected to fill the questioners. This study applies the primary data gathering through survey. Intention to use e-learning is dependent variables and the independent variables are Attitude (Att), Facilitating Condition (Fas) and Personalization (Per). Intention to use e-learning was developed by [7]. Further, Attitude is measured by four items which developed based on [11] and adapted by [8]. Thus, facilitating condition use the items developed by [6], [12]. In addition, personalization use a measurement adapted by [7]. Data is measured by five-points Likert scales. The data analysed by using the smart-PLS. There are two model in smart-PLS: the measurement model and structural model [13]. Four criteria are used for measurement model [14]: (i) indicator reliability, (ii) internal consistency, (iii) convergent validity, and (iv) discriminant validity. Indicator validity applies the outer loadings and the value must be greater than 0.60. In addition, internal consistency uses the composite reliability and Cronbach alpha and the value must be more than 0.70. Thus, convergent validity utilizes the AVE and it must be greater than 0.50. Finally, discriminant validity applied the Fornell-Lacker criterion and it should be more than its cross loading. Meanwhile, the structural model use two criteria [14]: predictive power, and predictive relevance. In addition, predictive power using R-square and higher value is better. Thus, predictive relevance uses Q-square and it must be greater than 0. To answer the hypotheses, this study uses the significance of structural path coefficient [13].

3 Result and discussion

Final sample of this study is seventeen respondents. Demographic analysis is demonstrated in Table 1. Based on gender, male respondents are teen academic staffs or 58.82% and the rest are female (41.18%). Respondent age is ranging from 31 to 53 years old. In addition, respondent with age 31 to 40 years old is about four respondents or 23.53%, followed by 11 to 40 years old (58.82%) and more than 50 years old (17.65%). Based on faculty, academic staffs from economics, law, humanity, education and postgraduate are 11.76%, 5.88%, 5.88%, 23.53%, 23.53%, and 29.41% respectively. Level of respondent education are 35.29% of doctorate and the rest are master holders. Academic position is tutor, lecturer and senior lecturer (29.41%, 52.94%, and 17.65%).

Table 1. Demographic. Analysis

Demographic	Category	Number	Percentage
Gender	Male	10	58,82
	Female	7	41,18
Age	31 to 40 years	4	23,53
	11 to 50 years	10	58,82
	> 50 years	3	17,65
Faculty	Economics	2	11,76
	Law	1	5,88
	Humanity	1	5,88
	Education	4	23,53
	Civil Engineering	4	23,53
Education Level	PostGraduate	5	29,41
	Doctor	6	35,29
	Master	11	64,71
Position	Tutor	5	29,41
	Lecturer	9	52,94
	Senior Lecturer	3	17,65

3.1 Analysis of Measurement model

As mention above, there are four criteria used to analysis the measurement model. The result can be seen in Table 2. First analysis show that there are items having outer loading less than 0.60 and these items are excluded (att-4, per-2 and per-3). The analysis run again and the result indicate all outer loadings are greater than 0.6. second criteria is internal consistency and the result show that composite reliability and Cronbach alpha are satisfied, except for Personalisation, but still greater than 0.5 and it also has high composite reliability. Third criteria is convergent validity and it uses AVE. The value of AVE for all variables are greater than 0.5 and satisfied the requirement.

Table 2. Measurement Model

Construct	**Item**	**Loadings**	**CA**	**CR**	**AVE**
Intention to Use	iu_1	0,84			
	iu_2	0,66	0,702	0.817	0,601
	iu_3	0,82			
Facility Condition	fc_1	0,95			
	fc_2	0,97	0.935	0.953	0,836
	fc_3	0,85			
	fc_4	0,88			
Attitude	at_1	0,64			
	at_2	0,90			
	at_3	0,83	0.758	0.836	0,635
Personalisation	pe_1	0,82			
	pe_4	0,83	0.515	0.805	0,674

Finally, discriminant validity uses the Fornel-Lacker criterion and the result satisfied the requirements (see Table 3). The coefficient correlation is created through square root of AVE (bold number). The value of coefficient correlation must be greater than the value of coefficient correlation with other construct. For example, coefficient correlation between Attitude (Att) is 0.797 and this value is greater than coefficient correlation between Attitude Facility (-0.257), Intention (0.560), and Personalization (0.351). therefore, it can be concluded that the construct has a good discriminant validity. The same explanation is also applied for other constructs.

Table 3.Fornell-Lacker Criterion

Construct	Att	Fas	Int	Per
Att	**0.797**			
Fas	-0.254	**0.914**		
Int	0.560	-0.568	**0.775**	
Per	0.351	-0.223	0.735	**0.821**

3.2 Analysis of Structural Model

The result of analysis of structural model could be seen in Table 4. There are two criteria in structural model analysis [14]: predictive power, and predicative relevance. Predictive power use R-square and its value should be higher, the higher the better. Meanwhile, the predictive relevance applies the Q-square and the cut-off value must be more than zero. The t value of structural path coefficient is applied to see if any significant relationship between independent variables and dependent variable. The result show that R-square is 0.774 which means that intention to use e-learning is explained 77.4% by independent variables and the rest is explained by other variables which does not include in this study. Predicative relevance also indicates the value (0.268) which is greater than zero and it, therefore, satisfy the requirement.

Table 4.Structural Model Analysis

Endogenous Construct	**R-Squared**	**Q-Squared**
Int	0.774	0.268
Relation	**Path Coefficient**	**T Value**
Att -> Int	0.269	4.895
Fas -> Int	-0.374	7.949
Per -> Int	0.554	13.594

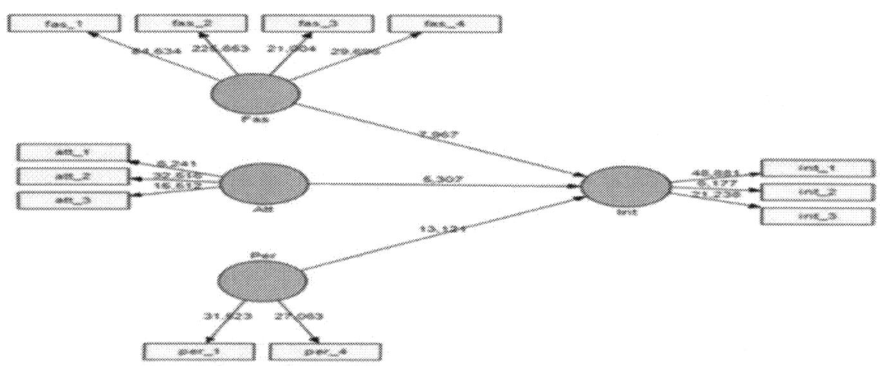

Fig 1. Structural model

The effect of Attitude, Facility condition and Personalization on intention to use e-learning could be seen from t-value and path coefficient. The Attitude has a positive relationship with intention to use e-learning due to the t value of Attitude is greater than 1.96 (α=1%). Thus, it means that the higher the attitude toward e-learning, the higher the intention to use e-learning. This finding is consistent with the finding of [8,9]. The second finding is the significant effect of facility condition on the intention to use e-learning due to the t value of this variable is greater than 1.96. Nevertheless, the relationship is negative which means that the better the facility condition, the lower the intention to use e-learning. This finding is not supported by [6] who conclude that there is a positive relationship between facility condition and intention to use e-learning. Personalization has a positive significant effect on the intention to use e-learning. this finding is alignment with the finding of [7]. The higher the Personalization, the higher the intention to use e-learning. the structural model is also demonstrated in Figure 2 below

4 Conclusion and Recommendation

Today higher education institution has been relying on e-learning to facilitate learning process. Many higher education institution has been used e-learning and other is starting to use it, such as Universitas Bung Hatta. However, it is not known how intention to use among academic staffs. Therefore, this study investigates the effect of attitude, facility condition and personalization on intention to use e-learning among academic staff in Universitas Bung Hatta. Using seventeen respondents and SEM-PLS, it is found that attitude and personalization have a positive effect on intention to use e-learning. However, facility condition has a negative relationship with intention to use e-learning. Theoretically, this study contributes to the theory of plan behaviour [11]. Practically, these findings could be used by higher education institution to formulate e-learning concept by considering Attitude, Facility condition and Personalization of user, such as academic staffs. Several limitations to this study need to be acknowledged. First, the sample size is too small. Second, this study use object in one higher education institution. Finally, the number of independent variables used in this study is limited to three variables. It is recommending that further research be undertaken in the following areas: (i) future research add sample size in order to gain more robust result, (ii) future research also can examine intention to use e-learning in other different higher education institution, and (iii) next study can add other variables from other perspective, such as from Technology Acceptance Model (TAM).

References

1 M. Ali, S. A. Raza, W. Qazi, C.-H. Puah, *Interact. Technol. Smart Educ.*, **15**, 1, (2018)
2 J. C. Roca, C. Chiu, F. J. Martinez, *Int. J. Human-Computer Stud.*, **64**, (2006).
3 K. Rui-Hsin, C.-T. Lin, *Polic. an Int. J. police Strateg. Manag.*, **41**, 1, (2018)
4 J. Fleming, K. Becker, C. Newton, *Educ. Train.*, **59**, 1, (2017)
5 Y. Zhang, Y. Fang, K. Wei, Z. Wang, *Inf. Technol. People*, **25**, 4, (2012)
6 T. Teo, *Campus-Wide Inf. Syst.*, **28**, 2, (2011).
7 Y.-M. Cheng, *Int. J. Web Inf. Syst.*, **10**, 1, (2014).
8 R. Arteaga Sánchez, A. Duarte Hueros, M. García Ordaz, *Campus-Wide Inf. Syst.*, **30**, 2, (2013)
9 R. Cheung, D. Vogel, *Comput. Educ.*, **63**, (2013)
10 F. D. Davis, *Int. J. Man-Machine Stud.*, **38**, (1993)
11 I. Ajzen, *Organ. Behav. Hum. Decis. Proccess*, 50, (1991)
12 T. Teo, C. B. Lee, *Campus-Wide Inf. Syst.*, **27**, 2, (2010).
13 J. F. Hair, G. T. M. Hult, C. M. Ringle, M. Sarstedt, *A Primer on Partial Least Squares Structural Equation Modeling (PLS-SEM).* (Sage, 2013).
14 V. E. Vinzi, W. W. Chin, J. Henseler, H. Wang, *Handbook of Partial Least Square: Concepts, Methods and Applications.* (Springer, 2010).

Application of Compound Bonding Based On Augmented Reality

Alexander Setiawan[1,*], *Silvia* Rostianingsih[2] , and *Timotitus* Reinaldo Widodo[3]

[1,2,3] Informatics Department, 60236 Petra Christian University, Indonesia

Abstract. Augmented reality technology is very important in visualizing real objects or concepts. The advancement of computing equipment enables 3D modelling to be performed virtually with relatively easy processes. The appearance of 3D models in virtual is very interesting because it can be combined with interactive multimedia applications so that users can control the course of the application as needed. This augmented reality application has been through process testing, system testing on different mobile devices, and marker testing. The results of process testing indicate that every process on the application goes according to existing model design. Test results show that applications created using the library vuforia and Unity can run well. High school students show that as many as 84.6% who agree with using this application is more interesting than the usual method of learning so as to be able to recognize compound bonding using this application well.

1 Introduction

The development of science and technology more rapidly, it can be seen and we feel around us directly. These developments are impacting on all aspects of human life. Augmented Reality is one example of reversible technological developments in the field of chemical education. Augmented Reality is a technology where the virtual world is brought into the real world for a specific purpose. In the field of education, can be made markers combined with AR technology, so that the possibility of the emergence of 3D objects based on images or markers. With the implementation of AR technology implemented can be obtained interesting and interactive learning.

Augmented Reality (AR) is a variation of Virtual Environment (VE), or better known as Virtual Reality [1]. Virtual Reality technology makes the user incorporated in a virtual environment as a whole. With the help of augmented reality technology, the real environment around will be able to interact in digital (virtual).

Chemical learning about compound bonding is usually boring by children. This is because children only know the theory of books and teacher explanations only. Therefore, with the creation of applications that utilize Augmented Reality, children are expected to be more interested and enthusiastic to learn compound bonding because with applications that utilize Augmented Reality children can see 3D models of these chemical elements and also see the merger between elements.

* Corresponding author: alexxander@petra.ac.id

2 Applications Theory

2.1 Augmented Reality

Augmented Reality (AR) technology can incorporate 3D objects into real environment using webcam media. Unlike Virtual Reality that completely replaces reality, Augmented Reality only adds or complements reality. Virtual objects that are merged into the real environment serve to display information that can't be received by man directly. This makes Augmented Reality useful as a tool to help the user's perception and interaction with the real world. The information displayed by the displayed object helps the user perform activities in the real world. There are three principles of Augmented Reality, the first of which is the combination of the real world and the vitality, the second runs interactively in real time, and the third is the integration between objects in three dimensions, virtual objects integrated in the real world. [2].

2.2 Android Programming

In general, application of "Compound Bonding" is used to introduce chemical engineering using mobile learning technology. This application is designed to facilitate the introduction of compound bonding, aimed at students and people to provide information to users. In building this application program using Android [3]. Android is an operating system developed by Google, which promises openness, affordability, open source, and quality framework, to meet the needs of the operating system that supports standard and publishing APIs and can be utilized as a whole with low cost. [4]. Android provides an open platform for developers to create their own applications. Initially developed by Android Inc., a firm new comers who make software for mobile phones which was later purchased by Google Inc. For its development, formed the Open Handset Alliance (OHA), a consortium of 34 hardware companies, software companies, and telecommunications, including Google, HTC, Intel, Motorola, Qualcomm, T-Mobile, and NVIDIA. [5]. The open nature of Android that has made a large number of communities have sprung up application developers to use Android as the basis for the project of making applications, by adding new features for Android on the devices officially released by using another operating system. [6]. Android is a software stack for mobile devices that includes an operating system, middleware and key applications. The Android SDK provides the tools and APIs necessary to begin developing applications on the Android platform using the Java programming language. [7]

2.3 AJAX

AJAX is a technique to make the display more quickly and dynamically. AJAX allows the display can perform updates by exchanging small amounts of data. The advantage of using AJAX is able to reload in a section on the display without having to refresh the entire page. The goal is to move most of the interaction on a computer web surfer, to exchange data with the server behind the scenes, so that the web page does not have to be read as a whole every time a user makes a change. This will increase interactivity, speed, and usability. [8].

3 Design Methods

The design for the system will be divided into 2 (two) parts, the overall system design, search process design and tabulation process design. Overall system design describe the design throughout the system, as in Figure 1. Design of the search process describes the steps taken by the system in searching and finding the desired data chemical. The schematic of search process design can be seen in Figure 2.

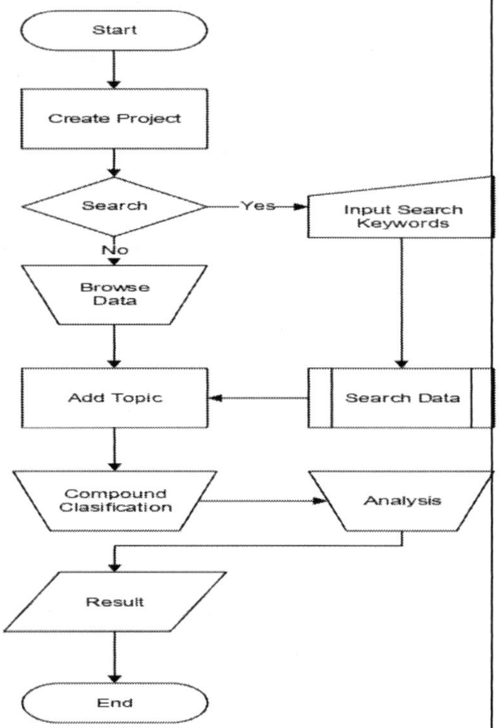

Fig 1. System Design Compound Bonding

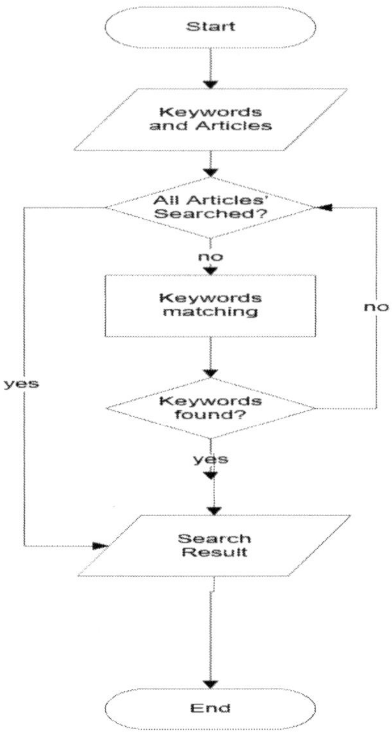

Fig. 2. Search Process Design

4 Implementation System

The Marker has been identified, an overlay image and a 3D model will appear, corresponding to the name of the element, the illustration of Augmented Reality implementation can be seen Figure 3.

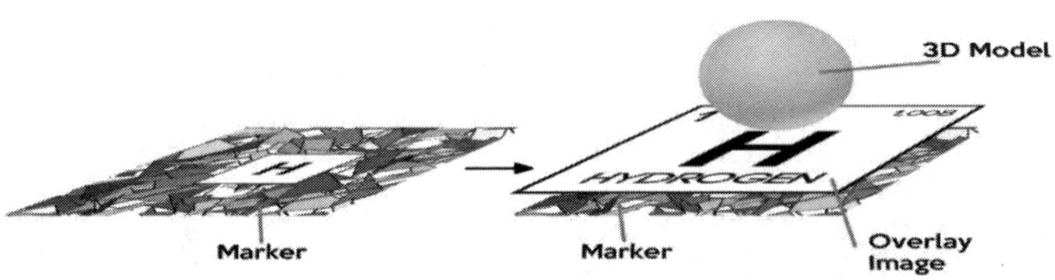

Fig 3. The Illustration of Augmented Reality Implementation

It appears that Marker is recognizable by software, so Overlay Image and 3D models can be placed just above the Marker. Closing 2 Hydrogen objects will produce H_2. The result of combining elements (H_2) can be seen in the figure 4.

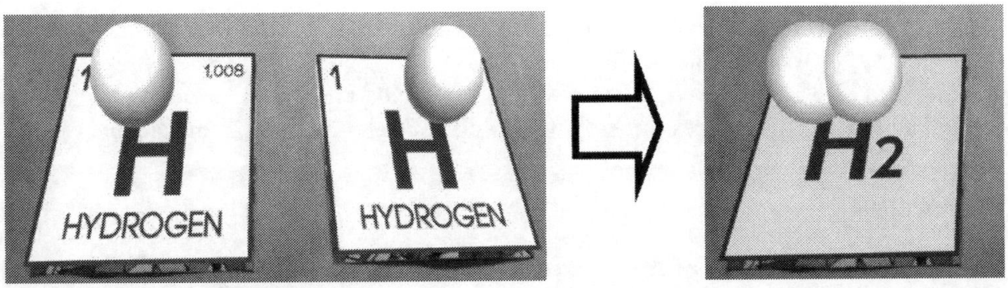

Fig 4. The Combining Elements H_2

If the result of element H_2 is added with oxygen (O_2) will produce H_2O. The result of combining elements (H_2O) can be seen in the figure 5.

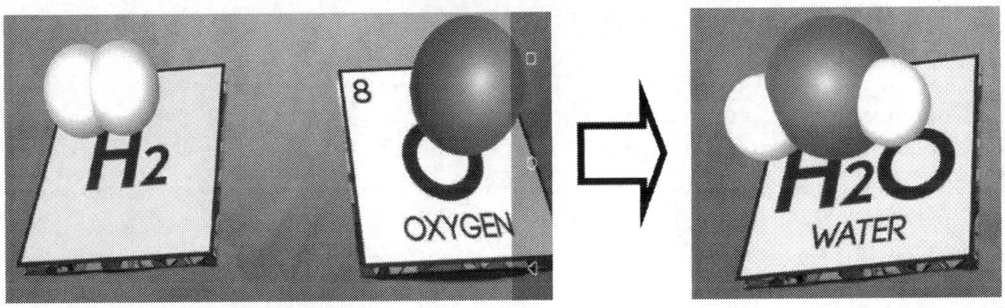

Fig 5. The Combining Elements H_2O

A covalent bond, also called a molecular bond, is a chemical bond that involves the sharing of electron pairs between atoms. These electron pairs are known as shared pairs or bonding pairs, and the stable balance of attractive and repulsive forces between atoms, when they share electrons, is known as covalent bonding. The following is the result of compound bonding can be seen on Figure 6.

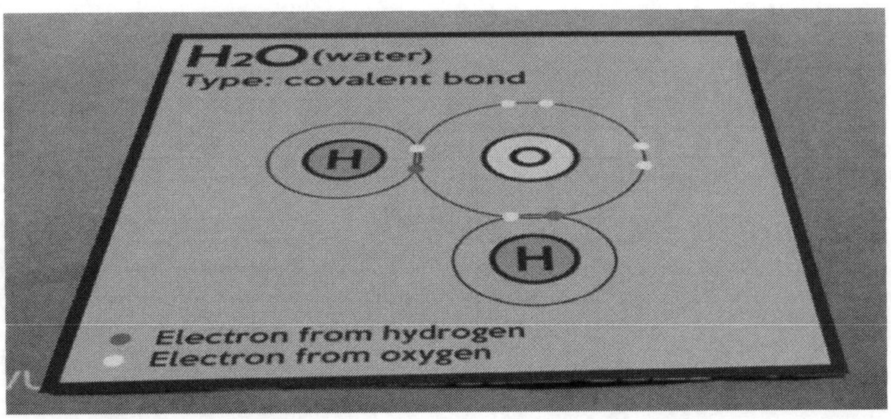

Fig 6. The Compound Bonding H_2O

5 Conclusion

At the end of the design and development of application program for the compound bonding application, some conclusions can be drawn through the test :

- Marker's detection by mobile camera are already work well. Marker can be detected and the application can shown the 3D object.
- Chemical reaction process is already well for combination 2 and 3 elements.
- Transformation 3D object is already work well, it can be enlarged, reduced, and rotated.

References

1. A. Ronald T.. "A Survey of Augmented Reality". Presence : Teleoperators and Virtual Environtments. (1997)
2. S. Pangestu..Jurnal Edukasi dan Penelitian Informatika, Vol.1-2 h.8, Universitas Tanjungpura, Pontianak. (2015)
3. Setiawan, A., Handojo, A., & Hadi, R. Indonesian Culture Learning Application Based on Android. *International Journal of Electrical and Computer Engineering*, *7*(1), 526. (2017).
4. Huda U. Design of Prototype Payment Application System with Near Field Communication (NFC) Technology based on Android. *Computer Engineering and Application Journal* (ComEngApp).; 1(1); 1-12 (2012)
5. Safaat, N. Pemrograman Aplikasi *Mobile Smartphone* dan *Table* PC Berbasis Android Revisi Kedua. Bandung: Informatika Bandung. (2014).
6. Salbino, S.. Buku Pintar Gadget Android untuk Pemula. Jakarta: Kunci Komunikasi. (2014)
7. Teddy M, Andri Z., Securing E-Mail Communication Using Hybrid Cryptosystem on Android-based Mobile Devices. *TELKOMNIKA Indonesian Journal of Electrical Engineering*.; 10(2); 827-834. (2012)
8. W3schools. AJAX Introduction, URL=http://www.w3schools.com/Ajax/ajax_intro.asp. (2016)

A Novel of Internal Corrosion Assessment Methods on Drinking Water Distribution Pipelines

Edi Septe[1,2,*], *Nizwardi* Jalinus[2], and *Refdinal* [2]

[1]Mechanical Engineering Dept., Industrial Engineering Faculty, Universitas Bung Hatta, Indonesia
[2]Doctoral Program Vocational Education and Technology Department, Engineering Faculty, Universitas Negeri Padang, Indonesia

Abstract. Various methods of corrosion assessment on drinking water distribution pipelines have been reviewed in this paper. Five methods of corrosion assessment consisting of sampling method, scanning method, ultrasonic testing method, electromagnetic inspection method and corrosion index assessments described provide valuable information regarding their implementation mechanisms. Since each method has its advantages and limitations, the application of methods on the pipeline must consider the environmental conditions of the pipeline to be inspected, as well as the equipment and resources available.

1 Introduction

The main pipeline consist of iron pipes such as gray cast-iron pipes, ductile cast iron pipes and steel pipes [1] is susceptible to internal corrosion [2] which causes the dissolution of iron into the water, and forms an iron oxide that produces particulate iron compounds in water and causes "red water" that affects the water quality [3]. Corrosion that occurs in pipes can vary significantly in one segment with another segment [4] and it is the key factor affecting failure pressure of the pipeline [5].

Drinking water distribution pipelines are generally planned for a period of more than 40 years to ensure water supplies to more than 90% of the city's population smoothly with good quality [6] [7]. To ensure the quality of the water supply, as well as to maintain the stability of drinking water supply in a sustainable manner, the drinking water company should rehabilitate the damaged pipes or old pipes systematically [3]. Rehabilitation of pipelines requires up to date data and information on pipeline conditions. Without proper information, it can lead to un optimal rehabilitation associated with cost, time and method [4]. The assessment of pipe corrosion is one way to obtain information: the condition of the pipeline which is necessary to avoid failure of equipment that may cause interruption of operation, about the relationship between the corrosion process and the operating variables, with the result that pipes can be used more efficiently [8].

* Corresponding author: edysepte@yahoo.com

The assessment of corrosion of drinking water distribution pipelines can be focused on three factors: physical, operational and environmental factors. Physical factor parameters are pipe mass loss within a certain time period, which can be converted into pipeline corrosion rate per year. The operational factor parameters are related to the performance of water flow in the pipeline, in the form of transient pressure, flow rate and flow capacity. Environmental factors that affect the occurrence of damage to the pipeline consists of external and internal environment. External environmental parameters are soil corrosivity or resistivity, soil moisture, pH, sulphide and chloride concentration, whereas internal environmental parameters are water aggressiveness affecting pipe corrosion, such as pH, hardness and oxygen dissolution [9].

Based on the corrosion assessment factor, corrosion assessment can be done directly and indirectly. There are several methods of corrosion assessment on applicable drinking water distribution pipelines. The intention of this article is to describe several methods of internal corrosion assessment of drinking water distribution pipelines as part of pipeline assessment technology.

2 Literature Review

To improve the reliability of the water supply system, it is necessary to conduct a pipeline inspection, because if the extent of damage to the pipe wall and its operating period is known, it can calculate the average corrosion rate as the basis for decision making on pipe replacement or rehabilitation [7].

The study to determine the condition and predict the life time of the pipe future of the ductile cast-iron pipes that are used as drinking water distribution pipes was conducted [3]. Assessment is done with the impact of elastic-wave method on the piece of pipe after receiving the mechanical loading at one end. Because pipeline stiffness will decrease due to decrease of pipe thickness caused by corrosion, the thickness of the pipe is known based on the peak vibration frequency data of the pipe [10]. While the hole on the surface of the pipe is known based on the frequency distribution of pipe vibration waves, because the frequency of pipe waves that have a hole in the distribution of a lower range than the massive pipe frequency waves [3].

Pipeline corrosion assessment can also be done through coupon testing [11]. Investigated the corrosion rate of reclaimed water distribution pipes through coupon testing is performed by making a closed test installation with a test chamber as a coupon rack, located in the middle of the installation. The weight changes that occur in each coupon over a period of time become the basis for calculating coupon corrosion rate, as an indicator of corrosion assessment [12].

Corrosion assessment on drinking water distribution pipelines can also conducted based on acoustic resonance technology method [4]. The assessment of corrosion is done using pipe scanner which has a number of ultrasound transducers, as well as data acquisition and data storage. Pipe scanners are inserted into the pipe through a hole made at the beginning of the pipe and moved along the inspected pipe. Acoustic resonance technology refers to the use of wide-band acoustic pulses to generate resonance modes in the pipe. This mode will be reflected from the pipe wall and received by the transducer for further analysis [13]. The area that goes in the resonance and from which the acoustic signal is reflected is called 'foot print'. The acoustic reflection signal contains information such as wall thickness and corrosion in 'foot print'. The analysis can determine the thickness of the pipe wall, although the pipe is covered by corrosion or sediment products. This method can determine the thickness of the remaining pipe wall in the range of 1.5 to 25 mm, distinguish between corrosion inside and outside the pipe, and indicate the position and area of rust, tuberculosis and sedimentation in the pipe [4][14].

Non-destructive techniques can be done to inspect water pipes made of steel and ductile cast iron are based on eddy current and ultrasonic technique [15] [16]. Ultrasonic testing method used the ultrasonic waves with high frequency above 20 Khz. The ultrasonic testing system consists of several functional units such as receiver waves, transducers, and display devices. Receiver waves are electronic devices that can generate high-voltage electrical energy that makes high-frequency transducers produce ultrasonic energy. Ultrasonic waves are generated by changes in electrical energy to the mechanical energy of a transducer or probe, through a piezoelectric effect. The piezoelectric effect is a reversible effect, which in the event of a change of electrical energy to a mechanical, then a change of mechanical energy to electrical energy also occurs.

Sound energy will spread in the form of waves through pipe material and when there are defects detected by waves, some energy will be reflected back from the defective surface. Ultrasonic scan, has the ability to detect metal thickness changes down to less than 0.2 mm [17]. Examination of thickness and presence of defects in the pipes with ultrasonic waves can be done through resonance techniques, transmission techniques, and echo techniques. A resonance technique that measures the thickness of the material through measurement of ultrasonic wave lengths that can lead to maximum resonance in the material and the determination of defects based on resonance changes. The transmission technique determines the defect based on the decrease in the intensity of the ultrasonic wave received by the receiver probe. The echo technique determines the thickness of the material, the location and the size of the defect based on the time of propagation and wave amplitude received by the probe [15].

Electromagnetic inspection method is done by application of remote field eddy current. This method is based on the measurement of the attenuation and phase delay of the electromagnetic signal as it passes through the thickness of the metal pipe wall. Pipe inspection is performed using an exciter coil that produces a direct (internal) electromagnetic field moving inside the pipe, but its power will weaken rapidly because of the circumferential eddy current induced in the wall of the conductor pipe. Simultaneously, the exciter produces another indirect (external) field that travels through the wall of the pipe with a small attenuation. The change of field strength and attenuation depends on the thickness of the wall of the pipe thus from this change allows the determination of the thickness of the pipe wall [15][18].

Corrosion assessment on drinking water distribution pipelines can also be performed based on the water condition in the pipelines. The water condition described as the corrosion index which can cause corrosion impact to the pipe walls. Langelier Saturation Index (LSI), Ryznar Stability Index (RSI), Larson-Skold (LS), Puckorius Index Scaling (PSI) and Aggressive Index (AI), are usually used to determine whether water flowing in a pipe affects corrosion or forms a water sediment on a pipe wall [19][20].

Physicochemical reactions that occur between metals and environments that have electrochemical properties will cause changes in the properties of metals. Corrosive water in the pipe will dissolve the pipe material during flow and this causes many problems [21].

3 Recommendation for Future Work

There are five methods that can be used to assess the corrosion of drinking water distribution pipes. In practice, the methods have their own advantages and disadvantages. The sampling method provides pipeline condition information at a particular location at which the sample was taken [11]. This method can't provide comprehensive information from the inspected pipeline. While the scanning method can provide more comprehensive information on the inner pipe surface conditions [4][14]. Ultrasonic testing and electromagnetic inspection may provide better pipeline condition information, but this method requires a skills operator and need a relatively larger cost. The corrosion index assessment method through the flowing water examination is a relatively easy method, but the information obtained from the application of this method does not directly describe the condition of the pipe surface.

Based on the parameters of corrosion assessment, the application of the sampling method with impact-elastic-wave method, scanning method, ultrasonic testing and electromagnetic inspection are performed for assessment the physical factors. Meanwhile the sampling method with coupon testing is done for operational parameters assessment, because in this test can be varied velocity, pressure and water temperature flowing in the pipe. Moreover, corrosion assessment by corrosion index method is an assessment based on the parameters of the pipe internal environment. To get better information on corrosion assessment of pipelines, two methods can be implemented on the pipelines to be inspected.

4 Conclusion

Several methods may be undertaken for assessment of corrosion of drinking water distribution pipelines, which provide information on pipeline conditions and environmental or water conditions that flow within the pipeline. Five methods of pipe corrosion assessment have been proposed in this article and each method has advantages and limitations. To conduct an effective corrosion assessment, the selection of methods to be applied to a pipeline must consider the environmental conditions of the pipeline to be inspected, as well as the equipment and resources available. The data and information obtained from the corrosion assessment can be considered for the decision of rehabilitation or replacement of drinking water distribution pipelines.

References

1. Y. Hou, D. Lei, S. Li, W. Yang, and C.Q Li, Experimental Investigation on Corrosion Effect on Mechanical Properties of Buried Metal Pipes, *International Journal of Corrosion*, Volume 2016, Article ID 5808372, 13 pages (2016)
2. A.C. Benjamin, J.L.F. Freire, R.D. Vieira, D.J.S. Cunha, Interaction of Corrosion Defects in Pipelines – Part 1: Fundamentals, *International Journal of Pressure Vessels and Piping* (2016)
3. Y. Kobayashi, Y. Takahashi, T. Kamada, A. Koizumi and M. Fujiwara, 2010, Corrosion Assessment of Aging Drinking Water Pipes by Impact Elastic-Wave Method, *Water Practice and Technology*, **5,** 3 (2010)
4. E. Kloosterman, Condition Assessment of Metallic Water Mains by Internal Pipe Inspection, (4th Pipeline Technology Conference, Hannover Germany, April 22-23, 2009)
5. Y. Chen, H. Zhang, J. Zhang, X. Liu, X. Li, J. Zhou, Failure Assessment of X80 Pipeline With Interacting Corrosion Defects, (Engineering Failure Analysis, 2014)
6. R. Gražulevičienė and G. Balčius, Assessment of Iron and Manganese Concentration Changes in Kaunas City Drinking Water Distribution System, *Environmental Research, Engineering and Management*, **4**, 50, 37–43 (2009)
7. V. Maslak, N. Nasonkina, V. Sakhnovskaya, M. Gutarova, S. Antonenko and D. Nemova, Evaluation of Technical Condition of Water Supply Networks on Undermined Territories, *Procedia Engineering* **117**, 980–989 (2015)
8. S.A. Protain, An Introduction to Corrosion Monitoring. Protection Anticorrosiva, (Master Document 0147 Issue E June 2005).

9. E.Septe, A. Naumar and A.H. Mohammed, A Review of Corrosion Assessment Model and Parameters of Drinking Water Distribution Pipelines, *Jurnal Teknologi UTM Malaysia*, **69**, 2 (2014).
10. S. Uchida, T. Kamada, T. Iwasaki, S. Iwano and H.S. Lee, A Study on Frequency Analysis Methods for Concrete Thickness Estimation by Impact Elastic-Wave Method, (Proceeding of Third International Conference on Sustainable Construction Materials and Technologies, Kyoto, Japan, August 18-21, 2013)
11. H. Lee, U. Rasheed and M. Kong, A Study on The Comparison of Corrosion in Water Supply Pipes Due to Tap Water (TW) and Reclaimed Water (RW), *Water*, **10**, 4 (2018)
12. C.E. Chuka, Odio. B.O, Chukwuneke J.L, Sinebe. J.E. Investigation of The Effect of Corrosion on Mild Steel in Five Different Environments, *International Journal Of Scientific & Technology Research*, 3, 7 (2014).
13. R. Urayama, T. Takagi, T. Uchimoto, S. Kanemoto, T. Ohira and T. Kikuchi, Implementation of Electromagnetic Acoustic Resonance in Pipe Inspection, *E-Journal of Advanced Maintenance*, 5, 1 (2013).
14. L. Djeddi, R. Khelif, S. Benmedakhene and J. Favergeon, Reliability of Acoustic Emission as a Technique to Detect Corrosion and Stress Corrosion Cracking on Prestressing Steel Strands, *International Journal of Electrochemical Science*, **8**, 8356 - 8370 (2013).
15. B. Rajani and Y. Kleiner, Non-Destructive Inspection Techniques to Determine Structural Distress Indicators in Water Mains,)Proceedings of Evaluation and Control of Water Loss in Urban Water Networks, Valencia, Spain, June 21-25, 2004)
16. N. Ulapane , A. Alempijevic, T.V. Calleja and J.V. Miro, Pulsed Eddy Current Sensing for Critical Pipe Condition Assessment, *Sensors*, **17**, 2208 (2017)
17. B.F. Al-Daajani, N.M. Al-Anazi, A.M. El-Sherik and A.I. AbdulHadi, Assessment of Advanced Corrosion Monitoring Technology for Pipeline Integrity, Saudi Aramco *Journal of Technology*, **2016**, 41-46 (2016)
18. X. Xu, M. Liu, Z. Zhang and Y. Jia, A Novel High Sensitivity Sensor for Remote Field Eddy Current Non-Destructive Testing Based on Orthogonal Magnetic Field,, *Sensors*, **14**, 24098–24115 (2014)
19. A. Dargahi, R. Shokri, M. Mohammadi, A. Azizi, L. Tabandeh, A. Jamshidi and S. Beidaghi, Investigating of The Corrosion and Deposition Potentials of Drinking Water Sources Using Corrosion Index: A Case Study of Dehloran, *Journal of Chemical and Pharmaceutical Sciences*, Special Issue **7**, 73-79 (2016)
20. A. Amouei, S.H. Fallah, H. Asgharnia, A.R. Yari and M. Mahmoudi, Corrosion and Scaling Potential in Drinking Water Distribution of Babol, Northern Iran Based on the Scaling and Corrosion Indices, *Archives of Hygiene Sciences*, **6**, 1 (2017).
21. S. Shahmohammadi, A. Noori, S. Karami, A. Amini, B. Shahmoradi, S.S. Ardakani. S.M. Lee and R.R. Pawar, A Study on Corrosion and Scaling Potential of Drinking Water Supply Resources in Rural Areas of Sarvabad, West of Iran, *Journal of Advances in* Environmental Health Research, **6**, 53-61 (2018).

Determinants of web-user satisfaction: using technology acceptance model

Zaitul [1], *Fanny* Ramadhani[2], and *Desi* Ilona[3*]

[1]Economics Faculty, Universitas Bung Hatta, Padang, Indonesia
[2]Master Student, Information Engineering, Universitas Sumatra Utara, Medan, Indonesia
[3]Economics Faculty, Universitas Putra Indonesia YPTK, Padang, Indonesia

Abstract: This study investigates the determinant of web-user satisfaction. User satisfaction is successful measurement for IT implementation. Technology Acceptance Model and Theory of Plan Behaviour are used to underpin the relationship. Two variables are applied as determinants of web-user satisfaction: perceived usefulness and perceived ease of use. Students from four faculties are used as research object and questioner were distributed online by giving URL-Link through WhatsApp. 53 Students filled the questioners. Smart-PLS use to analyse data. The finding show that perceived ease of use has a positive relationship with user satisfaction. However, perceived usefulness does not have a significant relationship with user satisfaction. This study has a practical and theoretical implication. Practically, the finding can be used to formulate the future university website development. Theoretically, this study contributes to the Technology Acceptance Model and Theory of Plan Behaviour.

1 Introduction

Information technology is being widely deployed throughout industry, education, government and other institution[1]. In fact, the use of information technology in education institutions have been growing recently, such as higher education institution[2]. Refernce[1]argue that IT services for satisfying personal needs are growing at a rate that will have substantial influence in the information infrastructure. The use of IT has been widely in almost of all management functions, including academics, financial, student affair and etc. For example, *Universitas Bung Hatta* has been developing a decade ago. It was started in 2005 when University used the Microsoft access as platform. Five years ahead, this university applied the *html* as platform and it has been moving forward using more advance html platform to satisfy all university's stakeholders. End-user satisfaction is critical to information system success[1].

There are bundle of studies investigating the system-user satisfaction, see for example, [1], [3], [4], [5], [6], [7] and [8]. [1] study about key success factors affecting of information technology through user's behavioral perspective. They found that perceived ease of use, and perceived usefulness are the most significant factors affecting an intention to use IT service and user-satisfaction. Whereas, refence[3] investigate nine variables: perceived attitude of

*Corresponding author email: desiilona@upiyptk.ac.id

top management toward the project, perceived usefulness, user expectation user experience, ease of use, user skills and user involvement in system development, organizational support, and user attitude toward information system. reference[3] conclude that the most significant variables affecting the end user satisfaction is involvement, perceived usefulness, user experience, organization support, and user attitude toward the IS. [4] validate the end user computing satisfaction instrument developed by [6] and conclude the instrument had its psychometric stability when tested for users of enterprise resources planning. Further, reference[5] study on end user information systems satisfaction formation and conclude that equitable work performance fulfillment and equitable relatedness fulfillment are significant factors affecting satisfaction of end users. Thus, reference[7] investigate the end user satisfaction of e-government services using quality-in-use satisfaction with advanced traveler information systems (ATIS). Finally, reference[8] review the previous studies that investigating factors affecting the system-user satisfaction.

To date, many attempts have been made to study factors affecting the IT user satisfaction. However, there is a lack of studies investigating the subject matters using the stakeholders of higher education institution (HEI). As mention before, HEI has been using the IT largely [2], but previous researchers paid less attention, especially study on user satisfaction. Therefore, there is a desired need to investigate the user-satisfaction in the context of HEI, such as University. Perceive ease of use is one of factor affecting the system user satisfaction [1][3] and [8]. End-user satisfaction is also determined by Perceived usefulness (i.e. [1]). However, the effect of these variables on end-user satisfaction is not conclusive. Therefore, this study aims to investigate the effect of perceived ease of use and perceived usefulness on end user satisfaction. This paper is organized as follow: first session discusses about background of study. Second session is talking about methods and followed by result and discussion. Final session is conclusion and recommendation.

2 Methods

Students in faculty of economics, industrial technology, education, and law are used as research object. These students were taken due to their experiences using the university's website.primary data used in this research and questioners were distributed to them through WhatsApp by sending the URL-link. There are two type of variables; dependent variable (website user-satisfaction), and independent variable (perceived ease of use and perceived usefulness). The end-user satisfaction (SAT) is measured by four items [7]. Example of item is "I felt very confident using this website". Further, perceived ease of use (PEU) and perceived usefulness (PU) using 7 and 4 items respectively. The instrument of PEU and PU were adopted from [7]. Example of item for PEU and PU are "I thought the system is easy to use" and "I found the information in website to be very useful". To measure the respondents' perception on the survey question, five-point Likert scale was applied. Data analyses was conducted using a software of Smart-PLS 2.0. SEM-PLSis a second-generation regression technique for complex causal modeling. This technique is also known as variance-based structural equation modeling [9]. There are two kinds of analysis using PLS: measurement model and structural model [10]. To assess the measurement model, there are four criteria[11]: outer loading, internal consistency, convergent validity and discriminant validity. Further, the structural model has two criteria: predictive power and predictive relevance [11].

3 Result and discussion

There are 53 students participating in this study. Demography data is shown in Table 1. Male students involved are 11 students or 20.75% and the rest is female (79.25%). Students with age of 20 years old are 11 or 20.75%, followed by 21 years old student (47.17%), 22 years old (28.30%) and more than 22 years old (3.77%). 39 Students are from faculty of economics (73.58%), education (18.87%), law (5.66%), and Technology industry (1.89%).

Table 1. Demographic Analysis

Demographic	Category	Number	Percentage
Gender	Male	11	20.75
	Female	42	79.25
Age	20 years	11	20.75
	21 years	25	47.17
	22 years	15	28.30
	>22 years	2	3.77
Faculty	Economics	39	73.58
	Education	10	18.87
	Law	3	5.66
	Technology Industry	1	1.89

3.1 Analysis of Measurement model

The final result for measurement model is demonstrated in Table 2. However, the first analysis of measurement model show that several outer loadings from three constructs are not valid due to the lower outer loading (less than 0.6)[12]. Therefore, non-valid items were excluded from analysis: one item for end-user satisfaction (item 3), and three items for Perceived ease of use (item 3,4, and 6). Having deleted all non-valid items, the algorithmanalysis was conducted again and the result show that all items have a loading greater than 0.6 for all constructs.the second criteria for measurement model is internal consistency. To see the internal consistency, Composite reliability (CR) and Cronbach Alpha (CA) are applied and the result show that the value of CR and CA for all construct are greater than 0.70. Third criteria of measurement model are the convergent validity that use AVE and the result show that the value of AVE is greater than 0.50.

Table 2. Measurement Model

Construct	Item	Loadings	CA	CR	AVE
SAT	sat_1	0.863	0.881	0.926	0.806
	sat_2	0.927			
	sat_4	0.903			
PEU	peu_1	0.923	0.826	0.886	0.662
	peu_2	0.864			
	peu_5	0.698			
	peu_7	0.749			
PU	pu_1	0.910	0.938	0.926	0.844
	pu_2	0.944			
	pu_3	0.914			
	pu_4	0.908			

The final criteria of measurement model (using Fornell-Lacker criterion) also indicate a good measurement model (see Table 3). The correlation coefficient was created through square root of the AVE (bold number). Further, this value is compared to coefficient of correlation with other construct. For example, square root of SAT's AVE (0.898) is compared to 0.864 and 0.765. if the square root of AVE is greater than coefficient correlation with other construct and it can be concluded that the measurement model for SAT is adequate. Therefore, it can conclude that all constructs have a good discriminant validity.

Table 3. Fornell-Lacker Criterion

Construct	SAT	PEU	PU
SAT	**0.898**		
PEU	0.864	**0.814**	
PU	0.765	0.635	**0.919**

3.2 Analysis of structural model

Second analysis is structural model. The result of structural model could be seen in Table 4 and Figure 2. To assess the structural model, there are two criterias[11] : predictive relevance and predictive power. The former use Q-square and the value must be more than zero. Whereas, the later apply the R-square and higher R square is required. The significance of structural path coefficient is used to see any significant relationship. The overall explanatory power (R-square) indicate that the model account for 58.7% for user satisfaction. In addition, the second requirement is Q-square and the result show that the Q-square (0.450) satisfied the cut off value (>0.00). thus, it can conclude that the model achieve the predictive validity [10]. The effect of perceived usefulness on user satisfaction is not significant due to its t value is less than 1.65 (α<10%). However, perceived ease of use has a significant relationship with user satisfaction. The higher the perceived ease of use, the higher the user satisfaction.

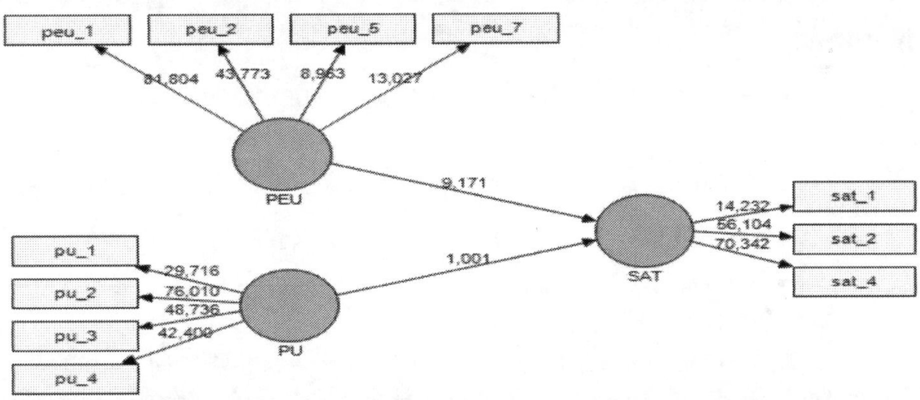

Fig 2. Structural model

Table 4. Structural Model Analysis

Endogenous Construct	**R-Squared**	**Q-Squared**
SAT	0.587	0.450
Relation	**Path Coefficient**	**T Value**
PU -> SAT	-0.104	1.001
PEU -> SAT	0.855	9.171***

Note; ***significant at 1%

The insignificant effect of perceived usefulness on user satisfaction is not consistent with the finding of [7][1]and suggestion of [8]. Further, the significant effect of perceived ease of use on user satisfaction is aligned with [3], [7], [8], [1]. If university intent to increase a website user satisfaction: (i) the website must be easy to use, (ii) there is a technical support to use it, (iii) no need to learn a lot thing before using it, and (iv) easy to learn to use the website quickly.

4 Conclusion and Recommendation

User satisfaction is a measurement of IT implementationsuccess. There are few theories used to understand why user satisfy or not, such as Technology acceptance model[14]. From Technology acceptance model (TAM), perceived ease of use and perceived usefulness. Theory of plan behaviour (TPB) [15] predict that these two variables influence the intention to use and finally user satisfaction. This study found that there is a significant relationship between user satisfaction and perceived ease of use. The finding from this study provide a new understanding of website user satisfaction from student's perspective and contribute to the TAM and TPB in the sense that website university user (student) satisfaction also can be explained by these two theories. Practically, the university could increase the perceived ease of use to website if they intent to increase user satisfaction. Finally, a number of important limitation need to be considered. First, this study uses the university students as respondent. Second, this study only applied two variables as factors affecting the user satisfaction. Finally, this study utilized a direct effect of these variables on user satisfaction. This study has thrown up many questions in need of further investigation. First, future researcher can study the user satisfaction but use the different respondents, such as lecturer or other university's stakeholders. Second, the future study also can use other independent variables suggested by previous researcher, such as coverage, customize access, and etc. [7]. Finally, future investigation also can consider the role of mediating or moderating variables.

References

1. Y. Liu, Y. Chen, C. Zhou, *international conference* (2006).
2. M. Fusilier, S. Durlabhji, *Campus-Wide Inf. Syst.*, **22**, 4 (2005)
3. M. Adam Mahmood, J. M. Burn, L. A. Gemoets, C. Jacquez, *Int. J. Hum. Comput. Stud.*, **52**, 4 (2000) .
4. T. M. Somers, K. Nelson, J. Karimi, *Decis. Sci.*, **34**, 3 (2003)
5. A. Au, E. W. T. Ngai, T. C. E. Cheng, *MIS Q.*,**32**, 1 (2008).
6. W. J. Doll, W. Xia, G. Torkzadeh,*MIS Q.*, 18, 4 (1994)
7. T. A. Horan, T. Abhichandani, R. Rayalu, *Conference on System Sciences* (2006)
8. M. Hidayat, W. W. Winarno, D. Adhipta, *Seminar Nasional Teknologi Informasi dan Multimedia*(2014)
9. J. F. Hair, C. M. Ringle, M. Sarstedt, *J. Mark. Theory Pract.*, **19**, 2 (2011)
10. J. F. Hair, G. T. M. Hult, C. M. Ringle, M. Sarstedt, *A Primer on Partial Least Squares Structural Equation Modeling (PLS-SEM)*, (Thousand Oaks: Sage, 2013).
11. V. E. Vinzi, W. W. Chin, J. Henseler, H. Wang, (Springer, 2010).
12. J. F. Hair, C. William, B. J. Babin, R. E. Anderson, *Multivariate Data Analysis*, 7th(Pearson Education Limited, 2014)
13. Y. Liu, Y. Chen, C. Zhou, international conference (2006).
14. F. D. Davis, *Int. J. Man-Machine Stud.*, **38** (1993)
15. I. Ajzen, *Organ. Behav. Hum. Decis. Proccess*, **50** (1991)

AUTHOR INDEX

AUTHOR INDEX

EDP Sciences – Web of Conferences
17, Avenue du Hoggar
Parc d'Activité de Courtabouf
BP 112
F-91944 Les Ulis Cedex A
France

ISBN 978-1-5108-7703-0

CURRAN ASSOCIATES INC.
proceedings.com